AF842868

Vegetable Breeding
Principles and Application

Vegetable Breeding

Principles and Application

Bandan Thapa
Genetics and Plant Breeding, Regional Research Station Kalimpong,
Uttarbanga Krishi Viswavidyalaya, West Bengal

Balagoni Maruthi
Department of Vegetable Science
Sri Konda Laxman Telangana State Horticultural University, Telangana

Sourav Roy
Department of Agriculture
Brainware University, Barasat, Kolkata

Umesh Thapa
Department of Vegetable Science
Bidhan Chandra Krishi Viswaidyalaya (BCKV), West Bengal

Pramod Kumar Pandey
College of Agriculture
Central Agricultural University (Imphal), Kyrdemkulai, Meghalaya

CRC Press
Taylor & Francis Group
Boca Raton London New York

CRC Press is an imprint of the
Taylor & Francis Group, an **informa** business

-EPH-

Elite Publishing House

First edition published 2026
by CRC Press
4 Park Square, Milton Park, Abingdon, Oxon, OX14 4RN

and by CRC Press
2385 NW Executive Center Drive, Suite 320, Boca Raton FL 33431

© 2026 Elite Publishing House

CRC Press is an imprint of Informa UK Limited

The right of Bandan Thapa, Balagoni Maruthi, Sourav Roy, Umesh Thapa and Pramod Kumar Pandey to be identified as author(s) of this work has been asserted in accordance with sections 77 and 78 of the Copyright, Designs and Patents Act 1988.

All rights reserved. No part of this book may be reprinted or reproduced or utilised in any form or by any electronic, mechanical, or other means, now known or hereafter invented, including photocopying and recording, or in any information storage or retrieval system, without permission in writing from the publishers.

For permission to photocopy or use material electronically from this work, access www.copyright.com or contact the Copyright Clearance Center, Inc. (CCC), 222 Rosewood Drive, Danvers, MA 01923, 978-750-8400. For works that are not available on CCC please contact mpkbookspermissions@tandf.co.uk

For Product Safety Concerns and Information please contact our EU representative GPSR@taylorandfrancis.com. Taylor & Francis Verlag GmbH, Kaufingerstraße 24, 80331 München, Germany.

Trademark notice: Product or corporate names may be trademarks or registered trademarks, and are used only for identification and explanation without intent to infringe.

British Library Cataloguing-in-Publication Data
A catalogue record for this book is available from the British Library

Print edition not for sale in India.

ISBN13: 9781041331292 (hbk)
ISBN13: 9781041340683 (pbk)
ISBN13: 9781003790976 (ebk)

DOI: 10.1201/9781003790976

Typeset in A Caslon Pro
by Elite Publishing House

-EPH-

Contents

Preface

Vegetable breeding, a specialized domain within plant breeding, has rapidly evolved over the years due to advances in genetics, molecular biology, and agricultural sciences. This book, Vegetable Breeding- Principles and Application, aims to provide a comprehensive yet accessible guide to the core principles and methodologies that drive successful vegetable breeding programs. Given the global demand for food security, the rising challenges of climate change, and the increasing consumer preference for nutrient-dense crops, vegetable breeding has taken on a more pivotal role than ever before.

The first unit of this book traces the historical significance of vegetable breeding, marking its divergence from cereal crop breeding and emphasizing its evolutionary aspects. Vegetable breeding holds a distinct place in agricultural sciences due to the vast diversity of vegetables in terms of species, reproductive biology, and genetic makeup. The opening chapter explores the importance of vegetable breeding in shaping global food security, improving nutrition, and adapting crops to varying environmental and market demands. Furthermore, it highlights the challenges breeders face due to the distinct evolutionary and genetic characteristics of vegetables compared to cereals.

Unit two transitions into the technical aspects of selection and hybridization procedures. This unit covers critical breeding techniques, such as selfing, crossing, and selection methods that are essential for improving desired traits in vegetables. Topics such as genetic architecture, breeding for disease resistance, and abiotic stress tolerance are also addressed, alongside strategies for improving water and nutrient use efficiency (WUE and NUE). This section is particularly valuable for breeders focusing on developing crops that are resilient to environmental stresses, including drought, high temperatures, and nutrient-poor soils.

Heterosis breeding, a crucial tool for producing high-yielding hybrid varieties, is explored in depth in unit three. This section examines the mechanisms and basis of heterosis (hybrid vigor), which is essential for improving vegetable productivity. Facilitating mechanisms such as male sterility, self-incompatibility, and different sex forms in vegetables are also discussed, providing breeders with valuable insights into maximizing hybridization success. These principles are vital in the production of vigorous, high-yielding, and adaptable hybrid vegetable varieties.

Unit four introduces mutation and polyploidy breeding, techniques that allow breeders to generate genetic variation and introduce novel traits into vegetable crops. This unit also addresses the improvement of asexually propagated vegetables and highlights breeding techniques for developing vegetables suited for protected environments, such as greenhouses. These approaches are becoming increasingly important in the context of controlled-environment agriculture, which is gaining prominence for vegetable production in regions with unfavorable climates. In unit five, the focus shifts to ideotype breeding, a method that involves designing crop varieties based on an ideal plant type for specific environmental conditions and agricultural practices. This unit outlines the varietal release procedures, including DUS (Distinctness, Uniformity, and Stability) testing for vegetable crops, ensuring that new varieties meet the required standards. Furthermore, the unit covers the application of in-vitro and molecular techniques, underscoring the role of biotechnology in modern vegetable breeding.

Unit six provides detailed information on breeding specific groups of vegetables, such as crucifers (Brassicaceae family), cucurbits (Cucurbitaceae family), solanaceous crops (Solanaceae family), and tuberous/bulbous vegetables. This unit presents practical examples and case studies, giving readers a clearer understanding of the unique breeding challenges and opportunities in these important vegetable families. Finally, the seventh unit offers a "vegetable breeder's kit," a collection of practical tools, techniques, and resources that breeders can use to facilitate the breeding process. Whether involved in academic research or commercial breeding, this kit provides essential information for implementing efficient breeding programs, from field trials to molecular techniques.

Vegetable Breeding- Principles and Application is crafted to serve as a comprehensive reference guide for students, researchers, and professionals in vegetable breeding. It aims to inspire innovations in crop improvement and foster the development of vegetable varieties that are productive, resilient, and nutritious. By integrating both classical breeding methods and modern biotechnological advances, this book equips the next generation of vegetable breeders with the tools needed to meet the agricultural challenges of the 21st century.

Authors

About the Authors

Dr Bandan Thapa is a distinguished academic with a solid educational background and extensive experience in plant genetics and breeding. Currently, Dr Thapa serves as an Assistant Professor at the Uttarbanga Krishi Viswavidyalaya, Regional Research Station in Kalimpong. He earned his B.Sc. Ag. from Bidhan Chandra Krishi Viswavidyalaya, his M.Sc. in Genetics and Plant Breeding from Banaras Hindu University, and his PhD from Uttarbanga Krishi Viswavidyalaya. . His research focuses on plant genetic resource management, breeding, conservation, and sustainable practices for high-altitude medicinal and horticultural plants. He has authored a book, contributed a chapter, published 26 research articles, and participated in over 30 conferences and trainings.

Dr. Balagoni Maruthi completed his B.Sc. (Hons) in Horticulture from SKLTSHU, Mulugu, Telangana in 2018, followed by an M.Sc. in Vegetable Science from the same institution in 2020. He earned his Ph.D. in Vegetable Science from BCKV, Mohanpur, West Bengal. Dr. Maruthi has qualified for the NET exam in Vegetable Science twice, in 2021 and 2023, conducted by ASRB, New Delhi. Currently, he serves as a Teaching Associate in Vegetable Science at SKLTSHU. His research portfolio includes 6 papers, 7 book chapters, numerous popular articles, 1 patent, 1 book, and 6 abstracts from various seminars and conferences.

Dr. Sourav Roy is an Assistant Professor and Academic Coordinator in the Department of Agriculture at Brainware University, Kolkata. He holds a Doctoral Degree in Vegetable Science from BCKV, an M.Sc. in Vegetable Science from BCKV with the 2nd highest OGPA, and a B.Sc. in Horticulture from Uttar Banga Krishi Viswavidyalaya, where he also ranked 87 in the ICAR-AIEEA-JRF exam. He has qualified the ASRB-NET exam and was awarded the National Fellowship for Other Backward Classes by UGC. His expertise includes crop improvement in vegetables like Okra, Tomato, and Sweet Pepper. Dr. Roy has published 4 books, 4 book chapters, and 8 research papers, and has participated in various national and international conferences. He also has experience as an extension worker.

Dr. Umesh Thapa is a highly esteemed academic and researcher specializing in Vegetable Science, with a career spanning over 27 years. He earned his B.Sc. (Hons) in Agriculture, M.Sc (Ag) in Horticulture, and Ph.D.in Vegetable Science from Bidhan Chandra Krishi Viswavidyalaya and Post-Doctoral Research in Vegetable Science in Hydroponic from the University of Turin, Italy. Dr. Thapa is currently a Professor at Bidhan Chandra Krishi Viswaidyalaya (BCKV), West Bengal, India, where he also serves as the Course Director for the Diploma in Agricultural Extension Services for Input Dealers (DAESI) and the In-Charge of the Farmers Academy Convention Centre.(2021-2023) At BCKV, Dr. Thapa has held several significant leadership roles: he was the Head of the Department of Vegetable Science from 2010 to 2014, Dean of Student Welfare from 2017 to 2018, Secretary of the Faculty Council for UG and PG Academic Programmes from 2019 to 2020, and Director of Extension Education from 2020 to 2023. His contributions have been instrumental in advancing the university's academic and extension activities. His areas of expertise include protected cultivation, seed production, and soilless/hydroponic technology for high-value vegetables.

Dr. Thapa has supervised 30 M.Sc. and 15 Ph.D. students and has led 20 research projects funded by various governmental and private organizations. He has published 100 research papers and authored 15 influential books and has 1 patent in Urban Farming. His accolades include the Distinguished Scientist Award (VIFRA-2015), Excellence in Research Award (EET CRS Research & Branding, 2016), Achiever Award (SADHNA, 2016), and Bharat Jyoti Award (India International Friendship Society, 2021). Additional honors include the Eminent Scientist Award (National Seminar on Farmer Centric Cinema, 2017), the Outstanding Scientist Award (International Scientist Award on Engineering, Science and Medicine, 2021), and the Life Time Achievement Award (SHUATS, 2023). He has also received the Eminent Scientist Award 2023 in the field of vegetable Science by AETDS at Sher-e-Kashmir University of Agricultural Sciences and Technology, Srinagar. Dr. Thapa is also a prominent member of several scientific societies, serves on various editorial boards, and has reviewed for numerous journals. His international academic missions have taken him to Switzerland, Germany, Italy, Malaysia, Thailand, Sri Lanka, Bhutan, Nepal, Bangladesh, and the Netherlands.

Dr. Pramod Kumar Pandey is working at the College of Agriculture, Kyrdemkulai, Meghalaya, which is part of the Central Agricultural University headquartered in Imphal, Manipur.

He was born in the village of Kondari, which is in the Ghazipur District of Uttar Pradesh. He did his undergraduate studies at PG College, Ghazipur, U.P., and obtained his postgraduate degree in Biotechnology from KIIT University, Bhubaneswar, Odisha. He obtained his Ph.D. in Biotechnology from Dibrugarh University, Dibrugarh, Assam. He has authored numerous scholarly research papers, and articles and contributed to various book chapters. He possesses about 10 years of professional and academic expertise in the field of Endophytes. He is currently involved in maize breeding activities and has more than 6 years of experience in the field of maize breeding. He serves as the In-Charge of AICRP on Maize, CAU Brapanai Centre.

Chapter - 1

Importance and History of Vegetable Breeding

A. Importance of Vegetable Breeding

Vegetable breeding holds immense significance for both modern agriculture and the global food system. It serves as a critical tool for developing improved varieties of vegetable crops, ensuring that they meet the demands of growing populations, changing climates, and the need for sustainable agricultural practices. One of the key objectives of vegetable breeding is to enhance productivity, quality, and resilience, making them more suitable for cultivation in diverse environmental conditions. By selecting and developing varieties with desirable traits, breeders can contribute to increased food production, better nutritional quality, and improved economic viability for farmers.

One of the primary benefits of vegetable breeding is its contribution to food security. As the global population continues to grow, so does the demand for food, particularly vegetables, a key source of essential vitamins, minerals, and fibre. Vegetable breeding enables the development of high-yielding varieties that are capable of producing more food per unit of land. This is particularly important in regions where arable land is limited, and the efficiency of land use must be maximized. By improving the yield potential of vegetable crops, breeders help to ensure a stable and sufficient food supply for future generations. Another critical aspect of vegetable breeding is enhancing the nutritional quality of vegetables. In recent years, there has been increasing recognition of the importance of diet quality in addressing global malnutrition and non-communicable diseases. Vegetables are rich in vitamins, minerals, and antioxidants that are essential for human health. Through breeding, it is possible to increase the concentration of these nutrients in vegetable crops by the development of nutrient-rich varieties, thus contributing to healthier diets. For example, biofortification efforts have focused on increasing the levels of provitamin A, iron, and zinc in staple vegetable crops to combat micronutrient deficiencies in vulnerable populations.

In addition to yield and nutrition, vegetable breeding plays a crucial role in improving the resilience of crops to biotic and abiotic stresses. Climate change poses significant challenges

DOI: 10.1201/9781003790976-1

to agriculture, with extreme weather events, droughts, heatwaves, and salinity becoming more common. Vegetable breeders work to develop varieties that are more tolerant of these adverse conditions, enabling farmers to maintain productivity even in the face of environmental stress. The development of drought-tolerant, heat-tolerant, and salt-tolerant vegetable varieties is crucial for sustaining agricultural production in regions most affected by climate change. Pest and disease resistance is another key focus of vegetable breeding. Crop losses due to pests and diseases are a major threat to food security and farmer livelihoods. Breeding for resistance to common pests and diseases can reduce the need for chemical pesticides, which can be harmful to both the environment and human health. By developing pest-resistant and disease-resistant varieties, breeders help to promote more sustainable and environmentally friendly farming practices.

B. History of Vegetable Breeding

The history of vegetable breeding is deeply intertwined with the history of agriculture itself. From the early domestication of wild plants to the modern era of scientific breeding techniques, vegetable breeding has undergone significant evolution, shaped by human ingenuity and advances in biological science. Understanding the historical development of vegetable breeding provides important insights into the progress and challenges that have shaped this field over millennia.

The origins of vegetable breeding can be traced back to the dawn of agriculture, approximately 10,000 years ago, when early human societies transitioned from hunting and gathering to settled farming. During this period, early farmers began to select and cultivate wild plants that exhibited desirable traits such as larger fruits, better taste, and higher yields. This process of selection and cultivation is considered the first form of plant breeding. Over time, the repeated selection of favourable traits led to the domestication of wild vegetable species, resulting in the development of the first cultivated vegetable crops.

Throughout the centuries, vegetable breeding remained largely informal, with farmers relying on traditional methods of selection and seed saving. However, significant advancements in vegetable breeding began in the 19th century with the emergence of modern scientific breeding techniques. One of the pivotal figures in the history of plant breeding was Gregor Mendel, whose work on inheritance patterns in pea plants laid the foundation for modern genetics. Mendel's experiments in the mid-19th century demonstrated that traits could be passed from one generation to the next in a predictable manner, following specific patterns of inheritance. This discovery revolutionized the field of plant breeding, providing breeders with a scientific basis for selecting and crossing plants to achieve desired traits.

The early 20th century saw further advancements in vegetable breeding with the development of hybridization techniques. Hybrid breeding involves crossing two genetically distinct parent plants to produce offspring that exhibit hybrid vigour, or heterosis. Hybrid vegetable varieties are often more vigorous, productive, and resistant to diseases than their parent lines. The introduction of hybridization led to the rapid development of high-yielding and disease-resistant vegetable varieties, which contributed to significant increases in agricultural productivity during the Green Revolution of the mid-20th century.

In the latter half of the 20th century, the application of biotechnology further transformed vegetable breeding. Techniques such as tissue culture, genetic engineering, and molecular marker-assisted selection have enabled breeders to manipulate the genetic makeup of vegetable crops with greater precision and efficiency. Genetic engineering, in particular, has allowed for the introduction of novel traits into vegetable crops, such as resistance to herbicides or enhanced nutritional content. Biotechnology has opened up new possibilities for improving vegetable crops, allowing breeders to address challenges such as climate change, pest resistance, and nutrient deficiencies more effectively.

The history of vegetable breeding is also closely linked to the broader history of agricultural development and food production. The domestication and breeding of vegetable crops have been central to the development of civilizations, enabling the growth of human populations and the establishment of settled societies.

S No	Period	Major Milestones in Vegetable Breeding
1	Prehistoric times (~10,000 years ago)	Domestication of wild plants. Early farmers selected plants with desirable traits such as larger fruits and better yields.
2	Ancient Civilizations (5000-3000 BCE)	Development of early agricultural practices in Mesopotamia, Egypt, and other regions. Continued selection for improved crops.
3	Middle Ages (5th - 15th century)	Use of simple selection techniques by farmers. Knowledge of grafting and basic hybridization techniques spread across Europe and Asia.
4	Renaissance Era (15th - 17th century)	Advancement in botanical knowledge, discovery of plant classification. Early scientific breeding techniques started to emerge.
5	18th Century Enlightenment	Increased understanding of plant biology, introduction of more organized breeding practices, first systematic breeding programs.
6	19th Century (Gregor Mendel's Experiments)	Gregor Mendel's work on inheritance patterns in pea plants laid the foundation of modern genetics and breeding practices.
7	Early 20th Century (Hybridization techniques)	Introducing hybridization techniques leads to development of more vigorous and productive vegetable varieties.

8	Mid-20th Century (Green Revolution)	Green Revolution led to the widespread adoption of high-yielding, disease-resistant varieties, boosting food production globally.
9	Late 20th Century (Biotechnology Era)	The advent of biotechnology. Genetic engineering and marker-assisted selection allowed precision breeding for desirable traits.
10	21st Century (Molecular Breeding and CRISPR)	Advances in molecular breeding techniques, including CRISPR gene editing. Increased focus on sustainability and climate resilience.

C. Evolutionary Aspects of Vegetable Breeding

Vegetable breeding, much like the breeding of other crops, follows the principles of evolution and natural selection. However, this process is human-directed and accelerated compared to the natural evolutionary changes in wild plant populations. Over time, the evolution of vegetable crops has been driven by the selection of specific traits deemed desirable by humans, such as improved yield, disease resistance, and better taste. The process of breeding, therefore, reflects an ongoing interaction between plants and their environment, in which breeders apply selective pressures that mimic or amplify the natural evolutionary forces acting on the plants.

Genetic variation (natural as well as artificial creation) is essential for any vegetable breeding program. In natural ecosystems, genetic diversity arises through mechanisms such as mutation, recombination, and gene flow, which provide a pool of genetic material that plants can draw from to adapt to their changing environments. Breeders harness this natural genetic variation by selecting and crossing plants with desired traits to create improved varieties. By continually selecting for favourable traits over multiple generations, breeders can significantly alter the genetic composition of a vegetable population within a relatively short time frame, compared to the slow pace of natural evolution.

One of the fundamental evolutionary processes utilized in vegetable breeding is artificial selection. This involves the intentional selection of plants that exhibit particular desirable traits, such as resistance to pests or a more appealing appearance. Through artificial selection, breeders can direct the evolution of vegetable crops to meet specific human needs. For example, in lettuce (*Lactuca sativa*), breeders have selected for traits such as non-bolting (delayed flowering), leaf color, texture, and head shape. In the case of carrots (*Daucus carota*), breeders have historically selected for larger root size, sweeter taste, and the now iconic orange color.

Another evolutionary aspect of vegetable breeding is the creation of hybrid varieties. Hybridization, the crossing of two genetically distinct parent plants, results in offspring that may exhibit hybrid vigor or heterosis. This means the hybrid plants often perform better than either of the parent plants, showing improvements in yield, growth rate, and resistance to environmental stressors. Hybridization mimics a natural evolutionary process where the crossing of different populations leads to genetic recombination, introducing new traits and improving genetic fitness. In vegetable breeding, hybridization has been particularly important for crops like tomatoes, peppers, and cucumbers, where hybrid varieties have dominated the market due to their superior performance.

Vegetable breeding also involves the management of evolutionary trade-offs. For instance, selecting for high-yield traits in some crops may come at the expense of other traits, such as disease resistance or nutritional quality. Breeders must carefully balance these trade-offs to ensure that the new vegetable varieties meet the needs of both producers and consumers. Additionally, the co-evolution of plants and their pests or pathogens plays a key role in breeding strategies. As breeders develop resistant varieties, pests and diseases may evolve new ways to overcome plant defenses, necessitating ongoing breeding efforts to keep pace with these evolutionary changes.

In the modern era, molecular breeding techniques have added a new dimension to the evolutionary process in vegetable breeding. Advances in genomics and biotechnology allow breeders to pinpoint specific genes associated with desired traits, making it possible to introduce or manipulate these genes with greater precision. Techniques like marker-assisted selection (MAS) and CRISPR-based gene editing have revolutionized vegetable breeding by accelerating the process of developing improved varieties, allowing breeders to achieve results in a fraction of the time compared to traditional breeding methods. These technologies also enable breeders to target complex traits, such as drought tolerance or disease resistance, which are controlled by multiple genes.

Through these evolutionary processes, vegetable breeding continues to evolve, producing crops that are better adapted to the challenges of modern agriculture, including climate change, resource constraints, and the need for sustainable production systems. The evolutionary aspects of vegetable breeding highlight the dynamic relationship between humans and plants, where the selective pressures imposed by breeders have shaped the development of vegetable crops for millennia.

D. Cereal Crop Breeding*vis a vis* Vegetable Crop Breeding

While vegetable breeding and cereal crop breeding share many similarities in terms of the basic principles of plant breeding, there are significant differences between the two, particularly in terms of the goals, methods, and challenges faced by breeders. These differences arise from the distinct biological and agricultural characteristics of vegetables versus cereals, as well as the different economic and environmental contexts in which these crops are grown.

One of the key distinctions between vegetable and cereal crop breeding lies in the diversity of species involved. Cereal crops, such as wheat, rice, and maize, are largely derived from a few domesticated species that have been cultivated on a massive scale for thousands of years. In contrast, vegetable breeding encompasses a much broader range of species, many of which have undergone relatively recent domestication. For example, vegetables like broccoli, kale, and Brussels sprouts all belong to the same species (*Brassica oleracea*), but they have been bred for different traits over the past few centuries. The diversity of vegetable crops presents unique challenges and opportunities for breeders, as each species may have different reproductive systems, growth habits, and environmental requirements.

Another important difference between vegetable and cereal breeding is the focus on quality traits. In cereal crops, breeding efforts have traditionally prioritized yield and agronomic traits, such as grain yield, number of branches per plant, number of tillers per plan, disease resistance and drought tolerance, given the role of cereals as staple food crops that provide a large portion of the world's calories. While yield is also important in vegetable breeding, there is a greater emphasis on traits related to consumer preferences, such as taste, texture, color, and nutritional content. This reflects the fact that vegetables are often consumed fresh or with minimal processing, so the appearance and flavor of the final product are critical factors in determining its marketability. Breeders of vegetable crops must therefore consider a wider range of traits, many of which are subjective and influenced by consumer trends.

Reproductive biology also plays a major role in the variation between vegetable and cereal crop breeding. Many vegetable crops are cross-pollinated, meaning they rely on pollinators or wind to transfer pollen from one plant to another. This creates greater genetic variation within a population and allows for more opportunities for breeders to select new traits. In contrast, cereal crops like wheat and rice are often self-pollinated, meaning that they reproduce by fertilizing themselves, leading to more uniformity in their genetic makeup. This uniformity can be an advantage in terms of stabilizing desirable traits, but it also limits the amount of genetic diversity available for breeding programs.

Hybridization techniques, which are commonly used in both vegetable and cereal breeding, differ in their application between the two crop types. In vegetable crops, hybrids are particularly valued for their ability to exhibit heterosis, or hybrid vigor, resulting in improved yield and performance. Many commercial vegetable varieties, especially those grown in industrial agriculture systems, are hybrids. Hybrid varieties dominate the market for vegetables such as tomatoes, peppers, and cucumbers, where they offer significant advantages in terms of yield, disease resistance, and uniformity. In cereal crops, hybrid breeding has also been used, particularly in maize, but it is less common in crops like wheat and rice, where open-pollinated varieties or self-pollinated lines are still widely grown.

Disease and pest resistance are critical objectives in both vegetable and cereal crop breeding, but the types of diseases and pests that affect these crops can differ significantly. Cereal crops are often targeted by fungal diseases, such as rusts and blights, while vegetable crops face a wider range of pathogens, including bacteria, viruses, and nematodes. The complexity of pest and disease management in vegetables requires breeders to focus on a broader spectrum of resistance traits, often involving multiple genes. In addition, because vegetables are often grown in diverse environments, from small-scale farms to large industrial operations, breeding for adaptability to local growing conditions is particularly important.

Finally, the economic and environmental context in which vegetable and cereal breeding takes place also varies. Cereal crops are typically grown on a large scale in monoculture systems, where the focus is on maximizing yield and efficiency. In contrast, vegetables are often grown in more diverse production systems, including organic farms, greenhouses, and urban agriculture settings. This diversity of production systems means that vegetable breeders must consider a wider range of environmental factors, such as soil health, water use, and climate variability when developing new varieties. Additionally, the shorter growing seasons of many vegetable crops allow for more rapid breeding cycles, enabling breeders to respond more quickly to emerging challenges and consumer demands.

Chapter - 2

Selection Procedures for Vegetable Crop Breeding

1. Selection Procedures – Techniques of Selfing and Crossing and Its Importance in Vegetable Breeding

1.1 Selfing (Self-Pollination)

Self-pollination, or selfing, is a technique wherein a plant's pollen is used to fertilize its ovules, forming seeds. This technique is common in plants that are naturally self-pollinated, such as wheat, barley, and beans. The objective of selfing in breeding programs is to increase the homozygosity of the plant population. When plants undergo repeated cycles of self-pollination, it results in the development of pure lines, which exhibit stable, fixed traits. Selfing plays an essential role in stabilizing desired characteristics such as disease resistance, yield, plant height, or specific fruit quality traits in crops. By reducing the genetic variation within a population, breeders can isolate and preserve traits that are crucial for the crop's performance and market value. Selfing provides a controlled way to study the inheritance of traits and allows breeders to fix certain traits in a plant population. For crops where hybrid vigor (heterosis) is a goal, selfing can help create inbred lines, which are then crossed to produce hybrids with superior performance. Though this process is time-consuming, selfing allows breeders to ensure consistency in the expression of traits. Additionally, it is a valuable tool for breeders who are working with crops that are naturally self-pollinating, as it can easily be integrated into their breeding strategies.

1.1.1 Stabilization of Traits

One of the principal benefits of selfing is its role in stabilizing desirable traits in crop populations. In the context of plant breeding, stabilization refers to the process by which certain phenotypic characteristics (such as yield, plant height, or disease resistance) become uniformly expressed across generations. During self-pollination, the offspring inherit two identical alleles from the same parent, leading to increased homozygosity. Over several

DOI: 10.1201/9781003790976-2

generations, the population becomes more genetically uniform, and breeders can reliably produce plants that express the desired traits.

In vegetable breeding, for example, selfing is widely used to fix traits such as fruit size, shelf life, taste, and nutritional quality in crops like tomatoes and lettuce. The more generations a plant undergoes selfing, the higher the likelihood that it will exhibit the same trait consistently across its entire population. This consistency is particularly valuable in commercial agriculture, where uniformity in crop production can reduce variability in market quality and increase economic returns. Stabilization of traits through selfing is also important in developing parent lines for hybrid seed production, as it ensures that the inbred parent plants will contribute specific, well-defined traits to their hybrid offspring.One notable example of trait stabilization through selfing can be seen in the development of improved tomato varieties. In tomatoes, breeders may aim to stabilize traits like uniform fruit ripening, disease resistance to pathogens such as Fusarium wilt, and higher lycopene content for improved nutritional value. By repeatedly self-pollinating these plants over several generations, breeders can produce lines of tomatoes that exhibit these traits consistently, making them more valuable for both farmers and consumers. Selfing thus becomes a powerful tool in ensuring the reliable performance of vegetable crops, even under different growing conditions.

1.1.2 Creation of Inbred Lines

Inbred lines are crucial in hybrid breeding programs because they provide a genetic base that ensures the predictable inheritance of traits. Inbred lines are developed by repeatedly self-pollinating a plant over multiple generations, usually six to ten until the plant becomes highly homozygous. Homozygosity refers to the presence of two identical alleles for a particular trait, which increases uniformity in the offspring produced from these inbred lines. Once homozygosity is achieved, breeders can use these inbred lines to create hybrid plants, which tend to perform better due to the phenomenon known as hybrid vigor or heterosis.For example, in vegetable breeding, the development of inbred lines through selfing is critical in crops like peppers, cucumbers, and melons. These inbred lines allow breeders to control the inheritance of traits such as fruit size,resistance to certain diseases, or tolerance to environmental stresses like drought or heat. The resulting hybrid plants often outperform both parent lines, showing improvements in yield, resistance to pests, and overall vigor. Hybrid varieties created from inbred lines dominate the commercial vegetable seed market, particularly in crops like maize, where hybrid vigor leads to significantly higher yields.In addition to yield improvements, inbred lines are also used to develop specific trait combinations that are desirable for certain markets or growing conditions. For instance, a breeder may develop an inbred line of cucumbers that is resistant to downy mildew, while another line may have superior fruit shape or size. Crossing these two inbred lines, the breeder can create a hybrid cucumber variety that combines the best of both traits. This approach is widely used in breeding programs for high-value vegetables, where both quality and resistance to biotic and abiotic stresses are critical.

1.1.3 Genetic Purity

Maintaining genetic purity is another major goal of selfing in plant breeding. Genetic purity ensures that the offspring produced from a given plant population are true-to-type, meaning they exhibit the same desirable traits as the parent plants. In the case of self-pollination, since the offspring inherit identical genetic material from both the male and female reproductive parts of the same plant, there is little to no introduction of foreign genetic material. This ensures that the resulting population maintains its genetic purity over successive generations.

In vegetable breeding, maintaining genetic purity is especially important for the production of high-quality seeds that farmers can use to grow consistent and reliable crops. For example, in crops like lettuce, selfing is used to maintain pure lines that express specific traits like crispness, head shape, and resistance to bolting (premature flowering). When these pure lines are used as parent material in breeding programs, they help ensure that the hybrids produced are uniform in their performance, which is essential for large-scale commercial production.Furthermore, genetic purity is critical in seed certification programs, where breeders must demonstrate that the seeds they produce meet specific genetic and performance standards. Selfing is often employed to develop genetically pure foundation seeds, which are then multiplied to produce commercial seed lots. Maintaining genetic purity is not only important for ensuring crop uniformity but also for protecting the intellectual property of plant breeders, as it prevents unintended cross-pollination and maintains the integrity of proprietary breeding lines.

1.1.4 Limitation of Genetic Diversity

While selfing offers many advantages in stabilizing traits and creating inbred lines, it also comes with the limitation of reducing genetic diversity. This reduction in diversity can make plants more susceptible to diseases and pests, as a lack of genetic variation means that all plants in a population may share the same vulnerabilities. In the wild, genetic diversity serves as a defense mechanism, allowing populations to adapt to environmental changes and resist emerging threats. By contrast, highly homozygous populations created through selfing may lack the genetic variability needed to cope with new challenges.

In crop production, this limitation can become a significant concern, especially when self-pollinated varieties are grown on a large scale. For example, if a disease or pest evolves that can overcome the resistance of a self-pollinated vegetable variety, the entire crop may be at risk. This was observed in the Irish potato famine of the mid-19th century, where the reliance on a single, genetically uniform potato variety made the crop vulnerable to the spread of late blight, resulting in catastrophic crop failures.To mitigate the risks associated with reduced genetic diversity, breeders often employ strategies such as crossing self-pollinated lines with more genetically diverse parent material. This helps to introduce new alleles into the population, enhancing its ability to respond to environmental pressures. Additionally, molecular breeding techniques such as marker-assisted selection can be used to identify and introduce beneficial traits into self-pollinated crops without compromising their genetic purity or uniformity.

1.2 Crossing (Cross-Pollination)

Cross-pollination, or crossing, involves the transfer of pollen from one plant to the ovule of a genetically distinct plant. This process leads to the production of offspring that are genetically diverse, inheriting traits from both parent plants. Cross-pollination can occur naturally through the action of wind, insects, or other pollinators, but it can also be controlled and directed by breeders to achieve specific breeding objectives. The primary goal of crossing is to combine desirable traits from different parent plants, resulting in offspring with improved characteristics. Cross-pollination is particularly important for crops that do not self-pollinate or that benefit from hybrid vigor when crossed.In vegetable breeding, crossing is used to create genetic variation into a population, which is essential for developing new varieties with improved traits. This process allows breeders to develop hybrids that exhibit traits such as increased yield, better resistance to diseases and pests, or enhanced tolerance to environmental stresses like drought or heat. Crossing is widely used in breeding programs for crops like tomatoes, peppers, and cucumbers, where hybrid varieties dominate the market due to their superior performance.

1.2.1 Introduction of Genetic Diversity

One of the most significant advantages of cross-pollination is the introduction of genetic diversity into a plant population. Genetic diversity is crucial for the adaptability and resilience of crops, as it provides a broader range of traits that can help plants survive in changing environments or under different stress conditions. By crossing plants with different genetic backgrounds, breeders can create offspring that are more diverse and possess a combination of traits from both parent plants.In vegetable breeding, genetic diversity is particularly important for developing varieties that can withstand a range of biotic and abiotic stresses. For example, breeders may cross a tomato variety that is resistant to certain pests with another variety that has better drought tolerance. The resulting hybrid may inherit both resistance and tolerance, making it more resilient and better suited to a wider range of growing conditions. Genetic diversity also plays a key role in disease resistance, as it reduces the likelihood that a single pest ordisease could wipe out an entire crop. This is particularly important in vegetable crops that are often grown in diverse environments, including greenhouses, fields, and urban gardens. By introducing genetic diversity through cross-pollination, breeders can develop varieties that are more adaptable and have a higher chance of thriving under varying conditions.Cross-pollination also allows for the introduction of new traits that may not be present in a self-pollinated population. For example, a breeder working with lettuce might cross a variety that has a superior flavor with one that has resistance to a particular pest. The resulting hybrid lettuce would ideally combine both the flavor and resistance traits, producing a more marketable and resilient crop. This introduction of genetic diversity helps breeders create varieties that meet both consumer demands and production challenges.

1.2.2 Hybrid Vigor (Heterosis)

One of the most well-known benefits of cross-pollination is the phenomenon of hybrid vigor, or heterosis. Hybrid vigor occurs when the offspring of two genetically distinct parent plants exhibit superior performance compared to either parent. This improvement can manifest in various ways, such as increased yield, faster growth, better resistance to diseases, or improved tolerance to environmental stresses. Hybrid vigor is a key reason why hybrid varieties often dominate the commercial seed market for many vegetable crops.In vegetable breeding, hybrid vigor is particularly important for crops like tomatoes, peppers, and cucumbers. For example, a hybrid tomato variety might have both higher fruit yield and better resistance to common diseases like Fusarium wilt, compared to either of its parent lines. The combination of traits from both parents leads to a plant that is more productive and resilient, making it highly valuable for both commercial growers and home gardeners.Hybrid vigor is also essential for ensuring consistent crop performance across different growing environments. By crossing two parent lines with complementary traits, breeders can create hybrids that perform well in a variety of conditions, from drought-prone regions to areas with high disease pressure. This versatility makes hybrid vegetables particularly attractive for large-scale commercial production, where uniformity and reliability are critical for economic success.

1.2.3 Combination of Desirable Traits

Another key advantage of cross-pollination is the ability to combine desirable traits from two different parent plants. In breeding programs, this is often referred to as trait stacking, where breeders aim to incorporate multiple beneficial traits into a single variety. For example, a breeder working with cucumbers might want to combine resistance to powdery mildew with superior fruit quality and higher yield. By crossing two parent plants that each possess one or more of these traits, the breeder can create offspring that combine all of the desired characteristics.This ability to combine traits is especially important in modern vegetable breeding, where consumer preferences and market demands are constantly evolving. For instance, consumers may prefer tomatoes with a certain flavor profile, while growers may prioritize disease resistance and shelf life. Cross-pollination allows breeders to address both of these needs by combining traits from different parent plants, resulting in a variety that meets the demands of both producers and consumers.The combination of traits through cross-pollination is also crucial for addressing the challenges posed by climate change and environmental stresses. For example, as temperatures rise and droughts become more frequent, breeders are increasingly focused on developing vegetable varieties that can tolerate heat and water stress. By crossing plants that have different tolerance levels to these stresses, breeders can create hybrids that are better equipped to handle the changing climate.

1.2.4 Challenges of Cross-Pollination

While cross-pollination offers many advantages, it also presents certain challenges for plant breeders. One of the main difficulties is the unpredictability of the offspring. Since cross-pollination involves the combination of genetic material from two different parent plants, the resulting offspring can exhibit a wide range of traits, some of which may not be desirable. This variability can make it difficult for breeders to achieve consistent results, especially in the early stages of a breeding program.

Additionally, cross-pollination can introduce unwanted traits, such as susceptibility to certain diseases or poor fruit quality. Breeders must carefully select parent plants to minimize the risk of undesirable traits being passed on to the offspring. This can be a time-consuming process, requiring multiple generations of crossing and selection to achieve the desired combination of traits.Another challenge is the need for controlled pollination. In natural settings, cross-pollination can occur through the action of wind, insects, or other pollinators, but in a breeding program, it is essential to control which plants are crossed to ensure that only the desired traits are passed on. This often requires hand-pollination or the use of controlled environments, such as greenhouses, to prevent unwanted pollination from external sources.Despite these challenges, cross-pollination remains a critical tool in plant breeding, allowing breeders to introduce genetic diversity, combine desirable traits, and produce hybrids with superior performance. The ability to control and direct the cross-pollination process is key to developing new and improved vegetable varieties that meet the needs of both growers and consumers.

2. Breeding Systems and Methods

Breeding systems and methods are the foundation of plant improvement and innovation in agriculture. These systems encompass a wide array of techniques aimed at enhancing crop performance, sustainability, and adaptability to environmental changes. Plant breeders utilize various breeding methods to achieve specific goals, such as increasing yield, improving disease resistance, enhancing nutritional content, and ensuring crop quality. The choice of breeding system and method depends largely on the crop species, breeding objectives, and the genetic architecture of the traits being improved.

2.1 Traditional Breeding Systems

Traditional breeding systems form the backbone of plant breeding, with methods that have been practiced for centuries. These systems primarily rely on natural processes such as selection and hybridization, without the use of modern biotechnological tools. The basic premise of traditional breeding is the selection of plants with desirable traits from a genetically diverse population and using these plants for further propagation.

2.1.1 Mass Selection

Mass selection is one of the oldest and simplest methods of breeding, where individual plants with desirable traits are selected on basis of phenotypic appearance from a diverse population. The seeds from these selected plants are then used to propagate the next generation. Over multiple generations, this process can lead to an increase in the frequency of the desirable trait within the population. Mass selection is particularly useful in crops where the goal is to improve traits like plant height, yield, or disease resistance.

In mass selection, no genetic control or testing of progeny is required, making it an effective method for improving crops that are naturally outcrossing, such as maize or some vegetable species. This method has been extensively used in the past to develop landraces, which are local varieties of crops that have adapted to specific environmental conditions over time. These landraces often possess a high degree of genetic diversity, making them resilient to environmental stresses.However, mass selection has limitations in achieving rapid genetic improvement. Because it does not involve controlled pollination or progeny testing, it may take several generations to fix a desirable trait in a population. Furthermore, in highly variable populations, the selection pressure may not be strong enough to achieve uniformity, especially in traits controlled by multiple genes.

2.1.2 Pure Line Selection

Pure line selection is a breeding method used primarily in self-pollinated crops, such as wheat, rice, and beans. In this method,superioe individual plants are selected based on their phenotype, and the progeny of each selected plant is grown separately to form pure lines. These pure lines are genetically uniform, as they arise from self-pollination and carry identical alleles at most loci. After several generations of self-pollination and selection, a pure line with desirable traits can be stabilized.This method is particularly effective in crops where uniformity in traits such as grain quality, disease resistance, and plant architecture is essential. For instance, pure line selection has been used to develop high-yielding varieties of wheat that are resistant to rust diseases. Because pure lines are genetically uniform, they are more predictable in their performance, making them ideal for large-scale commercial production. However, pure line selection is time-consuming, as it requires several generations of selfing to achieve genetic uniformity. Additionally, because pure lines are highly homozygous, they may lack the genetic diversity needed to respond to new environmental challenges, such as the emergence of new pests or diseases.

2.1.3 Pedigree Selection

Pedigree selection is a method that involves selecting plants based on their ancestry or pedigree, along with their phenotypic performance. In this system, breeders keep detailed records of the parent plants used in each cross and follow the performance of their offspring over several generations. By tracking the inheritance of traits through the pedigree, breeders can make informed decisions about which plants to select for further breeding. This method is widely used in crops where multiple traits are being improved simultaneously, such as in wheat, rice, and maize breeding programs. Pedigree selection allows breeders to combine traits from different parent lines, such as high yield from one parent and disease resistance from another, to create superior varieties. Because the pedigree of each plant is known, breeders can identify plants that are more likely to carry the desired combination of traits. One of the main advantages of pedigree selection is that it allows for the selection of complex traits controlled by multiple genes. By tracking the performance of progeny over multiple generations, breeders can make more precise selections, leading to faster genetic gains. However, this method requires careful record-keeping and is labor-intensive, as it involves evaluating a large number of plants over several generations.

2.2 Hybrid Breeding Systems

Hybrid breeding systems involve the crossing of two genetically distinct parent lines to produce offspring that exhibit hybrid vigor or heterosis. Hybrid breeding is widely used in both vegetable and cereal crops because hybrid plants often outperform their parents in terms of yield, disease resistance, and adaptability. The success of hybrid breeding relies on the selection of parent lines that complement each other in terms of desirable traits.

2.2.1 Development of Hybrid Varieties

The development of hybrid varieties involves several steps, starting with the creation of inbred parent lines through repeated self-pollination. These inbred lines are highly homozygous, meaning they carry identical alleles at most genetic loci. Once the inbred lines are established, they are crossed to produce hybrid seeds. The resulting hybrid plants often exhibit hybrid vigor, which can lead to increased growth rates, higher yields, and improved stress tolerance. In vegetable breeding, hybrid varieties dominate the commercial seed market for crops like tomatoes, peppers, and cucumbers. Hybrid tomatoes, for example, are known for their uniform fruit size, resistance to common diseases like Fusarium wilt, and improved shelf life. Hybrid vegetables are particularly valued by farmers because they offer consistent performance across different growing conditions, which is crucial for large-scale production. The development of hybrid varieties requires careful selection of parent lines to ensure that the desired traits are expressed in the hybrid offspring. This process can take several years, as breeders must test different combinations of parent lines to identify the best crosses. However, the benefits of hybrid vigor often outweigh the time and cost involved in developing hybrid varieties, as they offer significant advantages in terms of yield and disease resistance.

2.2.2 Three-Way and Double-Cross Hybrids

In addition to simple hybrids produced by crossing two parent lines, more complex hybrid breeding methods, such as three-way and double-cross hybrids, are used in some crops. A three-way hybrid is produced by first crossing two inbred lines to create a hybrid and then crossing this hybrid with a third inbred line. A double-cross hybrid, on the other hand, is produced by crossing two different hybrids, each of which was created by crossing two inbred lines.

Three-way and double-cross hybrids are often used in crops like maize, where the goal is to combine multiple traits from different parent lines. These more complex hybrid breeding methods allow breeders to stack traits from different parent lines, such as yield potential, disease resistance, and drought tolerance, into a single hybrid variety. Three-way and double-cross hybrids can offer even greater genetic diversity and hybrid vigor than simple hybrids, they are also more complex to develop and may require more extensive testing to ensure that the desired traits are expressed consistently in the offspring. Despite the added complexity, these hybrid breeding methods are valuable tools for developing high-performance crop varieties that can meet the demands of modern agriculture.

2.3 Modern Breeding Methods

Modern breeding methods incorporate advanced genetic and molecular tools to accelerate the breeding process and improve the precision of trait selection. These methods, such as marker-assisted selection (MAS), genomic selection, and genetic engineering, allow breeders to make more informed decisions and achieve faster genetic gains.

2.3.1 Marker-Assisted Selection (MAS)

Marker-assisted selection (MAS) is a modern breeding technique that uses molecular markers to identify plants carrying specific genes associated with desirable traits. These markers are sequences of DNA that are closely linked to the genes controlling traits like disease resistance, drought tolerance, or yield. By testing plants for the presence of these markers, breeders can select individuals with the desired genes, even before the traits are expressed phenotypically. MAS has revolutionized plant breeding by allowing breeders to make selections based on the genetic makeup of a plant, rather than relying solely on basis of phenotypic expression of observable traits. This is especially useful for traits that are difficult to evaluate in the field, such as resistance to soil-borne pathogens or tolerance to abiotic stresses like salinity or heat. By using MAS, breeders can accelerate the breeding process and reduce the number of generations needed to develop improved varieties.In vegetable breeding, MAS has been used to develop varieties of crops like tomatoes, peppers, and lettuce with improved resistance to diseases such as powdery mildew and bacterial spot. The use of molecular markers allows breeders to introduce these resistance traits into new varieties more quickly and with greater precision than traditional breeding methods.

2.3.2 Genomic Selection

Genomic selection is an advanced breeding method that uses genome-wide markers to predict the genetic potential of plants. Unlike MAS, which focuses on a few specific genes, genomic selection analyzes the entire genome of a plant to predict its performance based on the combined effect of many genes. This method is particularly useful for improving complex traits that are controlled by multiple genes, such as yield, drought tolerance, or nutritional content.In genomic selection, breeders first create a training population of plants with known genotypes and phenotypes. This population is used to develop a prediction model that can estimate the breeding value of new plants based on their genetic data. Once the model is established, breeders can use it to select plants with the highest predicted genetic potential, even before they are grown and tested in the field.Genomic selection is a powerful tool for improving crops because it allows breeders to make more accurate selections and reduces the time and cost involved in field testing. This method is particularly valuable for crops with long breeding cycles, such as fruit trees or perennial vegetables, where traditional breeding methods can take many years to produce results.

2.3.3 Genetic Engineering

Genetic engineering is a modern breeding technique that involves the direct manipulation of a plant's DNA to introduce or enhance specific traits. Unlike traditional breeding methods, which rely on crossing plants and selecting offspring over multiple generations, genetic engineering allows breeders to introduce desirable traits more quickly and precisely. This method can be used to insert genes from other species, modify existing genes, or remove undesirable traits. Genetic engineering has been instrumental in developing crops that are resistant to pests, diseases, and environmental stresses, as well as improving nutritional content.In vegetable breeding, genetic engineering has led to significant advancements in crops like tomatoes, potatoes, and eggplants. For instance, genetically modified (GM) tomatoes have been developed with enhanced shelf life and resistance to viral diseases. Similarly, genetically engineered potatoes resistant to late blight have been created, reducing the need for chemical pesticides. Genetic engineering allows breeders to introduce traits that are difficult or impossible to achieve through conventional breeding methods, such as resistance to specific pathogens or the ability to grow in saline soils.

A well-known example of genetic engineering in vegetables is the development of *Bt* crops, which contain a gene from the bacterium *Bacillus thuringiensis* that produces a protein toxic to certain insect pests. This trait has been widely used in crops like eggplants and corn to reduce insect damage and the need for chemical insecticides. Genetic engineering also enables the development of crops with enhanced nutritional profiles, such as biofortified vegetables that contain higher levels of essential vitamins and minerals. For example, genetically modified tomatoes have been developed with higher levels of lycopene, an antioxidant linked to reduced risks of certain diseases.Although genetic engineering offers tremendous potential for crop improvement, it is also subject to strict regulatory oversight and public scrutiny. Concerns about the safety of genetically modified organisms (GMOs) and their potential environmental impacts have led to stringent approval processes in many countries. Nonetheless, genetic engineering remains a powerful tool in modern plant breeding, allowing for the development of crops that can better withstand the challenges of modern agriculture, including climate change, pest pressures, and food security concerns.

2.3.4 CRISPR-Cas9 Technology

CRISPR-Cas9 is one of the most revolutionary developments in modern plant breeding. It is a genome-editing technology that allows breeders to make precise changes to a plant's DNA by cutting and modifying specific genes. Unlike traditional genetic engineering, which often involves introducing genes from other species, CRISPR-Cas9 can be used to edit a plant's own genome, making it a more natural and targeted approach to improving crops. The technology allows breeders to target specific genes responsible for traits like disease resistance, drought tolerance, or fruit quality and make precise modifications to enhance these traits.In vegetable breeding, CRISPR-Cas9 has been used to develop crops with improved resistance to diseases and pests, as well as enhanced nutritional content. For instance, CRISPR has been employed to edit the genes of tomatoes, making them more resistant to fungal diseases like powdery mildew. Similarly, the technology has been used to enhance the flavor and nutritional content of crops like lettuce and cucumbers. CRISPR-Cas9 offers a faster and more efficient way to introduce desirable traits compared to traditional breeding methods, as it eliminates the need for multiple generations of crossing and selection.

Another significant advantage of CRISPR-Cas9 is its ability to create crops that are free from transgenes, meaning that the edited plants do not contain foreign DNA. This addresses some of the public concerns associated with genetically modified organisms (GMOs), as CRISPR-edited crops are seen as more "natural" because they only involve changes to the plant's existing genome. As a result, CRISPR-Cas9 is increasingly being used in breeding programs for vegetables like peppers, eggplants, and potatoes to improve traits such as yield, disease resistance, and stress tolerance. CRISPR technology has the potential to revolutionize vegetable breeding by allowing for faster and more targeted improvements in crop traits. As technology continues to evolve, it is expected to play a central role in addressing the challenges posed by climate change, food security, and the need for sustainable agricultural practices. The precise nature of CRISPR-Cas9 editing makes it an ideal tool for breeding programs aimed at improving complex traits controlled by multiple genes, such as drought tolerance, disease resistance, and nutritional content.

3. Hybridization (Intra and Inter Specific Hybridization)

Selection procedures and hybridization are fundamental processes used by plant breeders to improve crop varieties, intending to enhance specific traits such as yield, disease resistance, stress tolerance, or quality. Selection procedures involve identifying and selecting individuals with favorable traits, while hybridization is the process of crossing two genetically distinct parent plants to create offspring with a combination of desirable characteristics from both parents. These methods are widely applied in both traditional and modern breeding programs and form the basis for genetic improvements in crop production.

3.1 Hybridization

Hybridization is a powerful breeding technique that involves crossing two genetically distinct parent plants to produce offspring with a combination of desirable traits from both parents. This method is widely used in both self-pollinated and cross-pollinated crops to create hybrids with enhanced characteristics, such as higher yield, improved disease resistance, or greater adaptability to environmental stresses. Hybridization can be classified into different types depending on the genetic relationship between the parent plants and the goals of the breeding program.

3.1.1 Intraspecific Hybridization

Intraspecific hybridization involves the crossing of two individuals from the same species but with different genetic backgrounds. This type of hybridization is commonly used in crops like maize, tomatoes, and peppers, where hybrid varieties are favored for their superior performance compared to open-pollinated varieties. Intraspecific hybridization allows breeders to combine complementary traits from both parents, resulting in offspring that are more vigorous, productive, and resilient.

For example, in maize breeding, intraspecific hybridization is used to combine traits such as drought tolerance, pest resistance, and high yield. The resulting hybrid plants often exhibit "hybrid vigor" or heterosis, which refers to the enhanced growth and productivity of the hybrid compared to its parents. This phenomenon is especially important in commercial agriculture, where hybrid maize varieties are widely grown for their consistent performance under different environmental conditions. Hybrid vigor is one of the main advantages of intraspecific hybridization, particularly in vegetable breeding. For instance, hybrid tomato varieties are known for their uniform fruit size, resistance to common diseases like Fusarium wilt, and improved shelf life. These traits make hybrid vegetables attractive to both growers and consumers, as they offer better market quality and higher yields. However, the production of hybrid seeds requires controlled pollination and careful selection of parent lines, which can be resource-intensive and costly.

3.1.2 Interspecific Hybridization

Interspecific hybridization involves crossing individuals from different species to introduce novel traits that may not be present within a single species. This method is less common than intraspecific hybridization but has been successfully used in some crops to introduce traits such as disease resistance, drought tolerance, or improved quality characteristics. Interspecific hybrids can provide valuable genetic diversity, especially for breeding programs focused on improving the adaptability of crops to changing environmental conditions.

A notable example of interspecific hybridization is the development of hybrid wheat varieties by crossing cultivated wheat species with wild relatives. These hybrids can introduce new genes for traits such as rust resistance, which may not be present in domesticated wheat.

Similarly, in vegetable breeding, interspecific hybridization has been used to introduce traits like resistance to specific pathogens or improved fruit quality in crops like tomatoes and eggplants. Despite the potential benefits, interspecific hybridization presents challenges due to genetic incompatibility between species. Hybrids produced from such crosses may have reduced fertility or vigor, and breeders may need to backcross them to the parent species to eliminate unwanted traits while retaining the desired characteristics. This process can be time-consuming and requires careful management to ensure the stability and performance of the resulting hybrids.

4. Genetic Architecture

Genetic architecture refers to the comprehensive genetic basis that underpins the variation observed in a plant's traits (phenotypes). It encompasses the number of genes influencing a particular trait, the location of these genes on the chromosomes, the magnitude of their effects, the interactions between genes (epistasis), and how these genetic elements interact with environmental factors. The concept of genetic architecture is crucial in plant breeding because it helps breeders understand how traits are inherited and how they can manipulate these traits to improve crop performance. In crop breeding, a thorough understanding of genetic architecture provides insights into the complexity of traits such as yield, disease resistance, plant morphology, and stress tolerance. This understanding helps breeders to predict how traits will respond to selection and to design effective breeding programs that maximize genetic gains. The genetic architecture of a plant can vary significantly depending on whether the traits in question are controlled by a single gene (monogenic traits) or multiple genes (polygenic traits). Breeding strategies are adjusted accordingly to manage the complexity of trait inheritance.

4.1 Monogenic Traits

Monogenic traits, also referred to as simply inherited traits, are traits that are controlled by a single gene or a very small number of genes. These traits typically follow Mendelian inheritance patterns, meaning that they are inherited according to the laws first described by Gregor Mendel in his experiments with peas. Monogenic traits are often easier to work with in plant breeding because they show clear patterns of inheritance, and their genetic control is relatively simple.

4.1.1 Mendelian Inheritance of Monogenic Traits

Monogenic traits exhibit Mendelian inheritance, which is characterized by predictable ratios of trait expression in the progeny based on the genotype of the parents. Mendel's laws of segregation and independent assortment describe how alleles (different versions of a gene) segregate during gamete formation and how they independently assort to create genetic variation in the offspring. In the context of genetic architecture, Mendelian inheritance allows breeders to easily track and select desirable monogenic traits, such as flower color, seed shape, or resistance to specific pathogens. For example, in tomato (*Solanum lycopersicum*), a single gene controls whether the fruit is resistant to the *Verticillium* wilt pathogen. This resistance gene can be selected and fixed within a breeding population using traditional selection methods, such as backcrossing or marker-assisted selection. Similarly, in rice (*Oryza sativa*), the *Xa21* gene confers resistance to bacterial blight, a serious disease that affects rice crops. The single-gene nature of these resistances makes them easier to incorporate into breeding programs compared to polygenic traits.

However, despite the simplicity of working with monogenic traits, their impact on crop improvement is somewhat limited because most economically important traits in crops are polygenic, meaning they are controlled by multiple genes. Monogenic traits are often qualitative in nature, meaning they result in distinct phenotypes that can be classified into discrete categories, such as resistant vs. susceptible or red vs. white flowers.

4.1.2 Advantages of Working with Monogenic Traits in Breeding

One of the main advantages of working with monogenic traits in plant breeding is the ease with which these traits can be identified, selected, and fixed in a population. Because monogenic traits follow predictable inheritance patterns, breeders can make rapid genetic progress by selecting individuals that carry the desired alleles. This is especially true when molecular markers are available for the gene of interest, allowing breeders to use marker-assisted selection (MAS) to track the gene through the breeding process.For instance, marker-assisted selection has been used to introgress (introduce) disease-resistance genes into commercial crop varieties. In wheat (*Triticum aestivum*), breeders have successfully used markers linked to the *Lr34* gene, which confers durable resistance to leaf rust, to develop resistant varieties. This approach allows breeders to accurately select the resistance gene without having to rely on visual assessments of disease symptoms, which can be affected by environmental conditions.

In addition to disease resistance, monogenic traits are also useful in improving traits related to plant morphology and quality. For example, in cucumbers (*Cucumis sativus*), the gene responsible for the bitter-free characteristic in fruit is monogenic. Breeders can select for this trait to produce cucumber varieties that are more palatable and marketable. However, the simplicity of monogenic traits also presents limitations. Many important traits, such as yield, drought tolerance, and nutritional content, are controlled by multiple genes and are influenced by complex interactions between these genes and the environment. Therefore, while monogenic traits are valuable for improving specific characteristics, they represent only one aspect of the broader genetic architecture that governs crop performance.

4.2 Polygenic Traits

Polygenic traits, also known as quantitative traits, are controlled by multiple genes, each of which has a small additive effect on the overall phenotype. These traits do not follow simple

Mendelian inheritance patterns and are influenced by both genetic and environmental factors. Polygenic traits are typically continuous in their expression, meaning that they result in a range of phenotypes rather than discrete categories. Examples of polygenic traits in crops include yield, plant height, drought tolerance, and nutrient efficiency.

4.2.1 Complexity of Polygenic Trait Inheritance

The inheritance of polygenic traits is much more complex than that of monogenic traits because the phenotype is influenced by the combined effects of multiple genes, each contributing a small amount to the overall trait expression. These genes are often located at different loci across the genome, and their effects can be additive or interact in non-additive ways. This complexity makes it more challenging for breeders to predict the outcome of crosses and select individuals with the best combination of alleles.

For example, in maize (*Zea mays*), yield is a highly polygenic trait that is controlled by numerous genes involved in processes such as photosynthesis, nutrient uptake, and stress tolerance. Improving yield through traditional selection methods can be slow because each gene contributes only a small amount to the overall phenotype. However, advances in molecular breeding techniques, such as genomic selection, have made it possible to accelerate the breeding process by analyzing the entire genome and selecting favorable combinations of alleles that contribute to yield improvement. Polygenic traits are also influenced by environmental factors, which further complicate their inheritance. For instance, drought tolerance in plants is controlled by a large number of genes involved in water uptake, stomatal regulation, and stress response pathways. The expression of these genes can vary depending on the environmental conditions, making it difficult to predict how a plant will perform under different stress scenarios. This interaction between genes and the environment is known as genotype-by-environment interaction (GxE), and it is a critical consideration in the breeding of polygenic traits.

4.2.2 Breeding Strategies for Polygenic Traits

Given the complexity of polygenic traits, breeders must use different strategies compared to those used for monogenic traits. One common approach is recurrent selection, which involves selecting individuals with the best performance in each generation and inter-crossing them to increase the frequency of favorable alleles in the population. Recurrent selection is particularly useful for improving traits that are difficult to assess visually, such as nutrient efficiency or resistance to multiple diseases. Marker-assisted selection (MAS) and genomic selection (GS) are also valuable tools for improving polygenic traits. MAS involves using molecular markers linked to specific genes that control a portion of the trait, while GS uses genome-wide markers to predict the overall genetic value of an individual based on the combined effect of all genes. These techniques have been used to improve complex traits in crops like rice, wheat, and tomatoes.

For instance, in rice (*Oryza sativa*), breeders have used MAS to improve traits such as yield, grain quality, and resistance to diseases like blast and bacterial blight. By selecting individuals with favourable alleles for multiple traits, breeders can develop varieties that combine high yield with improved resistance to biotic and abiotic stresses.Genomic selection takes this a step further by using high throughput genotyping to analyse thousands of markers across the genome. This approach allows breeders to make selections based on the genetic potential of individuals rather than relying solely on phenotypic evaluations. Genomic selection has been particularly successful in improving yield and drought tolerance in maize, where multiple genes contribute to the trait.

4.2.3 Gene Interactions and Epistasis

In addition to the additive effects of multiple genes, polygenic traits are also influenced by gene interactions, known as epistasis. Epistasis occurs when the expression of one gene is modified by one or more other genes. These interactions can complicate the breeding process because the effect of a gene may depend on the presence or absence of other genes in the genome.

For example, in wheat, certain genes that control grain size may interact with genes involved in nutrient uptake or disease resistance. The overall phenotype depends on how these genes work together, and the interaction between them can vary depending on the environment. Understanding epistatic interactions is crucial for breeders who aim to improve polygenic traits, as selecting for individual genes without considering their interactions may not lead to the desired outcomes.Modern breeding technologies, such as quantitative trait locus (QTL) mapping and genome-wide association studies (GWAS), have helped breeders identify epistatic interactions and understand how they contribute to complex traits. By mapping the loci that control these interactions, breeders can develop more effective strategies for improving polygenic traits in crops.

5. Breeding for Biotic Stress (Diseases, Insect Pests, and Nematodes)

Breeding for biotic stress resistance is a fundamental aspect of plant breeding, aimed at improving crop resilience against various biological threats such as diseases, insect pests, and nematodes. Biotic stresses pose significant challenges to agriculture, as they can drastically reduce crop yields, compromise quality, and increase production costs due to the need for chemical interventions. The goal of breeding for biotic stress resistance is to develop crop varieties that can naturally withstand these challenges, minimizing the need for pesticides and other inputs while improving productivity and sustainability.Biotic stress resistance is often controlled by a combination of genetic and environmental factors. Resistance can be conferred by specific genes that recognize and defend against pathogens, insects, and nematodes. Breeding efforts typically focus on introducing and stacking these resistance genes into elite varieties using conventional breeding, marker-assisted selection (MAS), and modern biotechnological tools like genetic engineering and genome editing. These breeding strategies are critical for developing crops that can endure biotic stress pressures in diverse agroecological environments.

5.1 Breeding for Disease Resistance

Diseases caused by fungi, bacteria, viruses, and other pathogens are among the most serious biotic stresses in crop production. Plant breeders have long focused on breeding for disease resistance as a way to reduce yield losses and ensure food security. Breeding for disease resistance can be achieved through the identification and incorporation of resistance genes, which can provide either vertical or horizontal resistance. Vertical resistance is usually controlled by single genes and provides strong, specific protection against a particular pathogen, whereas horizontal resistance involves multiple genes and offers more general, but often less intense, protection against a broad spectrum of diseases.

5.1.1 Vertical Resistance

Vertical resistance, also known as specific resistance, is controlled by single resistance (R) genes that recognize and respond to specific pathogen strains. When a plant with a vertical resistance gene encounters the corresponding pathogen, a hypersensitive response is triggered, leading to localized cell death at the site of infection and preventing the pathogen from spreading. This type of resistance is highly effective but is often race-specific, meaning it only protects against certain pathogen strains. Over time, pathogens can evolve to overcome vertical resistance by mutating and avoiding recognition by the plant's R genes.

In plant breeding, vertical resistance has been widely used to develop crop varieties resistant to particular diseases. For example, in potato (*Solanum tuberosum*), R_1 gene confers resistance to late blight of potato, in wheat (*Triticum aestivum*), the *Lr34* gene confers resistance to leaf rust, a fungal disease that can cause significant yield losses. Similarly, the *Xa21* gene in rice (*Oryza sativa*) provides resistance to bacterial blight, a major bacterial disease affecting rice production in many parts of the world. The single-gene nature of vertical resistance makes it relatively easy to incorporate into breeding programs through traditional backcrossing or marker-assisted selection.One of the advantages of vertical resistance is its strong and specific protection against the target pathogen, often leading to complete resistance. This makes it a valuable tool for managing diseases that have severe impacts on crop yields. However, the main drawback is that vertical resistance is vulnerable to breakdown due to pathogen evolution. As pathogens mutate, they can overcome the R genes, rendering the resistance ineffective. To mitigate this risk, breeders often combine multiple R genes (gene pyramiding) to create more durable resistance.

5.1.2 Horizontal Resistance

Horizontal resistance, also known as quantitative or partial resistance, is controlled by multiple genes, each contributing a small effect to the overall resistance. Unlike vertical resistance, horizontal resistance provides broader protection against a range of pathogen strains, making it less vulnerable to breakdown by evolving pathogens. However, the level of protection is typically not as strong as with vertical resistance, and plants with horizontal resistance may still experience some level of disease infection. Horizontal resistance is often more durable than vertical resistance because it does not rely on a single gene and is harder for pathogens to overcome. This type of resistance is especially important for managing diseases that are caused by highly variable pathogens, such as rust, powdery mildew, and late blight. For example, in potatoes (*Solanum tuberosum*), breeders have developed varieties with horizontal resistance to late blight, a devastating fungal disease caused by *Phytophthora infestans*. This resistance helps reduce the severity of the disease, although it may not provide complete immunity. Breeding for horizontal resistance typically involves selecting for polygenic traits, which can be more challenging than selecting for single-gene traits. Marker-assisted selection and genomic selection are valuable tools for identifying and selecting multiple resistance genes in breeding populations. By combining several minor genes that each contribute to disease resistance, breeders can create varieties with more durable and stable resistance, even in the face of evolving pathogens.

5.1.3 Gene Pyramiding

Gene pyramiding is a breeding strategy used to enhance disease resistance by combining multiple R genes into a single variety. This approach is designed to improve the durability of resistance by making it more difficult for pathogens to overcome the plant's defenses. By stacking several resistance genes, each targeting different pathogen strains or mechanisms, breeders can create varieties that are more resilient to a wide range of diseases. For instance, in cauliflower breeding, black rot and downy mildew are major devastating diseases reducing the yield and quality of the crop. To tackle these through host plant resistance, a marker-assisted backcross breeding method was followed to pyramid a black rot-resistant gene (*Xca1bo*) and a downy mildew-resistant gene (*Ppa3*) from donors BR-161 and BR-2, respectively, into the background of Pusa Meghna cauliflower cultivar. in rice breeding, gene pyramiding has been used to develop varieties with enhanced resistance to bacterial blight by combining several *Xa* genes, each providing resistance to different strains of the pathogen. Similarly, in wheat breeding, multiple rust resistance genes (e.g., *Lr*, *Sr*, and *Yr* genes) have been pyramided to create varieties that are resistant to leaf rust, stem rust, and stripe rust. The success of gene pyramiding relies on the availability of molecular markers linked to the resistance genes. The marker-assisted selection allows breeders to track the presence of each gene during the breeding process, ensuring that the final variety carries all of the desired resistance genes. While gene pyramiding can provide more robust and durable resistance, it is a time-consuming process that requires careful planning and multiple generations of breeding to achieve the desired results.

5.2 Breeding for Insect Pest Resistance

Insect pests pose a significant threat to crop production, causing direct damage to plants by feeding on leaves, stems, and fruits, as well as transmitting diseases. Breeding for insect pest resistance aims to develop crop varieties that can withstand or repel insect attacks, reducing the need for chemical insecticides and improving crop sustainability. Insect resistance can be conferred by specific genes that produce physical or chemical defenses, or by genes that affect the plant's ability to tolerate or recover from insect damage.

5.2.1 Host Plant Resistance

Host plant resistance refers to the natural ability of a plant to resist or tolerate insect pest attacks. This type of resistance can be conferred by various mechanisms, including physical barriers (such as thickened cell walls or trichomes) or chemical defenses (such as secondary metabolites that deter insects). Host plant resistance can be either constitutive, meaning it is always present in the plant, or inducible, meaning it is activated in response to insect feeding. For example, in tomatoes, a large number of biochemical and morphological characters have been implicated in resistance to tomato pests, including trichomes and trichome-associated compounds, resistance-related proteins, and a diverse assortment of secondary metabolites. Wild tomatoes often possess higher levels of these traits and thus demonstrate important genetic resources for resistance breeding in cotton (*Gossypium hirsutum*), breeders have developed varieties with resistance to bollworms, a major insect pest. These resistant varieties produce higher levels of gossypol, a chemical compound that is toxic to bollworms and deters their feeding. Similarly, in maize (*Zea mays*), breeders have selected for resistance to corn borers by identifying varieties with thicker stalks, which make it more difficult for the insects to penetrate and damage the plant. Breeding for host plant resistance is an important strategy for reducing reliance on chemical insecticides, which can have negative environmental and health impacts. By developing insect-resistant varieties, breeders can help farmers achieve more sustainable pest management while maintaining crop productivity. However, like disease resistance, insect resistance can be overcome by evolving pest populations, requiring ongoing breeding efforts to maintain effectiveness.

5.2.2 Transgenic Approaches to Insect Resistance

In addition to traditional breeding methods, genetic engineering has been used to introduce insect resistance into crops. One of the most well-known examples of transgenic insect resistance is the development of *Bt* crops, which contain genes from the bacterium *Bacillus thuringiensis*. These genes produce proteins that are toxic to specific insect pests, such as caterpillars and beetles but are harmless to humans and other non-target organisms. *Bt* crops, such as *Bt* maize and *Bt* cotton, have been widely adopted in many parts of the world due to their effectiveness in controlling major insect pests. The introduction of *Bt* genes into these crops has significantly reduced the need for chemical insecticides, leading to environmental and economic benefits. For instance, *Bt* cotton has been highly successful in reducing bollworm infestations, leading to higher yields and reduced pesticide use in cotton-producing regions. However, the widespread use of *Bt* crops has raised concerns about the potential for pests to develop resistance to the *Bt* proteins. To mitigate this risk, breeders and regulators have implemented strategies such as refuge planting, where non-*Bt* plants are grown alongside *Bt* crops to reduce the selection pressure for resistance in pest populations. In addition, new *Bt* genes are being developed and introduced into crops to provide multiple modes of action against insect pests, further reducing the likelihood of resistance development.

5.3 Breeding for Nematode Resistance

Nematodes are microscopic roundworms that can cause significant damage to crops by feeding on plant roots and interfering with nutrient uptake. Breeding for nematode resistance is a key component of integrated pest management (IPM) strategies aimed at reducing the impact of these pests on crop productivity. Resistance to nematodes can be conferred by specific genes that limit nematode reproduction or inhibit their ability to establish feeding sites in the plant's roots.

5.3.1 Resistance to Root-Knot Nematodes

Root-knot nematodes (*Meloidogyne* spp.) are among the most destructive nematode pests, causing root galls that impair the plant's ability to take up water and nutrients. This leads to stunted growth, reduced yield, and, in severe cases, plant death. Breeding for resistance to root-knot nematodes is critical in crops like tomatoes, peppers, and cucumbers, where these pests are prevalent. Resistance to root-knot nematodes is often conferred by specific resistance genes, such as the *Mi-1* gene in tomatoes (*Solanum lycopersicum*). The *Mi-1* gene provides effective resistance to several species of root-knot nematodes, preventing them from establishing feeding sites in the roots. Plants carrying the *Mi-1* gene exhibit little to no gall formation, which helps maintain root health and overall plant vigor. The identification of resistance genes like *Mi-1* has allowed breeders to incorporate nematode resistance into commercial tomato varieties, reducing the need for chemical nematicides. Breeding for nematode resistance is particularly challenging because nematode populations are highly variable, and some species can evolve to overcome plant defenses. To address this, breeders often stack multiple resistance genes to increase the durability of nematode resistance. Marker-assisted selection (MAS) is used to track these genes in breeding populations, ensuring that resistant varieties carry the desired genetic traits. In addition to single-gene resistance, quantitative resistance to root-knot nematodes has also been identified in crops like sweet potatoes and beans. This type of resistance is controlled by multiple genes, each contributing to the plant's overall ability to tolerate nematode infections. Although quantitative resistance may not provide complete immunity, it helps reduce nematode reproduction and minimizes damage to the plant.

5.3.2 Resistance to Cyst Nematodes

Cyst nematodes, including species such as soybean cyst nematode (*Heteroderaglycines*) and potato cyst nematode (*Globodera pallida*), are also major pests of crops. These nematodes form cysts on the roots of their host plants, disrupting nutrient uptake and causing significant yield losses. Breeding for resistance to cyst nematodes is a priority in crops like soybeans and potatoes, where infestations can lead to severe economic losses.Resistance to cyst nematodes is often polygenic, meaning that multiple genes contribute to the plant's ability to resist nematode infection. In soybeans (*Glycine max*), resistance to soybean cyst nematode is conferred by a combination of resistance genes, including *Rhg1* and *Rhg4*. These genes prevent the nematodes from successfully establishing feeding sites, reducing their ability to reproduce and minimizing root damage. Soybean breeders have used marker-assisted selection to introgress these resistance genes into commercial soybean varieties, resulting in improved yield and nematode resistance.

In potato breeding, resistance to cyst nematodes is controlled by major resistance genes such as *H1*, which protects*Globoderarostochiensis*. The development of potato varieties with resistance to cyst nematodes has been a major focus in regions where these pests are endemic, helping to reduce yield losses and the need for chemical control measures.One of the challenges in breeding for cyst nematode resistance is the genetic variability of nematode populations. Like root-knot nematodes, cyst nematodes can evolve to overcome plant resistance, making it necessary for breeders to continually monitor nematode populations and develop new resistance strategies. This ongoing effort often involves identifying new resistance genes and incorporating them into existing breeding programs through gene pyramiding and marker-assisted selection.

5.3.3 Breeding for Nematode Tolerance

In addition to breeding for resistance, breeders also focus on developing crop varieties that can tolerate nematode infections without suffering significant yield losses. Tolerance is the ability of a plant to endure nematode infestations and continue growing and producing, even in the presence of the pest. Tolerant plants may not prevent nematodes from infecting their roots, but they can withstand the damage and maintain productivity.

Breeding for nematode tolerance is particularly important in regions where nematode infestations are widespread and complete resistance is difficult to achieve. In crops like bananas (*Musa spp.*), for example, breeders have developed varieties that can tolerate high populations of root-knot nematodes without experiencing severe yield reductions. These tolerant varieties are valuable in integrated pest management (IPM) systems, where they are used in combination with cultural practices, crop rotation, and biological control to manage nematode populations. Tolerance breeding involves selecting plants that exhibit minimal yield losses despite nematode presence. In some cases, this tolerance is due to the plant's ability to regenerate new roots after nematode damage or its ability to limit the impact of nematode feeding on nutrient uptake. By selecting for these traits over multiple generations, breeders can develop varieties that are better equipped to handle nematode infestations under field conditions.

6. Abiotic Stress Resistance (Temperature, Moisture, and Salt) and Quality Improvement

Abiotic stresses, such as extreme temperatures, moisture stress (drought and flooding), and soil salinity, are major environmental factors that negatively affect crop productivity. These stresses can result in significant yield losses and impact the overall quality of the produce. With the increasing incidence of climate change, breeding for resistance to abiotic stresses has become a critical focus area in plant breeding programs worldwide. In addition to improving stress resistance, breeders also aim to maintain or enhance the quality of the crop, ensuring that improvements in stress tolerance do not compromise the market value or nutritional quality of the produce. Abiotic stress resistance is typically controlled by complex polygenic traits, making it more challenging to breed compared to single-gene traits. Advances in molecular breeding, marker-assisted selection (MAS), genomic selection (GS), and biotechnological tools such as genetic engineering and CRISPR-Cas9 genome editing have enabled breeders to make significant progress in developing stress-resistant varieties. These efforts not only enhance the resilience of crops but also help in maintaining or improving the quality of agricultural produce.

6.1 Temperature Stress Resistance

Temperature extremes, whether too high or too low, can severely affect plant growth, development, and yield. High temperatures can cause heat stress, leading to reduced photosynthesis, pollen sterility, and poor fruit set, while low temperatures can result in chilling or frost damage, impacting seed germination and vegetative growth. Breeding for temperature stress resistance aims to develop crops that can maintain productivity and quality under both heat and cold stress conditions.

6.1.1 Heat Stress Resistance

Heat stress occurs when temperatures exceed the optimal range for crop growth, causing damage to cellular structures, reducing photosynthesis, and leading to premature senescence. Crops such as maize, wheat, and rice are particularly vulnerable to heat stress during critical growth stages like flowering and grain filling. Heat stress can also impair pollen viability and reduce seed set, leading to lower yields. Breeding for heat stress resistance involves identifying and selecting genotypes that can tolerate high temperatures without compromising yield or quality. Heat-tolerant varieties often exhibit traits such as increased leaf cooling capacity through enhanced transpiration, improved membrane stability, and the ability to maintain photosynthesis at higher temperatures. For example, in wheat (*Triticum aestivum*), breeders have developed heat-tolerant varieties that maintain grain filling and yield under elevated temperatures.

Molecular breeding approaches, including quantitative trait locus (QTL) mapping and marker-assisted selection, have been used to identify genes associated with heat tolerance.

For instance, the *heat shock protein (HSP)* genes play a key role in protecting plants from heat stress by stabilizing proteins and membranes under high temperatures. Breeders have used markers linked to HSP genes to select heat tolerance in crops like rice (*Oryza sativa*) and tomatoes (*Solanum lycopersicum*). In addition to maintaining yield, quality traits such as fruit size, flavor, and nutritional content must be preserved in heat-tolerant varieties. In tomato breeding, for example, the challenge is to develop varieties that can set fruit under high temperatures without compromising fruit quality. This involves selecting for traits such as improved pollen viability and fruit set under heat stress, as well as maintaining desirable fruit characteristics like colour, texture, and taste.

6.1.2 Cold and Frost Resistance

Cold stress, which includes chilling (0°C to 15°C) and freezing (below 0°C) temperatures, can severely damage crops by disrupting cellular processes, leading to poor germination, stunted growth, and tissue damage. Crops like rice, maize, and tomatoes are particularly sensitive to low temperatures, while temperate crops such as wheat and barley have more natural tolerance to cold conditions. Breeding for cold and frost resistance aims to develop varieties that can withstand low temperatures without experiencing significant damage. Cold-tolerant varieties typically exhibit traits such as improved cell membrane integrity, accumulation of osmoprotectants (compounds that protect cells against freezing), and increased expression of antifreeze proteins. In crops like wheat, cold tolerance is critical during the early stages of growth, as low temperatures can affect seedling establishment and tillering. Breeders have developed cold-tolerant wheat varieties by selecting for traits such as enhanced root growth, early vigor, and the ability to accumulate soluble sugars, which act as cryoprotectants to protect cells from freezing.

Frost resistance is particularly important in fruit crops such as apples (*Malus domestica*) and grapes (*Vitis vinifera*), where late spring frosts can damage flowers and reduce fruit set. Breeding for frost resistance in these crops involves selecting late-flowering varieties or varieties that can tolerate freezing temperatures without damage to the flowers or young fruits. Breeding for cold stress resistance also focuses on maintaining crop quality. For example, in vegetables like lettuce (*Lactuca sativa*), cold-tolerant varieties must retain their crisp texture and flavor even when grown under low temperatures. This requires breeders to carefully balance traits related to stress tolerance with those that influence marketable quality.

6.2 Moisture Stress Resistance (Drought and Flooding)

Moisture stress, which includes both drought and flooding, is a major limiting factor for crop production in many parts of the world. Drought stress occurs when water availability is limited, leading to reduced plant growth, poor seed set, and lower yields. Flooding, on the other hand, results in oxygen deprivation in the root zone, which can lead to root damage and plant death. Breeding for moisture stress resistance aims to develop crops that can tolerate both water-deficit and water-excess conditions.

6.2.1 Drought Resistance

Drought resistance is a complex trait controlled by multiple genes, and it involves a variety of physiological and morphological mechanisms that allow plants to survive and produce under limited water availability. Drought-resistant varieties often exhibit traits such as deep root systems for accessing water, reduced stomatal conductance to minimize water loss, and increased water-use efficiency. In crops like maize (*Zea mays*) and wheat (*Triticum aestivum*), breeding for drought resistance has been a major focus due to the increasing frequency of droughts in key production regions. Drought-tolerant maize varieties have been developed by selecting for traits such as root architecture that allow the plant to access deeper soil moisture and maintain yield under water-limited conditions.

Molecular breeding techniques, such as marker-assisted selection and genomic selection, have been used to identify and select drought-resistance genes in various crops. For example, the *DREB (dehydration-responsive element-binding)* genes play a key role in regulating drought response pathways in plants. By using markers linked to these genes, breeders can select drought-tolerant lines that exhibit better water-use efficiency and improved yield stability under drought conditions. In addition to yield stability, maintaining or improving crop quality under drought stress is a key challenge for breeders. Drought stress can negatively affect fruit size, texture, and flavor in crops like tomatoes and melons. Therefore, breeding programs focus not only on enhancing drought tolerance but also on preserving quality traits that are important for market acceptance.

6.2.2 Flooding Tolerance

Flooding, or waterlogging, is another major abiotic stress that affects crops, particularly in low-lying areas and regions prone to heavy rainfall. Flooding deprives plant roots of oxygen, leading to root rot, reduced nutrient uptake, and, ultimately, plant death. Crops like rice, which are adapted to wet conditions, have evolved mechanisms to tolerate flooding, while most upland crops, such as maize and wheat, are highly sensitive to waterlogging.

Breeding for flooding tolerance focuses on selecting for traits that allow plants to survive and recover from extended periods of waterlogging. In rice, breeders have developed submergence-tolerant varieties that can survive complete submergence for several days. For example, the *Sub1* gene, which confers tolerance to submergence, has been introduced into high-yielding rice varieties through marker-assisted selection, enabling farmers in flood-prone areas to achieve better yields.For non-rice crops, flooding tolerance is often achieved by selecting for traits such as improved root porosity, which enhances oxygen diffusion to the roots, or the ability to form adventitious roots that grow above the waterlogged soil. In maize, for example, breeders have selected for varieties that can tolerate short-term waterlogging by developing stronger root systems and mechanisms for maintaining nutrient uptake under flooded conditions. Maintaining crop quality under flooding conditions is also a priority for breeders. In crops like potatoes and sweet potatoes, flooding can lead to poor tuber quality, affecting both yield and marketability. Breeding efforts focus on developing varieties that can tolerate waterlogging while maintaining desirable tuber characteristics such as size, texture, and storage quality.

6.3 Salt Stress Resistance

Soil salinity is a growing problem in many agricultural regions, particularly in arid and semi-arid areas where irrigation practices can lead to the accumulation of salts in the soil. High salt concentrations in the soil disrupt plant water uptake and ion balance, leading to reduced growth, leaf chlorosis, and lower yields. Salt stress can also affect the quality of the produce, leading to poor marketability. Breeding for salt stress resistance aims to develop varieties that can tolerate high soil salinity without compromising yield or quality.

6.3.1 Mechanisms of Salt Tolerance

In crops like rice, wheat, and barley, salt tolerance is controlled by multiple genes, each contributing to the plant's ability to withstand saline conditions. Salt-tolerant varieties typically exhibit traits such as the ability to regulate ion uptake and compartmentalize excess sodium (Na^+) and chloride (Cl^-) ions into vacuoles, preventing them from accumulating in the cytoplasm where they can interfere with cellular processes. This process helps maintain cellular function and water balance under high-salinity conditions.

Breeding for salt tolerance has been particularly important in regions where irrigation-induced salinity is a significant problem. For example, in rice (*Oryza sativa*), breeders have developed salt-tolerant varieties by selecting for traits such as improved sodium exclusion and enhanced root growth under saline conditions. The *Saltol* QTL (quantitative trait locus) is one such genetic marker that has been used to breed salt-tolerant rice varieties, helping farmers grow rice in coastal areas and other salt-affected regions. In addition to selecting physiological traits that enhance salt tolerance, breeders also focus on maintaining the quality of the crop. For example, in tomatoes, salinity can affect fruit quality by altering flavor, texture, and sugar content. Therefore, breeding programs aim to develop salt-tolerant varieties that can maintain desirable fruit quality traits while growing under saline conditions. This involves selecting for plants that can accumulate solutes like sugars and organic acids in the fruits, enhancing their taste and marketability even when grown in salty soils.

6.3.2 Marker-Assisted Breeding for Salt Tolerance

Marker-assisted breeding has become an important tool in developing salt-tolerant crop varieties. By identifying and using molecular markers linked to genes or QTLs associated with salt tolerance, breeders can more efficiently select salt-tolerant genotypes. This approach speeds up the breeding process and increases the precision of selecting for complex traits like salt tolerance, which are often controlled by multiple genes.

For example, in wheat (*Triticum aestivum*), breeders have identified QTLs linked to salt tolerance, such as those controlling sodium exclusion and osmotic regulation. These QTLs have been used in marker-assisted breeding programs to develop wheat varieties that can tolerate saline soils without significant yield losses. Similarly, in crops like barley (*Hordeum vulgare*) and maize (*Zea mays*), breeders have used markers linked to salt-tolerance genes to select varieties that can perform well in salt-affected regions. In addition to traditional marker-assisted breeding, genomic selection is increasingly being used to improve salt tolerance in crops. Genomic selection involves using genome-wide markers to predict the genetic potential of a plant for salt tolerance, allowing breeders to select the best-performing genotypes early in the breeding process. This technique has been applied in crops like rice and wheat, where salt tolerance is a critical trait for improving yields in saline environments.

6.3.3 Enhancing Quality in Salt-Tolerant Varieties

While improving salt tolerance is the primary goal of breeding programs, maintaining, or enhancing the quality of the crop under saline conditions is equally important. Salt stress can negatively affect various quality traits, such as flavor, texture, and nutritional content. Therefore, breeders aim to develop salt-tolerant varieties that can maintain high-quality produce even when grown in challenging environments. For example, in cucumbers (*Cucumis sativus*), high soil salinity can lead to poor fruit quality, affecting both taste and marketability. Breeding programs focus on developing salt-tolerant varieties that maintain desirable fruit characteristics such as crispness, flavor, and size. This involves selecting traits that enhance osmotic adjustment and ion compartmentalization in the fruit tissues, allowing the plant to cope with salt stress while preserving fruit quality. Similarly, in potatoes (*Solanum tuberosum*), salinity can reduce tuber size and quality, making the crop less marketable. Breeding efforts in potatoes focus on selecting for salt-tolerant varieties that can maintain tuber quality under saline conditions, ensuring that farmers can continue to produce high-quality potatoes even in areas affected by salinity.

6.4 Quality Improvement under Abiotic Stress

Breeding for abiotic stress resistance often involves trade-offs between stress tolerance and crop quality. While improving stress tolerance is crucial for maintaining yields under challenging conditions, it is equally important to ensure that the quality of the produce is not compromised. In many cases, abiotic stresses such as drought, salinity, and heat can negatively impact quality traits such as flavor, texture, color, and nutritional content. Therefore, breeders must carefully balance the selection for stress tolerance with the need to maintain or improve quality traits that are important for consumer acceptance and marketability.

6.4.1 Maintaining Nutritional Quality

One of the key challenges in breeding for abiotic stress resistance is maintaining the nutritional quality of the crop. Abiotic stresses can affect the accumulation of important nutrients, such as vitamins, minerals, and antioxidants, in fruits, vegetables, and grains. For example, drought stress in tomatoes can lead to reduced levels of lycopene, an antioxidant that is important for human health. Similarly, heat stress in leafy greens like spinach can reduce the concentration of essential vitamins such as vitamin C.

To address this challenge, breeders focus on selecting genotypes that can maintain or even enhance their nutritional content under stress conditions. This involves identifying and selecting genetic markers associated with both stress tolerance and nutritional quality traits. For instance, in tomatoes, breeding programs aim to develop drought-tolerant varieties that can maintain high levels of lycopene, ensuring that the nutritional quality of the fruit is preserved even under water-limited conditions. In rice, efforts to breed for drought and heat tolerance have been accompanied by a focus on maintaining or improving grain quality. Breeders select varieties that maintain grain size, starch content, and aroma even under stress conditions, ensuring that the rice produced is not only resilient but also meets the quality standards expected by consumers.

6.4.2 Enhancing Flavor and Marketability

Flavor is a critical quality trait for many fruits and vegetables, and abiotic stresses can have a significant impact on the flavor profile of crops. For example, drought stress in melons can lead to reduced sweetness, while heat stress in tomatoes can negatively affect fruit flavor by altering the balance of sugars and organic acids.

Breeding programs that focus on improving abiotic stress tolerance must also prioritize the maintenance of flavor and marketability. This involves selecting genotypes that can withstand environmental stress while maintaining desirable flavor characteristics. In crops like melons and strawberries, breeders select drought-tolerant varieties that can maintain high sugar content and flavor under water-limited conditions. Similarly, in tomatoes, breeding efforts aim to develop heat-tolerant varieties that preserve the balance of sugars, acids, and volatile compounds that contribute to the fruit's flavor. Marketability is also influenced by other quality traits, such as appearance, texture, and shelf life. Abiotic stresses can lead to poor fruit sets, smaller fruit sizes, or changes in texture that affect the market value of the crop. Therefore, breeding programs focus on selecting genotypes that can maintain good fruit set, size, and texture under stress conditions. This is particularly important in crops like cucumbers, where heat or drought stress can lead to bitter or misshapen fruits, reducing their marketability.

7. Breeding for Water Use Efficiency (WUE) and Nutrient Use Efficiency (NUE)

Breeding for water use efficiency (WUE) and nutrient use efficiency (NUE) is crucial in modern agriculture to ensure sustainable crop production in the face of limited natural resources, such as water and essential nutrients (e.g., nitrogen, and phosphorus). These efficiencies help crops to maximize productivity under constrained resource availability, reduce the environmental footprint of agricultural practices, and improve the sustainability of farming systems. With the growing challenge of climate change and the increasing global population, improving WUE and NUE through breeding is a major goal in crop improvement programs worldwide. Water use efficiency (WUE) refers to the ratio of biomass or yield produced per unit of water consumed by the crop. Improving WUE allows plants to maintain high productivity under water-limited conditions. Similarly, nutrient use efficiency (NUE) refers to the ability of plants to effectively utilize available nutrients for growth and development, minimizing losses due to leaching or volatilization. Both WUE and NUE are controlled by complex genetic and environmental factors, making their improvement challenging but critical for sustainable agriculture.

7.1 Breeding for Water Use Efficiency (WUE)

Water use efficiency is a key trait in regions where water is scarce or where drought conditions prevail. Breeding for WUE aims to develop crop varieties that can produce high yields with less water, either by improving water uptake, reducing water loss, or enhancing the plant's ability to maintain productivity under water-limited conditions. This involves selecting for traits related to drought tolerance, water-use optimization, and physiological adaptations to water scarcity.

7.1.1 Physiological Traits for Improved WUE

Improving WUE involves selecting a range of physiological traits that enhance the plant's ability to use water more efficiently. These traits include deeper root systems for accessing water, reduced stomatal conductance to minimize water loss through transpiration, and improved leaf cooling mechanisms to maintain photosynthesis under heat stress. One of the key physiological traits linked to WUE is stomatal regulation. Plants with better control over their stomatal opening can reduce water loss while still allowing sufficient gas exchange for photosynthesis.

In wheat (*Triticum aestivum*), for instance, breeders have selected varieties with more efficient stomatal regulation, allowing them to maintain photosynthesis and yield under drought conditions. Similarly, in crops like sorghum (*Sorghum bicolor*) and maize (*Zea mays*), breeding programs have focused on selecting genotypes with deeper root systems that can access water stored in deeper soil layers, enhancing the plant's ability to survive drought periods. Another critical physiological trait is water-use partitioning, where water is preferentially allocated to reproductive organs rather than vegetative growth. In crops like rice (*Oryza sativa*), breeders have developed varieties that divert a higher proportion of available water to grain production, thus increasing grain yield under water-limited conditions. This approach maximizes the efficiency of water use for economically important traits such as grain or fruit yield.

7.1.2 Genetic Basis of WUE

WUE is a polygenic trait, meaning it is controlled by multiple genes that influence various aspects of water uptake, transpiration, and biomass production. The genetic basis of WUE has been studied extensively in several crops, leading to the identification of key genes and quantitative trait loci (QTLs) associated with WUE. In rice, for example, several QTLs linked to WUE have been identified, including those related to root architecture and osmotic adjustment. These QTLs have been incorporated into breeding programs through marker-assisted selection (MAS) to develop rice varieties that can maintain productivity with less water. Similarly, in wheat, QTLs associated with WUE and drought tolerance have been used to improve the crop's performance under water-limited conditions. Advances in genomics and molecular biology have enabled breeders to better understand the genetic regulation of WUE and to develop molecular markers linked to key WUE genes. This has allowed for a more precise selection of genotypes with improved WUE in breeding programs. For example, molecular markers linked to genes involved in stomatal regulation, osmotic adjustment, and root development are now used to select for drought-tolerant varieties of crops like wheat, maize, and barley.

7.1.3 Trade-offs between WUE and Yield

One of the challenges in breeding for WUE is the potential trade-off between water use efficiency and yield. While improving WUE often involves reducing water loss through transpiration, this can also limit the amount of carbon dioxide available for photosynthesis, potentially reducing yield. Therefore, breeders must carefully balance the selection for WUE with the need to maintain high productivity. In some cases, breeding for WUE has involved selecting traits that improve water use without compromising yield. For example, in maize, breeders have developed varieties that maintain high biomass production and yield under drought conditions by selecting deeper root systems and improved water-use partitioning. Similarly, in rice, breeders have focused on improving grain yield under water-limited conditions by selecting for varieties that can maintain efficient photosynthesis even under drought stress. To mitigate potential trade-offs, breeders are increasingly using genomic selection to identify genotypes that combine high WUE with high yield potential. Genomic selection allows breeders to predict the performance of genotypes based on genome-wide markers, enabling the selection of individuals that excel in both WUE and yield.

7.2 Breeding for Nutrient Use Efficiency (NUE)

Nutrient use efficiency refers to the ability of plants to effectively absorb, assimilate, and utilize essential nutrients, such as nitrogen (N), phosphorus (P), and potassium (K), to maximize growth and yield. Improving NUE is critical for reducing the reliance on chemical fertilizers, minimizing nutrient losses to the environment, and enhancing the sustainability of agricultural systems. Breeding for NUE focuses on improving the efficiency of nutrient uptake, assimilation, and remobilization in crops.

7.2.1 Nitrogen Use Efficiency (NUE)

Nitrogen is one of the most important nutrients for plant growth and development, but it is also one of the most inefficiently used. Large amounts of nitrogen fertilizers are applied to agricultural fields, but a significant proportion is lost through leaching or volatilization, contributing to environmental pollution. Improving nitrogen use efficiency (NUE) is a major breeding objective, especially in crops like wheat, rice, and maize that require substantial nitrogen inputs. Breeding for NUE involves selecting traits that enhance nitrogen uptake, assimilation, and remobilization within the plant. In wheat, for example, breeders have developed varieties that are more efficient at taking up nitrogen from the soil and assimilating it into proteins and other essential compounds. These varieties can produce higher yields with lower nitrogen inputs, reducing the environmental impact of nitrogen fertilization. Key traits associated with improved NUE include enhanced root architecture for better nitrogen uptake, increased nitrogen assimilation efficiency in the leaves, and improved nitrogen remobilization during grain filling. In maize, breeders have selected varieties with more extensive root systems that allow the plants to capture more nitrogen from the soil, improving their ability to maintain high yields with lower nitrogen inputs. Molecular breeding tools such as marker-assisted selection and genomic selection have been used to identify QTLs and genes associated with NUE. In rice, for example, several QTLs linked to nitrogen uptake and utilization have been identified, allowing breeders to develop nitrogen-efficient varieties that can maintain high productivity with reduced nitrogen fertilization.

7.2.2 Phosphorus Use Efficiency (PUE)

Phosphorus is another essential nutrient for plant growth, but its availability in soils is often limited due to its tendency to become fixed in the soil and unavailable for plant uptake. Phosphorus use efficiency (PUE) refers to the plant's ability to acquire and utilize phosphorus effectively. Breeding for PUE aims to develop crop varieties that can thrive in phosphorus-deficient soils or make better use of applied phosphorus fertilizers. Improving PUE involves selecting for traits such as enhanced root architecture, increased root exudation of organic acids that solubilize phosphorus, and improved internal phosphorus remobilization. For example, in common bean (*Phaseolus vulgaris*), breeders have selected varieties with enhanced root exudation of organic acids, which increases phosphorus availability in low-phosphorus soils. This allows the plant to take up more phosphorus and maintain higher productivity under nutrient-limited conditions.In wheat and maize, breeders have also focused on selecting varieties with root systems that can better explore the soil and access phosphorus reserves. Additionally, genes involved in phosphorus remobilization from older leaves to developing grains have been targeted in breeding programs to improve phosphorus efficiency during grain filling. Molecular markers associated with PUE traits, such as root length, root surface area, and phosphorus uptake efficiency, have been identified in crops like rice, maize, and barley. These markers are used in marker-assisted breeding programs to develop phosphorus-efficient varieties that can maintain yield with reduced phosphorus inputs.

7.2.3 Potassium and Other Nutrient Use Efficiencies

Micronutrients such as zinc (Zn), iron (Fe), and boron (B) play critical roles in plant metabolic processes and are also essential for human nutrition. Deficiencies in these micronutrients, both in plants and in human diets, can lead to poor crop performance and malnutrition. Breeding for improved micronutrient use efficiency (MUE) involves selecting plant varieties that can thrive in soils with low micronutrient availability and accumulate sufficient levels of these essential nutrients in edible parts like grains, fruits, and leaves. In wheat and rice, breeding efforts have focused on improving zinc and iron uptake efficiency to address both agronomic and nutritional deficiencies. For instance, high-zinc and high-iron varieties of rice have been developed through biofortification, a process that enhances the micronutrient content of crops. These varieties not only perform well in micronutrient-deficient soils but also provide nutritional benefits to consumers by increasing the levels of essential nutrients in the grain.

For potassium (K) use efficiency (KUE), breeders have selected varieties with enhanced potassium uptake and utilization, which play crucial roles in plant water regulation, enzyme activation, and stress tolerance. Potassium-efficient varieties are important for maintaining yield under low-potassium conditions, especially in crops like rice, maize, and wheat, where potassium is critical for overall crop health and productivity. To improve the breeding process for nutrient use efficiency, molecular markers linked to traits such as root architecture, nutrient uptake, and nutrient translocation have been developed. These markers are used in marker-assisted selection (MAS) to accelerate the identification and selection of nutrient-efficient genotypes in various crops.

7.3 Genetic and Molecular Tools for Improving WUE and NUE

Advances in genomics, molecular biology, and breeding technologies have greatly enhanced the ability of plant breeders to improve water and nutrient use efficiency in crops. Genetic tools such as quantitative trait locus (QTL) mapping, genome-wide association studies (GWAS), and genomic selection (GS) are widely used to identify key genes and loci associated with WUE and NUE. These tools allow breeders to make more informed and precise selections based on the genetic potential of plants to perform well under resource-limited conditions.

7.3.1 Marker-Assisted Selection (MAS)

Marker-assisted selection (MAS) involves using molecular markers linked to desirable traits, such as WUE and NUE, to select for individuals with improved performance. MAS is particularly useful for traits that are difficult to assess visually, such as root architecture or nutrient uptake efficiency. By using molecular markers, breeders can track the presence of key genes in breeding populations and make more precise selections. In rice, for example, markers linked to WUE and drought tolerance have been used to develop drought-tolerant varieties that can maintain productivity with less water. Similarly, in wheat and maize, markers associated with nitrogen and phosphorus use efficiency have been identified and used in breeding programs to improve nutrient uptake and utilization in resource-limited environments.

7.3.2 Genomic Selection (GS)

Genomic selection (GS) is a more advanced breeding tool that uses genome-wide markers to predict the genetic potential of individuals based on their entire genome. This approach is particularly valuable for improving complex traits like WUE and NUE, which are controlled by multiple genes. By analyzing the genetic information of large populations, breeders can identify individuals with the best genetic potential for WUE and NUE and select them early in the breeding process.GS has been successfully applied in crops like maize and wheat to improve both WUE and NUE. For example, genomic selection has allowed breeders to develop maize varieties that perform well under drought conditions and maintain high yields with reduced nitrogen inputs. Similarly, in wheat, GS has been used to improve both water and nitrogen use efficiency, leading to the development of varieties that are more resilient and sustainable.

7.3.3 CRISPR-Cas9 and Genetic Engineering

The CRISPR-Cas9 genome-editing technology is revolutionizing plant breeding by allowing precise modifications to be made to specific genes associated with WUE and NUE. With CRISPR-Cas9, breeders can directly target and edit genes responsible for water uptake, root development, nutrient transport, and assimilation, leading to the creation of crop varieties with enhanced efficiency in resource use. For example, CRISPR-Cas9 has been used to edit genes involved in drought tolerance and nitrogen use efficiency in rice, leading to the development of varieties that can better withstand water-limited conditions and require less nitrogen fertilizer. Similarly, in maize, CRISPR-Cas9 has been used to improve root architecture, enhancing the plant's ability to access water and nutrients from deeper soil layers. Genetic engineering has also played a significant role in improving nutrient use efficiency. For instance, transgenic approaches have been used to introduce genes that enhance phosphorus uptake and utilization in crops like maize and soybeans, reducing the need for phosphorus fertilizers. Similarly, nitrogen-fixing bacteria genes have been explored for their potential to improve nitrogen uptake in non-leguminous crops.

7.4 Combining WUE and NUE for Sustainable Agriculture

Breeding for both WUE and NUE simultaneously is critical for developing crop varieties that can perform well under conditions of resource scarcity. Water and nutrients are interrelated in plant growth, and improving one without considering the other may not lead to optimal outcomes. For instance, improving nitrogen use efficiency without improving water use efficiency may result in suboptimal growth under drought conditions, and vice versa. To address this challenge, breeders aim to develop varieties that can efficiently use both water and nutrients. This involves selecting genotypes that can optimize resource allocation, such as by improving root architecture for better water and nutrient uptake or enhancing physiological mechanisms that allow plants to maintain productivity under stress conditions. For example, in wheat and maize breeding programs, efforts are being made to develop varieties that can maintain high yields under both drought and low-nitrogen conditions. By selecting traits that enhance both WUE and NUE, breeders can create varieties that are more resilient to climate change and require fewer inputs, contributing to more sustainable agricultural systems. Additionally, integrated breeding strategies that combine traditional breeding methods with modern molecular tools and genomic technologies are being used to accelerate progress in improving WUE and NUE. These strategies allow breeders to make more informed decisions and create crop varieties that are not only more efficient in using water and nutrients but also capable of maintaining high quality and yield under challenging environmental conditions.

Chapter - 3

Heterosis Breeding and Sterility

1. Heterosis Breeding

Heterosis, commonly referred to as hybrid vigor, is a genetic phenomenon that results in offspring exhibiting superior traits compared to either of their parental lines. These traits may include enhanced growth rates, increased yield, better resistance to diseases, greater tolerance to environmental stress, and overall improved vigor. In plant breeding, heterosis is used to develop hybrids that perform better in both yield and stress resilience, leveraging the genetic diversity between parent lines. The study and application of heterosis are critical in crops like maize, rice, sorghum, and vegetables, where hybrid vigor has been key to improving agricultural productivity. Heterosis can be understood and categorized based on various genetic mechanisms and the relationship between parent lines. A deeper look into these mechanisms helps in understanding how heterosis can be utilized effectively in breeding programs. The primary types of heterosis include dominance heterosis, overdominance heterosis, and epistatic heterosis. These forms of heterosis arise from different genetic interactions within the hybrid and provide a basis for selecting parent lines in breeding programs.

A. Types of Heterosis

There are several types of heterosis based on the comparison of hybrid performance to either the average of the parents, the better parent, or a commercial standard. The classification of heterosis plays a critical role in plant breeding, especially in determining the efficiency and effectiveness of hybrid development. The three main types of heterosis are **Average Heterosis, Heterobeltiosis**, and **Economic Heterosis**. Each of these types of heterosis is quantified using specific formulas that help assess the extent to which a hybrid surpasses its parents or other benchmarks.

DOI: 10.1201/9781003790976-3

1. Average Heterosis

Average Heterosis, also called mid-parent heterosis, refers to the performance of the hybrid being superior to the average performance of the two parents. This type of heterosis compares the hybrid's performance to the mid-parent value, which is the mean of the traits of both parents. In simpler terms, if the hybrid's trait, such as yield or growth, exceeds the average of the parents, it is considered to exhibit average heterosis. The extent of this heterosis is expressed as a percentage relative to the mid-parent value.

1.1 Calculation of Average Heterosis

Average heterosis is quantified using the following formula:

$$\textbf{Average Heterosis} = \{(F1-MP)/MP\}\times 100$$

Where:

- **F1** = Mean of the hybrid or the performance of the hybrid offspring.
- **MP** = Mid-parent value, calculated as the mean of the values of the two parents.

The mid-parent value is derived from the following formula:

$$MP = (P1+P2)/2$$

Where:

- **P1** = Value of the first parent.
- **P2** = Value of the second parent.

1.2 Practical Implications of Average Heterosis

In practical terms, average heterosis provides insight into the overall genetic advantage that a hybrid may have over its parental lines. It is particularly useful in assessing the general vigor and performance of the hybrid in traits such as growth rate, biomass production, or early-stage development. However, average heterosis is typically considered to have limited utility in commercial agriculture, especially when it comes to traits like yield or resistance to stress. This is because average heterosis does not guarantee that the hybrid will surpass the better-performing parent; it only ensures that the hybrid is better than the average performance of the parents.

For instance, in a crop like maize, if one parent has a yield of 5 tons per hectare and the other has a yield of 4 tons per hectare, the mid-parent value would be 4.5 tons per hectare. If the hybrid yields 4.8 tons per hectare, it would exhibit average heterosis, but this increase may not be significant enough for commercial purposes if the goal is to achieve a yield higher than the better parent.

1.3 Limitations of Average Heterosis in Agriculture

While average heterosis indicates that the hybrid performs better than the average of its parents, it is generally not of high significance in agricultural production. This is because farmers and breeders are typically interested in hybrids that outperform the best-performing parent, which average heterosis does not necessarily achieve. In commercial agriculture, the goal is to produce hybrids that show substantial improvements in economically important traits, such as yield, disease resistance, or stress tolerance. If a hybrid only exceeds the mid-parent value but does not surpass the better parent, it may not offer enough of an advantage to warrant its commercial use.

2. Heterobeltiosis

Heterobeltiosis is a form of heterosis where the performance of the hybrid surpasses that of the better parent. This type of heterosis is of particular importance in agricultural breeding because it demonstrates the hybrid's ability to outcompete the best-performing parent, making it highly desirable for commercial production. Heterobeltiosis is considered a more useful measure of hybrid vigor in terms of improving economically important traits such as yield, disease resistance, or environmental stress tolerance.

2.1 Calculation of Heterobeltiosis

Heterobeltiosis is calculated using the following formula:

$$\textbf{Heterobeltiosis} = \{(F1 - BP)/BP\} \times 100$$

Where:

- **F1** = Mean of the hybrid or the performance of the hybrid offspring.
- **BP** = Performance of the better parent.

In this calculation, the hybrid's performance is directly compared to that of the better-performing parent. If the hybrid surpasses the better parent in terms of yield, disease resistance, or another critical trait, it is considered to exhibit heterobeltiosis.

2.2 Practical Applications of Heterobeltiosis

Heterobeltiosis is highly valuable in breeding programs because it identifies hybrids that not only surpass the mid-parent value but also outperform the better parent. This is particularly significant in crops where breeders aim to maximize yield or improve resistance to diseases and pests. For instance, in rice breeding, a hybrid that produces more grains per panicle than the better parent is considered to exhibit heterobeltiosis and is likely to be selected for commercial production.

In maize breeding, heterobeltiosis has been widely exploited to develop high-yielding hybrids. For example, if one parent produces 6 tons per hectare the other parent produces

4 tons per hectare, and the hybrid produces 7 tons per hectare, the hybrid would exhibit significant heterobeltiosis. This superiority over the better parent makes such hybrids valuable for large-scale agricultural production. Heterobeltiosis is also important for improving traits such as disease resistance. In wheat, for instance, if one parent is resistant to rust diseases and the other parent is susceptible, a hybrid that exhibits enhanced resistance compared to the better parent would be highly desirable for breeding programs aimed at developing rust-resistant wheat varieties.

2.3 Benefits of Heterobeltiosis in Agriculture

The primary benefit of heterobeltiosis is its ability to identify hybrids that are superior to the best-performing parent. This makes it a powerful tool in crop improvement, as breeders are constantly looking for ways to increase productivity, improve quality, and enhance stress tolerance in crops. Heterobeltiosis allows breeders to select hybrids that combine the best traits of both parents and amplify them to a level beyond the better parent. In crops like sorghum and sunflower, heterobeltiosis has been used to improve drought tolerance and oil content, respectively. Hybrids that exhibit heterobeltiosis for these traits are able to produce higher yields under water-limited conditions or generate more oil per seed compared to their parent lines. This makes such hybrids valuable for both commercial production and ensuring food security in regions prone to environmental stresses.

3. Economic Heterosis

Economic Heterosis, also referred to as useful heterosis, compares the performance of the hybrid to that of a high-yielding commercial variety in a particular crop. This type of heterosis is crucial in determining the economic viability of a hybrid in real-world agricultural production. Economic heterosis is used to evaluate whether the hybrid offers a significant improvement over existing commercial varieties, making it suitable for large-scale production and market adoption.

3.1 Calculation of Economic Heterosis

Economic heterosis is calculated using the following formula:

$$\textbf{Economic Heterosis} = \{(F1 - CV)/CV\} \times 100$$

Where:

- **F1**= Mean of the hybrid or the performance of the hybrid offspring.
- **CV** = Mean performance of a high-yielding commercial variety.

In this case, the performance of the hybrid is compared to the best commercial variety available, which serves as the benchmark for determining whether the hybrid provides an economic advantage in terms of yield, quality, or other critical traits.

3.2 Importance of Economic Heterosis

Economic heterosis is critical for determining whether a new hybrid is worth adopting on a commercial scale. Even if a hybrid outperforms its parents, it must also offer a significant improvement over existing commercial varieties to be considered viable for farmers. For example, if a commercial variety of maize yields 10 tons per hectare, but a newly developed hybrid yields 12 tons per hectare, the hybrid would exhibit economic heterosis. This makes it a potentially valuable replacement for the commercial variety.

In addition to yield, economic heterosis can also be used to assess improvements in traits such as pest resistance, drought tolerance, or fruit quality. For instance, in tomato breeding, a hybrid that exhibits better disease resistance and longer shelf life compared to the most popular commercial variety would be considered to exhibit economic heterosis and would be favored for large-scale production.

3.3 Economic Benefits in Crop Production

In crops such as rice, maize, and wheat, economic heterosis provides a critical measure of whether a new hybrid can outcompete established commercial varieties in terms of profitability and efficiency. The adoption of hybrids that exhibit economic heterosis can lead to significant increases in crop productivity, ultimately benefiting farmers through higher yields per hectare, improved crop resilience, and reduced losses due to pests and diseases. For example, in maize production, a hybrid that shows economic heterosis by outperforming a widely grown commercial variety in yield per hectare would be more attractive to farmers, especially in regions where maximizing yield is crucial for both food security and profitability. Additionally, hybrids that exhibit superior traits, such as increased tolerance to drought or resistance to common pests like the fall armyworm, can provide substantial economic benefits by reducing the need for costly inputs such as irrigation and pesticides.

Similarly, in rice cultivation, a hybrid that outperforms a popular commercial variety in grain yield, grain quality, or adaptability to adverse environmental conditions (e.g., flooding or drought) would offer farmers a significant economic advantage. Hybrid rice varieties that exhibit economic heterosis are especially valuable in regions like Southeast Asia, where rice is a staple crop, and increased production is essential for feeding growing populations.In the case of vegetables such as tomatoes, cucumbers, and peppers, hybrids that demonstrate economic heterosis by producing better-quality fruits with longer shelf life, improved flavor, or resistance to diseases are more likely to be adopted in commercial farming. These hybrids can fetch higher prices in the market due to their superior quality, leading to increased profitability for growers. Additionally, hybrids with higher resistance to pathogens, such as late blight in tomatoes or powdery mildew in cucumbers, can reduce crop losses, making them economically valuable for farmers and producers.

B. Mechanisms, and Basis of Heterosis

1.1 Dominance Heterosis

Dominance heterosis, or pseudo-dominance, occurs when dominant alleles at several loci mask the expression of deleterious recessive alleles from both parents. The result is a hybrid that exhibits superior traits because the negative effects of the recessive alleles are minimized or suppressed by the presence of dominant alleles. This type of heterosis is crucial in many crops where improved yield, disease resistance, and stress tolerance are essential.

1.1.1 Mechanism of Dominance Heterosis

Dominance heterosis is driven by the concept that inbreeding in parental lines tends to expose deleterious recessive alleles. These alleles, which may negatively affect growth, yield, or other traits, become more prevalent in homozygous conditions. When two inbred lines are crossed to produce a hybrid, dominant alleles from one parent mask the expression of harmful recessive alleles from the other, leading to improved performance in the hybrid.

This masking effect is particularly evident in crops that suffer from inbreeding depression, such as maize. Inbred maize lines often exhibit reduced vigor, lower yields, and greater susceptibility to diseases because of the accumulation of deleterious recessive alleles. However, when these lines are crossed to form a hybrid, the dominance of beneficial alleles from both parents can eliminate the negative effects of inbreeding depression. This results in hybrids with enhanced plant vigor, increased kernel development, and greater overall productivity. The dominance hypothesis of heterosis proposes that hybrid vigor arises primarily from the accumulation of dominant alleles that contribute positively to traits such as yield, height, and resistance to environmental stresses. This hypothesis has been substantiated by decades of plant breeding experiments, particularly in crops like maize and sorghum, where dominance heterosis plays a pivotal role in hybrid breeding programs. In maize, for instance, hybrid vigor resulting from dominance heterosis has led to substantial increases in grain yield, which is essential for feeding growing populations. Recent molecular studies have identified specific genes that contribute to dominance heterosis in maize and other crops. For example, genes associated with stress tolerance, such as those involved in heat and drought resistance, have been found to exhibit dominance effects in hybrid combinations. These genes enhance the hybrid's ability to withstand environmental stresses, allowing for improved yield stability in challenging growing conditions.

1.1.2 Applications of Dominance Heterosis in Crop Breeding

Dominance heterosis has been widely applied in the development of hybrid varieties in several staple crops, including maize, sorghum, and sunflower. These crops benefit from the significant yield improvements and resilience to environmental stresses that dominance heterosis offers. By selecting parent lines that complement each other genetically, breeders can create hybrids with superior performance in terms of yield, drought tolerance, and disease resistance. One of the most notable applications of dominance heterosis is in maize breeding. Hybrid maize varieties consistently outperform their inbred counterparts due to the dominance heterosis effect. These hybrids exhibit increased biomass, improved photosynthetic capacity, and better resistance to diseases like maize leaf rust and corn earworm. The introduction of hybrid maize has revolutionized maize production worldwide, allowing farmers to achieve higher yields per hectare and adapt to changing environmental conditions.

In sorghum breeding, dominance heterosis has been exploited to develop hybrids that are more drought-tolerant and heat-resistant. Sorghum is a staple crop in arid and semi-arid regions, where water scarcity is a significant challenge. By leveraging dominance heterosis, breeders have been able to produce sorghum hybrids that are better suited to these harsh environments, providing farmers with crops that can maintain productivity even under drought stress. Similarly, in sunflower breeding, dominance heterosis has been utilized to improve oil yield and disease resistance. Sunflower hybrids exhibit better resilience to fungal diseases like downy mildew and show enhanced oil content, making them more valuable in the commercial market. The ability to improve both yield and quality through dominance heterosis has made hybrid sunflower varieties the preferred choice for commercial production.

1.1.3 Future Prospects of Dominance Heterosis

While dominance heterosis has been highly successful in many crops, ongoing research aims to further improve its application in plant breeding. Advances in molecular genetics and genomics are providing breeders with tools to better understand the genetic basis of dominance heterosis. By identifying specific genes and molecular pathways involved in heterosis, breeders can more precisely select parent lines that will maximize hybrid vigor.

One promising area of research involves using marker-assisted selection (MAS) and genomic selection to identify and track dominant alleles that contribute to heterosis. These tools allow breeders to predict the performance of hybrids based on their genetic makeup, making the breeding process more efficient and reducing the time required to develop high-performing hybrids. In maize, for example, MAS has been used to select for dominant alleles associated with improved drought tolerance and disease resistance, leading to the development of hybrids that are better suited to climate change. Additionally, advances in gene editing technologies, such as CRISPR-Cas9, offer new possibilities for enhancing dominance heterosis. By editing specific genes involved in stress tolerance or disease resistance, breeders can create hybrids with even greater levels of vigor and resilience. These technologies have the potential to revolutionize heterosis breeding by enabling breeders to directly manipulate the genetic factors that contribute to hybrid vigor.

1.2 Overdominance Heterosis

Overdominance heterosis, or true heterosis, is a phenomenon where the heterozygous state at certain loci produces superior performance compared to both homozygous parents. In this case, the hybrid exhibits better traits because the combination of different alleles at a locus results in enhanced expression of a particular trait. This type of heterosis is especially relevant in traits controlled by a small number of genes, where heterozygous genotypes outperform both parental homozygous genotypes.

1.2.1 Mechanism of Overdominance Heterosis

The overdominance hypothesis suggests that when different alleles at the same locus combine in a heterozygous state, they create a synergistic effect that enhances the expression of beneficial traits. This interaction between the alleles leads to superior performance in the hybrid compared to either parent, which may be homozygous for one of the alleles. Unlike dominance heterosis, which relies on masking deleterious alleles, overdominance heterosis is based on the enhancement of positive trait expression through heterozygosity. In crops such as maize and rice, overdominance heterosis is often observed in yield-related traits. For example, hybrids between two genetically diverse rice lines may produce plants with greater tiller numbers, larger panicles, and higher grain yield than either parent. This superior performance is due to the heterozygous advantage at key loci controlling yield-related traits.

Overdominance heterosis is not limited to yield traits. It has also been observed in traits related to stress tolerance, such as drought or heat resistance, where the hybrid exhibits better adaptability to environmental stress compared to the parental lines. In these cases, the heterozygous condition allows the hybrid to maintain physiological functions, such as photosynthesis and water uptake, under stress conditions, leading to improved survival and productivity.

1.2.2 Overdominance Heterosis in Crop Improvement

Overdominance heterosis has been widely used in the development of hybrid varieties for crops where yield and stress tolerance are critical. Hybrid rice is one of the most successful examples of overdominance heterosis in action. The introduction of hybrid rice varieties has significantly increased rice production in Asia, where rice is a staple food for billions of people. These hybrids consistently outperform inbred rice varieties in terms of grain yield, tillering capacity, and resistance to both biotic and abiotic stresses. Hybrid rice varieties that exhibit overdominance heterosis have been key to addressing food security challenges in regions with high population density and limited arable land. The ability of these hybrids to produce higher yields under a range of environmental conditions has made them a valuable tool for increasing rice production while reducing the need for additional land and inputs. Vegetable crops such as tomatoes, peppers, and cucumbers also benefit from overdominance heterosis. In hybrid tomato breeding, for example, hybrids exhibit improved fruit size, color, flavor, and shelf life compared to inbred lines. This has made hybrid tomatoes the preferred choice for both commercial growers and consumers, as they provide higher market value and better-quality produce.

1.2.3 Challenges and Future Directions in Overdominance Heterosis (continued)

One of the main challenges with overdominance heterosis is the complexity involved in identifying specific loci where overdominance occurs. Since overdominance heterosis relies on the interaction between different alleles at individual loci, breeders must carefully identify these loci and maintain heterozygosity at the key loci responsible for the superior trait expression. This can be a time-consuming and resource-intensive process, especially in polygenic traits where multiple loci contribute to overall performance. Furthermore, hybrids that exhibit overdominance heterosis may show variability in performance across different environments due to genotype-by-environment interactions. For example, a hybrid that performs exceptionally well under one set of environmental conditions may not exhibit the same level of performance under different environmental conditions. This variability presents a challenge for breeders, as it requires extensive multi-location and multi-environment testing to ensure that hybrids consistently exhibit superior performance across a range of growing conditions.

Another challenge with overdominance heterosis is maintaining hybrid vigor in subsequent generations. Like dominance heterosis, the overdominance effect is lost if the hybrid is self-pollinated or crossed with other hybrids, leading to a breakdown of heterozygosity. This means that hybrid seeds must be produced anew each generation, which can increase production costs and require sophisticated seed production systems to maintain genetic purity. Despite these challenges, the future of overdominance heterosis is promising, especially with advances in molecular breeding technologies. Marker-assisted selection (MAS) and genomic selection (GS) have been instrumental in identifying the loci associated with overdominance heterosis. These tools allow breeders to more efficiently select parent lines and predict hybrid performance based on genetic markers linked to heterozygous advantage. In addition, advances in gene editing technologies, such as CRISPR-Cas9, have the potential to further enhance overdominance heterosis by allowing precise manipulation of specific alleles involved in heterozygous advantage. By targeting and editing key genes associated with overdominance, breeders can develop hybrids with even greater levels of hybrid vigor and improved adaptability to environmental stresses. As climate change continues to impact agricultural systems worldwide, breeding for hybrids that can maintain high yields and quality under diverse environmental conditions will become increasingly important. Overdominance heterosis offers a powerful tool for achieving this goal, as it enables the development of hybrids that combine the best traits of both parents and outperform in a wide range of growing conditions.

1.3 Epistatic Heterosis

Epistatic heterosis occurs when the interaction between different genes at multiple loci results in superior performance in the hybrid compared to the parents. Unlike dominance and overdominance heterosis, which involve interactions between alleles at the same locus, epistatic heterosis involves the interaction between alleles at different loci. These gene-gene interactions, known as epistasis, can lead to enhanced trait expression in the hybrid, contributing to heterosis. Epistatic heterosis is particularly important for traits controlled by multiple genes, such as yield, disease resistance, and stress tolerance. The combined effect of favorable interactions between genes at different loci can result in hybrids that show superior performance in these complex traits.

1.3.1 Mechanism of Epistatic Heterosis

Epistasis refers to the interaction between genes at different loci, where the expression of one gene can be influenced or modified by the presence of alleles at other loci. In the context of heterosis, epistatic interactions can result in enhanced trait expression in the hybrid, leading to better performance compared to either parent. These interactions are particularly important in polygenic traits, where multiple genes work together to control the trait.

In crops like maize and rice, epistatic heterosis has been observed in yield-related traits. For example, in hybrid maize, the interaction between genes controlling photosynthesis, water-use efficiency, and nutrient uptake can lead to higher biomass production and improved grain yield. Similarly, in hybrid rice, epistatic interactions between genes controlling tillering, grain filling, and disease resistance result in hybrids that outperform both parent lines in these key traits. The epistatic model of heterosis suggests that the combination of favorable alleles at different loci can create a synergistic effect that enhances the hybrid's overall performance. This model is supported by molecular studies that have identified specific QTLs (quantitative trait loci) involved in epistatic interactions that contribute to heterosis. For example, in rice, QTL mapping has revealed that epistatic interactions between loci involved in grain size and number significantly contribute to the superior yield of hybrid rice varieties.

1.3.2 Applications of Epistatic Heterosis in Breeding

Epistatic heterosis has been widely utilized in breeding programs for crops where complex traits such as yield, disease resistance, and stress tolerance are important. In maize breeding, for instance, hybrids that exhibit epistatic heterosis are known for their superior biomass production, drought tolerance, and nutrient-use efficiency. By selecting parent lines that complement each other genetically, breeders can develop hybrids that benefit from favorable epistatic interactions between genes controlling these traits. In rice breeding, the use of epistatic heterosis has been instrumental in the development of hybrid varieties that produce higher yields and show better adaptability to varying environmental conditions. Hybrid rice varieties that benefit from epistatic interactions between genes controlling grain number, grain size, and plant architecture have consistently outperformed inbred varieties in terms of yield and resilience to environmental stresses. Epistatic heterosis is also important in vegetable breeding, particularly in crops like tomatoes, peppers, and cucumbers. Hybrids that show epistatic heterosis often exhibit improved fruit quality, disease resistance, and stress tolerance compared to their inbred parents. This makes hybrid varieties more desirable for commercial production, as they provide higher yields, better-quality produce, and greater adaptability to different growing conditions.

1.3.3 Challenges and Future Prospects for Epistatic Heterosis

One of the main challenges in exploiting epistatic heterosis is the complexity involved in identifying and selecting favorable epistatic interactions. Unlike dominance and overdominance heterosis, where specific alleles at individual loci can be targeted, epistatic heterosis involves interactions between multiple genes at different loci. This makes it more difficult to predict which parent lines will produce hybrids that exhibit strong epistatic heterosis. Another challenge is the stability of epistatic heterosis across different environments. Since epistatic interactions can be influenced by environmental factors, the expression of heterosis may vary depending on the growing conditions. This variability requires extensive testing of hybrids across multiple locations and environments to ensure that they consistently exhibit superior performance.

Despite these challenges, the future of epistatic heterosis is promising, particularly with the advent of modern breeding technologies. Advances in genomic selection and marker-assisted selection are helping breeders identify the genetic regions involved in epistatic interactions, making it easier to select parent lines that will produce hybrids with enhanced heterosis. Additionally, advances in CRISPR-Cas9 and other gene-editing technologies offer new opportunities to manipulate epistatic interactions and further enhance hybrid vigor. In the coming years, as the demand for high-yielding, resilient crop varieties continues to grow, the use of epistatic heterosis will become increasingly important. By leveraging gene-gene interactions, breeders can develop hybrids that not only outperform their parents but also show greater adaptability to the challenges posed by climate change and resource limitations.

C. Male Sterility in Vegetable Breeding

1. Male sterility in vegetable breeding plays a crucial role in the production of hybrid varieties, which are widely used for their superior traits, such as higher yields, better resistance to diseases, and improved adaptability to environmental stresses. Male sterility refers to the inability of a plant to produce viable pollen, making it incapable of self-pollination and seed production through its pollen. This feature is especially important in hybrid seed production, where controlled cross-pollination is needed to combine desirable traits from different parent lines.

In vegetable breeding, male sterility is an essential tool used to ensure efficient and cost-effective hybrid seed production. By eliminating the need for manual emasculation (the removal of male reproductive organs to prevent self-pollination), male sterility simplifies the process of producing hybrid seeds on a large scale. Vegetables such as onions, cabbages, carrots, and peppers have benefited from breeding programs that utilize male sterility to develop high-performing hybrid varieties.

Male sterility can be classified into several types based on the genetic and cytoplasmic factors that control it: **Cytoplasmic Male Sterility (CMS), Genetic Male Sterility (GMS),** and **Cytoplasmic-Genetic Male Sterility (CGMS).** Each of these types has distinct mechanisms and applications in vegetable breeding.

1.1 Cytoplasmic Male Sterility (CMS) in Vegetable Breeding

Cytoplasmic Male Sterility (CMS) is one of the most widely used mechanisms in vegetable breeding for hybrid seed production. CMS is a maternally inherited trait, meaning it is passed down from the female parent through the cytoplasm of the egg cell. It is controlled by mitochondrial genes in the cytoplasm of the plant, which affect the development of pollen. CMS is especially valuable in breeding programs because it eliminates the need for labor-intensive manual emasculation to prevent self-pollination, facilitating easier and more efficient hybrid seed production.

1.1.1 Mechanism of CMS in Vegetables

In CMS, mutations or alterations in the mitochondrial genome disrupt the normal development of the male reproductive organs (anthers), leading to the production of non-viable pollen or the complete absence of pollen. The cytoplasmic control of this trait ensures that male sterility is inherited maternally, meaning that all offspring of a CMS plant will also be male-sterile, provided that no fertility-restoring nuclear genes are present. This ensures that male-sterile plants can be used as female parents in hybrid seed production.

The mechanism of CMS can vary depending on the vegetable species. In some cases, CMS is linked to defects in mitochondrial functions related to respiration or energy production, which in turn affects pollen development. For example, in onion breeding (*Allium cepa*), CMS is used to prevent self-pollination in male-sterile plants, enabling breeders to cross-pollinate them with pollen from selected male parents. This ensures that the resulting seeds are hybrids with traits from both parents.CMS is particularly useful in vegetables where emasculation (removal of the male organs) is difficult or impractical. For instance, in crops like carrots (*Daucus carota*), where the flowers are small and delicate, CMS allows breeders to produce hybrid seeds without having to manually remove the male reproductive parts. Similarly, in brassica crops such as cabbage and broccoli, CMS facilitates the production of hybrids that are more uniform and vigorous compared to open-pollinated varieties.

1.1.2 Role of CMS in Hybrid Vegetable Breeding

In hybrid vegetable breeding, CMS is employed to ensure that the female parent plant does not produce viable pollen, thus preventing self-pollination. This enables controlled cross-pollination with a chosen male parent, allowing breeders to combine desirable traits from both parents into a single hybrid variety. CMS is particularly advantageous in crops where hybrid vigor (heterosis) is sought, as hybrids often outperform their parent lines in terms of yield, disease resistance, and stress tolerance. For example, in the breeding of hybrid onions, male-sterile plants are used as female parents, and they are pollinated by male-fertile lines with desirable traits such as disease resistance or improved bulb quality. The resulting hybrid seeds produce onions that are more uniform in size, have a higher yield, and possess enhanced resistance to pests and diseases. Similarly, CMS is widely used in carrot breeding, where hybrid varieties with improved root quality, color, and flavor are developed through controlled cross-pollination facilitated by CMS.

Another key application of CMS is in the breeding of hybrid brassica crops, such as cabbage (*Brassica oleracea var. capitata*) and cauliflower (*Brassica oleracea var. botrytis*). These hybrids exhibit superior traits, such as uniform head formation, earlier maturity, and better resistance to environmental stresses like cold or heat. The use of CMS in these crops ensures that breeders can efficiently produce hybrid seeds without the need for costly and labor-intensive emasculation techniques. The widespread use of CMS in vegetable breeding has revolutionized the production of hybrid varieties, making it possible to produce high-quality seeds on a commercial scale. CMS not only simplifies the process of hybrid seed production but also ensures that the resulting hybrids exhibit the desired combination of traits from both parents, leading to improved productivity and profitability for farmers.

1.1.3 Limitations and Challenges of CMS in Vegetable Breeding

While CMS is a powerful tool in vegetable breeding, it is not without its limitations. One of the primary challenges associated with CMS is the need for **restorer-of-fertility (Rf) genes** to ensure that male fertility can be restored in subsequent generations of hybrids if desired. In some vegetable crops, breeding lines with the appropriate restorer genes may not be readily available, making it difficult to produce hybrids with both male sterility and fertility restoration capabilities. Additionally, CMS systems can sometimes be unstable, with environmental factors such as temperature or humidity influencing the expression of male sterility. In certain cases, male-sterile plants may produce small amounts of viable pollen under specific environmental conditions, which can complicate the production of true hybrids. This environmental sensitivity can affect the reliability of CMS systems in some vegetable crops, particularly in regions with fluctuating climatic conditions. Despite these challenges, the advantages of CMS in vegetable breeding far outweigh the drawbacks, particularly in crops where hybrid vigor is a key driver of productivity. Advances in molecular breeding techniques, such as marker-assisted selection and genomic selection, are helping breeders to overcome some of the limitations of CMS by enabling the more precise selection of parent lines with stable CMS and fertility restoration genes.

1.2 Genetic Male Sterility (GMS) in Vegetable Breeding

Genetic Male Sterility (GMS) refers to male sterility that is controlled by nuclear genes rather than cytoplasmic factors. Unlike CMS, which is maternally inherited, GMS is governed by recessive or dominant nuclear genes that affect pollen production. In GMS systems, male sterility can be inherited in a simple Mendelian fashion, making it easier to manipulate and select for in breeding programs.

1.2.1 Mechanism of GMS in Vegetables

In GMS, male sterility is caused by mutations or genetic alterations in nuclear genes that control the development and function of the male reproductive organs. These genes can either be recessive or dominant, depending on the specific genetic system involved. In the case of recessive GMS, male sterility only occurs when the plant is homozygous for the recessive allele, meaning that both copies of the gene are mutated. In dominant GMS, male sterility can occur in heterozygous individuals, where only one copy of the gene is affected. The genetic basis of GMS makes it more flexible for breeders, as they can use traditional Mendelian breeding techniques to select male-sterile plants. In some vegetable crops, GMS is used in combination with other breeding tools to produce hybrids with enhanced traits.

1.2.2 Applications of GMS in Vegetable Breeding

GMS is used in crops such as tomatoes (*Solanum lycopersicum*), where breeders aim to develop hybrids with improved fruit quality, disease resistance, and yield. By selecting male-sterile plants in GMS systems, breeders can control cross-pollination more efficiently and produce hybrids that exhibit heterosis. GMS also plays a role in the breeding of cucurbit crops such as cucumbers (*Cucumis sativus*) and melons (*Cucumis melo*), where it facilitates the production of hybrid seeds for improved fruit size, flavor, and disease resistance.

1.2.3 Limitations and Challenges of GMS in Vegetable Breeding

While **Genetic Male Sterility (GMS)** offers flexibility due to its nuclear inheritance and the potential for traditional breeding techniques, it presents some challenges. One of the primary limitations of GMS is the segregation of sterile and fertile plants within breeding populations. Since GMS is often controlled by recessive genes, breeders must carefully manage breeding populations to ensure that male-sterile plants can be maintained in sufficient numbers while also allowing for the restoration of fertility when necessary. Additionally, GMS may not always be stable across different environmental conditions. Just as with CMS, certain environmental factors such as temperature fluctuations or soil conditions may influence the expression of sterility. In some cases, GMS plants that are expected to be sterile may produce viable pollen under specific environmental conditions, complicating the breeding process. Ensuring the stability of GMS expression across varied climates and environments remains a key challenge for vegetable breeders. Another challenge lies in the propagation of male sterile lines. In many vegetable crops, seed multiplication must be carefully managed to maintain the male-sterile trait across generations. This requires the use of marker-assisted selection or other molecular tools to ensure that the correct genetic combinations are maintained. While traditional breeding methods can be used to select male sterile plants, the process can be labor-intensive, requiring multiple generations to achieve desired results. Despite these challenges, GMS is a valuable tool in vegetable breeding, particularly in crops where hybrid vigor is sought. By employing GMS, breeders can facilitate controlled cross-pollination and the production of hybrid seeds, enabling the development of high-yielding, disease-resistant vegetable varieties.

1.3 Cytoplasmic-Genetic Male Sterility (CGMS) in Vegetable Breeding

Cytoplasmic-Genetic Male Sterility (CGMS) is a hybrid system that combines elements of both CMS and GMS. In CGMS, male sterility is controlled by cytoplasmic factors inherited from the female parent (similar to CMS), but fertility can be restored by nuclear restorer genes (Rf genes) from the male parent. This system offers the advantages of CMS in terms of efficient hybrid seed production, while also allowing breeders to restore fertility in subsequent generations through the use of Rf genes.

1.3.1 Mechanism of CGMS

In CGMS, male sterility is induced by mitochondrial mutations or alterations, which are passed down maternally through the cytoplasm. However, unlike CMS, where fertility can only be restored through nuclear restorer genes, CGMS systems allow for the restoration of fertility by introducing specific Rf genes from the male parent. These Rf genes interact with the cytoplasmic factors to restore normal pollen production in the hybrid or subsequent generations. The flexibility of CGMS makes it particularly valuable in crops where both male sterility and fertility restoration are required. This system allows breeders to produce male sterile plants for hybrid seed production while ensuring that fertility can be restored in the next generation if desired.

1.3.2 Applications of CGMS in Vegetable Breeding

CGMS is used in a variety of vegetable crops, including **peppers** (*Capsicum annuum*), **cabbages** (*Brassica oleracea*), and **carrots** (*Daucus carota*), where hybrid seed production is essential for achieving higher yields and improved quality. For instance, in **pepper breeding,** CGMS is employed to produce hybrid seeds that offer superior disease resistance, better fruit quality, and improved adaptability to different environmental conditions. The use of CGMS in pepper breeding eliminates the need for manual emasculation and allows for the production of uniform, high-performing hybrids.

In **carrot breeding**, CGMS facilitates the development of hybrid varieties that exhibit enhanced root color, texture, and flavor. By using male-sterile plants as female parents and crossing them with male-fertile lines carrying Rf genes, breeders can produce hybrid seeds that combine the best traits of both parents while ensuring that fertility can be restored in future generations if needed. Similarly, in **cabbage breeding,** CGMS is used to develop hybrid varieties with improved head formation, disease resistance, and stress tolerance. The ability to control male sterility and fertility restoration through CGMS enables breeders to produce high-quality hybrid seeds that are more consistent and productive than open-pollinated varieties.

1.3.3 Advantages of CGMS in Vegetable Breeding

The primary advantage of CGMS is its ability to combine the benefits of both CMS and GMS. CGMS allows for the efficient production of male-sterile plants, which are essential for hybrid seed production, while also providing the option to restore fertility through nuclear restorer genes. This flexibility makes CGMS an attractive system for breeders working with vegetables where hybrid vigor is critical for improving yield, quality, and resistance to biotic and abiotic stresses.In crops like **onions**, **cabbages**, and **carrots**, the use of CGMS has led to the development of hybrid varieties that are more uniform, vigorous, and resistant to diseases compared to their open-pollinated counterparts. CGMS systems also reduce the labor and cost associated with manual emasculation, making hybrid seed production more efficient and scalable.

1.4 Molecular Advances in Male Sterility Systems

Recent advances in **molecular genetics** and **genomics** have greatly enhanced the understanding and manipulation of male sterility systems in vegetable breeding. Molecular markers and **marker-assisted selection (MAS)** are now used to identify and select male-sterile plants more efficiently, reducing the time and effort needed to develop male-sterile lines. In addition, **genomic selection** and **CRISPR-Cas9 gene editing** technologies are being explored to directly modify genes responsible for male sterility, allowing for the precise control of this trait in breeding programs.

1.4.1 Marker-Assisted Selection and Genomic Tools

Molecular markers associated with CMS, GMS, and Rf genes have been identified in several vegetable crops, allowing breeders to track the inheritance of male sterility and fertility restoration traits. Marker-assisted selection enables breeders to select parent lines with the desired combination of sterility and fertility genes, ensuring that hybrid seed production can be carried out more efficiently. In **onion breeding**, for example, molecular markers linked to CMS have been developed, making it easier to identify and select male-sterile plants for hybrid production. Similarly, in **pepper and carrot breeding**, molecular tools are being used to track the inheritance of male sterility and fertility restoration traits, enabling breeders to produce high-quality hybrid seeds with improved efficiency.

1.4.2 Gene Editing for Male Sterility

Gene-editing technologies like **CRISPR-Cas9** offer new possibilities for developing male-sterile lines with greater precision and control. By directly targeting and editing the genes responsible for male sterility or fertility restoration, breeders can create customized male-sterile plants that are more stable and adaptable to different environments. For instance, CRISPR-Cas9 has the potential to be used in **tomato** and **pepper breeding** to precisely manipulate genes involved in pollen development, enabling the production of hybrids with enhanced traits such as disease resistance, fruit quality, and yield.The use of molecular tools and gene-editing technologies in male sterility systems is expected to revolutionize hybrid vegetable breeding by making the process more efficient, reliable, and scalable.

1.4.3 Future Prospects of Male Sterility in Vegetable Breeding

The use of male sterility in vegetable breeding is expected to continue expanding, particularly with the advent of molecular breeding tools and biotechnological advancements. The ability to produce hybrids more efficiently and with better control over sterility and fertility restoration will likely lead to significant improvements in vegetable crop performance and yield in the coming years. As breeders gain a deeper understanding of the genetic basis of male sterility and fertility restoration, they will be able to fine-tune these traits to meet the specific needs of different crops and growing environments.

In particular, the use of **molecular markers** for identifying male sterile plants and **gene-editing technologies** such as CRISPR-Cas9 for inducing male sterility or restoring fertility at precise stages of development will revolutionize hybrid seed production. These tools will not only enhance the efficiency of hybrid breeding programs but will also make it easier to develop male-sterile lines that are stable across diverse environmental conditions, which is a key challenge in many vegetable crops.

The combination of **marker-assisted selection, genomic selection**, and **gene-editing technologies** will open up new possibilities for creating hybrids that are more productive, disease-resistant, and resilient to environmental stresses. Moreover, these advances will allow breeders to address specific challenges, such as improving water-use efficiency, nutrient-use efficiency, or tolerance to extreme temperatures in vegetables, ensuring that new hybrid varieties can thrive in changing climatic conditions. In addition, the development of **precision breeding techniques** and **high-throughput phenotyping** platforms will enable breeders to screen large populations of male sterile plants more efficiently, selecting those with the optimal combination of traits for hybrid seed production. This will further reduce the time and cost associated with developing high-quality hybrids, making it possible to respond more quickly to market demands and emerging agricultural challenges.

D. Self-Incompatibility in Vegetable Breeding

1. Self-incompatibility (SI) is a genetic mechanism that prevents self-pollination and promotes cross-pollination in plants. In self-incompatible plants, the pollen from the same plant or genetically similar individuals is unable to fertilize the ovules, thus preventing self-fertilization. This mechanism is crucial for maintaining genetic diversity in plant populations, as it encourages outcrossing and ensures that offspring are the result of a genetic exchange between different individuals.

In vegetable breeding, self-incompatibility plays a significant role, especially in crops where hybrid seed production is desirable. By preventing self-pollination, self-incompatibility ensures that cross-pollination occurs, which can result in the production of hybrid seeds with superior traits. The use of self-incompatible plants in breeding programs allows breeders to exploit heterosis (hybrid vigor) and develop varieties with enhanced yield, improved resistance to diseases, better quality, and greater adaptability to environmental stresses. Self-incompatibility is typically controlled by specific genes located at the **S-locus** in the plant genome. There are two primary types of self-incompatibility found in plants: **Gametophytic Self-Incompatibility (GSI)** and **Sporophytic Self-Incompatibility (SSI)**. Each type operates through different mechanisms and has unique applications in vegetable breeding.

1.1 Gametophytic Self-Incompatibility (GSI) in Vegetable Breeding

Gametophytic Self-Incompatibility (GSI) is the most common form of self-incompatibility in many vegetable crops, particularly in the **Solanaceae** and **Rosaceae** families. In GSI, the compatibility of pollen is determined by the genotype of the pollen grain itself (the gametophyte). When pollen from the same plant or genetically similar plants lands on the stigma, the pollen tube fails to grow, preventing fertilization from occurring. This ensures that only pollen from genetically distinct plants can fertilize the ovules, leading to cross-pollination.

1.1.1 Mechanism of GSI

The GSI system is controlled by a single genetic locus known as the **S-locus**, which contains multiple alleles. The S-alleles in both the pollen and the pistil determine whether pollen germination and fertilization will occur. When pollen carrying an S-allele matches one of the S-alleles in the pistil, the pollen tube is prevented from growing through the style, thus blocking fertilization. This ensures that pollen from the same plant or closely related plants (sharing similar S-alleles) cannot fertilize the ovules.

For example, in **tomato** (*Solanum lycopersicum*) and **eggplant** (*Solanum melongena*), GSI prevents self-pollination by inhibiting the growth of pollen tubes when genetically similar pollen grains land on the stigma. The pollen that fails to fertilize the ovules is eventually degraded, allowing only pollen from genetically diverse plants to succeed in fertilizing the ovules.

1.1.2 Applications of GSI in Vegetable Breeding

GSI is particularly useful in hybrid seed production for vegetables. By ensuring that self-pollination does not occur, breeders can efficiently produce hybrid seeds through controlled cross-pollination. This is important in crops like **tomatoes** and **eggplants**, where hybrid varieties often outperform open-pollinated varieties in terms of yield, disease resistance, and overall vigor.

In **pepper breeding** (*Capsicum annuum*), GSI allows breeders to maintain genetic diversity and prevent inbreeding depression. The use of GSI helps breeders develop hybrids that combine the best traits from different parental lines, leading to hybrids with improved fruit size, color, flavor, and resistance to environmental stress. GSI also plays a role in the production of hybrid seeds for **cabbages** (*Brassica oleracea*), where it helps facilitate outcrossing between genetically distinct plants. The resulting hybrid cabbages exhibit superior traits such as earlier maturity, uniform head formation, and increased resistance to pests and diseases. These hybrid varieties are more profitable for farmers due to their higher yields and better quality compared to self-pollinated varieties.

1.1.3 Challenges and Management of GSI in Vegetable Breeding

While GSI is advantageous in preventing self-pollination, it can pose challenges in breeding programs. For example, the presence of multiple S-alleles in a population can make it difficult for breeders to control the specific crosses that occur. Additionally, environmental factors such as temperature and humidity can affect the expression of self-incompatibility, leading to variations in pollen tube growth and fertilization success. To overcome these challenges, breeders often use **marker-assisted selection** (MAS) to identify and select specific S-alleles in parent lines. This allows them to control the genetic combinations that result from cross-pollination, ensuring that the desired traits are incorporated into the hybrid offspring. Advances in molecular genetics have also provided tools for identifying and manipulating the S-locus, enabling breeders to develop more stable self-incompatible lines.

1.2 Sporophytic Self-Incompatibility (SSI) in Vegetable Breeding

Sporophytic Self-Incompatibility (SSI) is another form of self-incompatibility, primarily found in the **Brassicaceae** family, including important vegetable crops like **cabbages, broccoli**, and **cauliflower**. In SSI, the compatibility of the pollen is determined by the genetic makeup of the plant that produces the pollen (the sporophyte), rather than the individual pollen grains. This system prevents pollen from fertilizing the ovules of plants with the same or similar S-alleles, promoting cross-pollination.

1.2.1 Mechanism of SSI

In SSI, the genotype of the pollen-producing plant determines the success or failure of fertilization. The S-locus controls the interaction between the pollen and the pistil. If the S-alleles in the pollen-producing plant match the S-alleles in the pistil, the pollen is rejected at the surface of the stigma, and fertilization does not occur. This mechanism ensures that only pollen from genetically distinct plants can successfully fertilize the ovules. For example, in **cabbage breeding**, SSI prevents self-pollination by rejecting pollen from plants with the same S-alleles as the pistil. This ensures that cross-pollination occurs between genetically diverse plants, leading to greater genetic variation in the offspring. This genetic diversity is essential for improving traits such as head size, leaf texture, and resistance to diseases like **black rot**.

1.2.2 Applications of SSI in Hybrid Vegetable Breeding

SSI is widely used in the production of hybrid seeds for brassica vegetables, including **cabbages, broccoli**, and **cauliflower**. By preventing self-pollination, SSI allows breeders to control cross-pollination between parent lines with complementary traits. This results in hybrid varieties that exhibit superior performance compared to open-pollinated or self-pollinated varieties.

In **broccoli breeding**, for example, SSI facilitates the development of hybrids with uniform head formation, improved tolerance to cold temperatures, and better disease resistance. Hybrid broccoli varieties are preferred by commercial growers due to their higher yields, better quality, and resistance to common pathogens like **downy mildew**. In **cauliflower breeding**, SSI is used to produce hybrids with improved curd quality, earlier maturity, and better adaptability to different growing conditions. Hybrid cauliflower varieties are more productive and consistent in their performance, making them more profitable for farmers in both temperate and tropical regions.

1.2.3 Overcoming Challenges in SSI

While SSI provides many benefits for hybrid seed production, it can present challenges for breeders. The main challenge with SSI is the complexity of the S-locus, which contains many alleles that can interact in different ways. Managing the large number of possible S-allele combinations can make it difficult for breeders to control which plants successfully cross-pollinate. To manage SSI in breeding programs, breeders often use **genetic markers** to track S-alleles and select parent lines with the desired allelic combinations. Advances in **genomic selection** and **molecular breeding** techniques have made it easier to manage SSI and ensure successful hybrid seed production.

1.3 Molecular Approaches to Enhance Self-Incompatibility Systems

Molecular tools such as **marker-assisted selection (MAS)** and **CRISPR-Cas9** gene editing have revolutionized the study and application of self-incompatibility in vegetable breeding. By using these tools, breeders can more efficiently identify and select parent lines with specific S-alleles, ensuring successful cross-pollination and hybrid seed production.

1.3.1 Marker-Assisted Selection and S-Allele Management

Marker-assisted selection (MAS) is a powerful tool for managing self-incompatibility in vegetable breeding. By using molecular markers linked to the S-locus, breeders can track S-alleles in breeding populations and select parent lines that are compatible for cross-pollination. This allows breeders to overcome some of the challenges associated with managing large populations of self-incompatible plants.In crops like **tomatoes** and **peppers**, MAS is used to select for specific S-alleles that promote successful cross-pollination and hybrid seed production. By identifying the S-alleles present in each parent line, breeders can ensure that only genetically distinct plants are crossed, resulting in hybrids with superior traits.

1.3.2 Gene Editing and CRISPR-Cas9 in Self-Incompatibility

CRISPR-Cas9 and other gene-editing technologies provide plant breeders with the ability to precisely manipulate the genetic elements responsible for self-incompatibility. This allows for targeted modifications of the **S-locus**, where the S-alleles that regulate pollen-pistil interactions are located. By editing specific genes that control self-incompatibility, breeders can either enhance or suppress the expression of self-incompatibility, depending on the breeding objectives. For example, in crops like **tomatoes** or **peppers**, where hybrid vigor is highly sought after, CRISPR-Cas9 can be used to enhance the expression of self-incompatibility to ensure more controlled cross-pollination. Breeders can engineer plants with highly stable self-incompatibility systems, minimizing the risk of unwanted self-pollination and ensuring that cross-pollination occurs with genetically distinct plants. This approach enables the creation of more vigorous hybrid offspring with improved yields, disease resistance, and quality. Conversely, in crops where self-incompatibility can occasionally pose a challenge, CRISPR-Cas9 can be used to suppress or modify the function of certain S-alleles to facilitate self-pollination in specific breeding lines. This flexibility in managing self-incompatibility can help overcome barriers to breeding programs where controlled self-pollination is necessary.

One of the key advantages of using gene-editing technologies is the ability to create stable self-incompatible lines that are less influenced by environmental factors, such as temperature and humidity. Traditional self-incompatibility systems can be affected by these external conditions, which may lead to the breakdown of self-incompatibility under specific environmental stresses. With CRISPR-Cas9, breeders can modify the self-incompatibility system to ensure consistent expression, regardless of environmental fluctuations, thus improving the reliability of hybrid seed production. The potential for CRISPR-Cas9 in managing self-incompatibility is vast, and as gene-editing technologies continue to advance, they are likely to become a standard tool in vegetable breeding. This will help breeders optimize self-incompatibility systems for a wide range of crops, ensuring that hybrid seed production remains efficient and scalable.

1.4 Importance of Self-Incompatibility in Commercial Hybrid Production

Self-incompatibility systems are particularly important in the context of **commercial hybrid seed production**, where the goal is to produce high-quality seeds that exhibit heterosis (hybrid vigor). Self-incompatible plants provide a natural mechanism to ensure cross-pollination, which is essential for creating hybrids that combine the best traits of two parent lines. This results in offspring with superior characteristics, such as increased yield, disease resistance, and adaptability to different environmental conditions. Hybrid seeds produced through the use of self-incompatibility systems are highly valued in commercial agriculture due to their uniformity, improved performance, and economic viability. Farmers benefit from hybrid crops that are more resistant to pests and diseases, require fewer inputs, and yield more per unit of land. In crops such as **cabbages**, **broccoli**, **tomatoes**, and **peppers**, hybrids developed through self-incompatibility systems are favored by both small-scale and large-scale growers because of their consistency in performance and higher returns on investment.

Self-incompatibility also reduces the labor and cost associated with manual emasculation in crops where hybrid seed production is labor-intensive. For example, in crops like **onions** and **cabbages**, manual emasculation to prevent self-pollination can be costly and time-consuming. The use of self-incompatible lines eliminates the need for such practices, making the process of hybrid seed production more efficient and affordable for breeders and seed companies.

1.5 Challenges and Future Directions for Self-Incompatibility in Vegetable Breeding

While self-incompatibility systems offer significant advantages in vegetable breeding, they are not without challenges. One of the primary challenges is the management of **S-alleles** within breeding populations. The presence of multiple S-alleles can make it difficult to control cross-pollination, as breeders must ensure that plants with compatible S-alleles are crossed. This requires careful planning and monitoring, which can complicate the breeding process, especially in large populations. Environmental factors also play a role in the expression of self-incompatibility. Conditions such as temperature fluctuations, humidity, and soil fertility can affect the success of pollen tube growth, potentially leading to the breakdown of self-incompatibility. This instability poses challenges for breeders aiming to maintain consistent expression of self-incompatibility across diverse environments. Developing more environmentally stable self-incompatibility systems through gene-editing technologies like **CRISPR-Cas9** offers a promising solution to this challenge.

Looking ahead, **molecular breeding tools** such as **genomic selection** and **high-throughput phenotyping** will likely enhance the management of self-incompatibility systems in vegetable breeding. By identifying the specific genetic factors that regulate self-incompatibility and developing markers associated with these traits, breeders can more precisely select and control S-allele combinations. This will facilitate the development of self-incompatible lines that are more predictable and stable across a range of environments. Moreover, as **molecular biology** continues to advance, the integration of **biotechnology** into traditional breeding programs will further optimize self-incompatibility systems. This will enable breeders to develop hybrid vegetable varieties with even greater levels of genetic diversity, resilience, and productivity. By leveraging both conventional breeding techniques and cutting-edge molecular tools, the future of self-incompatibility in vegetable breeding looks promising for improving food security and meeting the growing global demand for high-quality vegetables.

E. Sex Forms in Vegetable Breeding

In vegetable breeding, understanding the **sex forms** or sexual expression of plants is essential for developing strategies that promote effective pollination, facilitate controlled breeding, and ultimately improve crop yields. The term "sex forms" refers to the different ways in which plants express their male and female reproductive organs. In vegetable crops, plants can exhibit various sexual forms, including **monoecious, dioecious, gynoecious, andromonoecious, gynodioecious**, and **hermaphroditic** forms. These forms have important implications for breeding programs, particularly in the production of hybrid seeds and the manipulation of flowering characteristics to enhance yield and quality. Different sex forms offer unique advantages and challenges in vegetable breeding, especially when it comes to controlling pollination, ensuring genetic diversity, and producing hybrids with desirable traits such as higher yield, improved resistance to pests and diseases, and better adaptability to different environments. The manipulation of plant sex forms through breeding techniques, such as selection and hybridization, is critical for maximizing crop performance.

1.1 Monoecious Plants in Vegetable Breeding

Monoecious plants are those that have both male (staminate) and female (pistillate) flowers on the same plant. This characteristic allows monoecious plants to produce both types of reproductive organs, making it possible for them to self-pollinate or cross-pollinate with other plants. Monoecious sex forms are commonly found in many vegetable crops, including **cucumbers** (*Cucumis sativus*), **pumpkins** (*Cucurbita pepo*), **squash** (*Cucurbita maxima*), and **melons** (*Cucumis melo*).

1.1.1 Breeding Monoecious Vegetables

In vegetable breeding, monoecious plants are important for ensuring a reliable source of pollen and for promoting cross-pollination between plants. Monoecious crops are advantageous in breeding programs because they allow for both self-pollination and cross-pollination, providing breeders with flexibility in choosing breeding strategies. For example, in **cucumber breeding**, monoecious varieties are often used as parent lines in hybrid seed production because they produce a consistent supply of both male and female flowers, which can be used to facilitate controlled pollination. However, while monoecious plants can self-pollinate, breeders often aim to promote cross-pollination to enhance genetic diversity and produce hybrids that exhibit heterosis (hybrid vigor). Cross-pollination between monoecious plants from different genetic backgrounds can result in hybrid offspring with superior traits, such as higher yield, improved disease resistance, and better fruit quality.

1.1.2 Manipulating Monoecious Forms in Breeding Programs

In monoecious vegetable crops, breeders can manipulate flowering patterns to optimize pollination and seed production. For instance, in **melon breeding**, breeders may select lines that produce a higher proportion of female flowers to increase fruit set and yield. Techniques such as the application of growth regulators, selection for specific genotypes, and environmental management (e.g., temperature and light control) can influence the expression of male and female flowers, allowing breeders to adjust the sex ratio in favor of female flowers. In some cases, breeders may also develop **gynoecious** lines from monoecious plants to further enhance yield potential. Gynoecious plants produce only female flowers, which can significantly boost fruit production, as all flowers are capable of setting fruit if pollinated by male plants. Gynoecious lines are particularly valuable in crops like cucumbers, where maximizing fruit production is a key breeding objective.

1.2 Dioecious Plants in Vegetable Breeding

Dioecious plants have separate male and female plants, meaning that individual plants produce either male or female flowers, but not both. This strict separation of sexes requires the presence of both male and female plants for pollination and seed production. Dioecious sex forms are less common in vegetable crops but can be found in species such as **asparagus** (*Asparagus officinalis*) and **spinach** (*Spinacia oleracea*).

1.2.1 Breeding Dioecious Vegetables

In dioecious vegetable crops, breeding programs must manage the presence of both male and female plants to ensure successful pollination. In **asparagus breeding**, for example, male and female plants must be planted in proximity to facilitate pollination. However, breeders often prefer to select for all-male hybrids because male asparagus plants tend to be more productive and vigorous than female plants, which allocate more energy to seed production rather than spear growth. All-male hybrids are produced by crossing male plants with female plants, followed by selecting offspring that exhibit only male flowers. These hybrids offer several advantages, including higher yields, better spear quality, and greater resistance to environmental stresses. In addition, the absence of female plants in all-male asparagus hybrids eliminates the production of seeds, which can compete with spear production and reduce overall yield.

In **spinach breeding**, dioecy is used to control pollination and promote cross-pollination between genetically diverse lines. Male and female plants are planted together, and breeders can select for traits such as disease resistance, bolting resistance, and improved leaf quality in the resulting hybrids. Dioecious spinach varieties provide breeders with opportunities to enhance genetic diversity and develop hybrids that perform well under a variety of growing conditions.

1.2.2 Challenges of Dioecious Plants in Breeding Programs

One of the main challenges of working with dioecious plants in breeding programs is the need to maintain both male and female plants to ensure successful pollination. This can complicate field management and increase the cost of seed production, as separate male and female plants must be grown and managed together. Additionally, breeders must carefully control the proportion of male and female plants to optimize pollination and seed set. However, the advantages of dioecious sex forms, particularly in terms of promoting genetic diversity and enabling hybrid seed production, often outweigh these challenges. By selecting desirable traits in both male and female plants, breeders can develop hybrids that combine the best characteristics of both parents, leading to improved performance in terms of yield, quality, and resistance to pests and diseases.

1.3 Gynoecious Plants in Vegetable Breeding

Gynoecious plants produce only female flowers, making them highly valuable in vegetable breeding programs aimed at maximizing fruit production. Gynoecious lines are commonly used in the breeding of cucurbit crops, such as **cucumbers** and **melons**, where increased fruit set is a key breeding objective.

1.3.1 Benefits of Gynoecious Plants in Hybrid Breeding

The primary advantage of gynoecious plants in breeding programs is their ability to produce a higher proportion of female flowers, which leads to increased fruit production. In crops like cucumbers, where each female flower has the potential to develop into a fruit, gynoecious plants are preferred for their high yield potential. Gynoecious lines are often used as female parents in hybrid seed production, while male flowers are provided by pollinator plants that produce both male and female flowers (monoecious plants). Gynoecious plants are particularly beneficial in controlled environments, such as greenhouses, where breeders can manipulate pollination by introducing male plants at specific times to ensure efficient pollination of the female flowers. This controlled pollination results in uniform, high-quality fruit that meets the market's demands for size, shape, and color.

1.3.2 Developing Gynoecious Lines in Vegetable Breeding

Breeders develop gynoecious lines through selection and hybridization techniques. In cucumber breeding, for example, breeders select lines that consistently produce only female flowers. These gynoecious lines are then crossed with monoecious or andromonoecious plants to create hybrids that exhibit a high degree of fruit set and yield. The use of gynoecious plants in hybrid breeding has revolutionized the production of high-yielding cucumber varieties that are uniform in size and shape, making them ideal for commercial production. Additionally, the development of **parthenocarpic** (seedless) varieties in cucumbers and other cucurbit crops has further enhanced the value of gynoeciouslines. Parthenocarpic varieties can produce fruit without the need for pollination, making them particularly useful in greenhouse production, where pollination can be challenging.

1.4 Andromonoecious and Gynodioecious Forms in Vegetable Breeding

Other sex forms, such as **andromonoecious** (plants that produce both male and hermaphroditic flowers) and **gynodioecious** (plants that produce both female and hermaphroditic flowers), are also important in certain vegetable breeding programs.

1.4.1 Andromonoecious Plants

In crops like **tomatoes** (*Solanum lycopersicum*), andromonoecious plants are used in breeding programs to promote flexibility in pollination. Andromonoecious plants produce both male and hermaphroditic flowers, allowing for both self-pollination and cross-pollination. This dual capability is useful in hybrid breeding programs, where breeders aim to balance the benefits of self-pollination (to maintain desirable traits) with the advantages of cross-pollination (to promote genetic diversity).

1.4.2 Gynodioecious Plants

In **gynodioecious** systems, plants produce both female and hermaphroditic flowers, providing an opportunity to balance the benefits of female-only reproductive structures with the potential for cross-pollination from hermaphroditic flowers. This sex form allows breeders to maintain a population of female plants that are more productive, while still ensuring that pollination occurs through the hermaphroditic flowers. In crops like **spinach** (*Spinacia oleracea*), gynodioecious plants are used to enhance cross-pollination and ensure genetic diversity in breeding programs. The presence of both female and hermaphroditic flowers allows for more controlled pollination strategies, ensuring that breeders can develop hybrids with desired traits such as increased yield, improved resistance to diseases like downy mildew, and better tolerance to environmental stresses like bolting or extreme temperatures. The combination of hermaphroditic and female flowers in gynodioecious plants ensures efficient pollination without the need for large numbers of male plants, making it easier for breeders to manage breeding populations. This system also allows breeders to focus on selecting and propagating female plants that exhibit superior traits, ensuring that the best characteristics are passed on to the next generation of hybrids.

1.5 Hermaphroditic Plants in Vegetable Breeding

Hermaphroditic plants, also known as perfect-flowered plants, have both male (stamens) and female (pistils) reproductive organs within the same flower. This structure allows for self-pollination as well as cross-pollination, depending on the environmental conditions and the breeding goals. Hermaphroditic flowers are common in many vegetable crops, including **tomatoes** (*Solanum lycopersicum*), **peppers** (*Capsicum annuum*), **eggplants** (*Solanum melongena*), and **beans** (*Phaseolus vulgaris*).

1.5.1 Breeding with Hermaphroditic Plants

Hermaphroditic plants are advantageous in breeding programs because they offer breeders the flexibility to either self-pollinate or cross-pollinate plants. In crops like **tomatoes** and **peppers**, self-pollination is commonly used to maintain desirable traits within a specific line. This allows breeders to produce stable, uniform varieties that consistently express favorable characteristics such as fruit size, shape, color, and disease resistance. In contrast, breeders can also encourage cross-pollination between hermaphroditic plants to promote genetic diversity and produce hybrids with enhanced traits. For example, in **tomato breeding**, cross-pollination between two hermaphroditic plants can result in hybrids that combine the best traits from both parent lines, leading to hybrids with improved fruit quality, higher yield, and better resistance to pests and diseases.

1.5.2 Challenges of Hermaphroditic Plants in Breeding

One of the challenges of breeding with hermaphroditic plants is managing the balance between self-pollination and cross-pollination. While self-pollination is useful for maintaining desirable traits, it can also lead to inbreeding depression over time, reducing vigor and increasing susceptibility to diseases and environmental stresses. To mitigate this, breeders often use techniques such as manual emasculation (removal of the male reproductive organs) or the introduction of pollinators to encourage cross-pollination.In some cases, breeders may also select for partial sterility in hermaphroditic plants to reduce the likelihood of self-pollination and promote outcrossing. This strategy is particularly useful in hybrid seed production, where breeders want to ensure that the resulting offspring are the product of cross-pollination between genetically diverse parent lines.

1.6 Applications of Sex Forms in Hybrid Vegetable Breeding

The manipulation of sex forms in vegetable breeding is crucial for producing high-quality hybrid seeds that exhibit superior traits such as higher yields, better disease resistance, and improved adaptability to different environmental conditions. Breeders use different sex forms strategically to control pollination and maximize the potential for heterosis (hybrid vigor) in the resulting hybrids. For example, in **cucumber breeding**, breeders may use **gynoecious** lines (which produce only female flowers) as the female parent and **monoecious** or **hermaphroditic** plants as the male parent to ensure efficient cross-pollination. The resulting hybrid seeds produce plants that are more productive, uniform, and resistant to diseases such as powdery mildew or downy mildew. In **asparagus** breeding, breeders select **all-male hybrids** because male plants are generally more vigorous and productive than female plants. By crossing dioecious male and female plants and selecting for male offspring, breeders can develop hybrids that yield higher-quality spears and are more resistant to environmental stresses. Similarly, in **broccoli** and **cauliflower** breeding, breeders manipulate sex forms to ensure that cross-pollination occurs between genetically diverse parent lines, resulting in hybrids with improved head formation, earlier maturity, and better resistance to diseases like clubroot and black rot.

1.7 Future Directions for Sex Form Manipulation in Vegetable Breeding

As advances in **molecular genetics** and **biotechnology** continue to shape the future of vegetable breeding, the manipulation of sex forms will become even more precise and efficient. Molecular tools such as **marker-assisted selection (MAS)** and **CRISPR-Cas9** gene-editing technology offer breeders the ability to modify the expression of genes that control sex determination in plants. These tools enable breeders to create custom sex forms that are optimized for specific breeding goals, such as increasing the number of female flowers to boost yield or promoting cross-pollination for greater genetic diversity.

For example, **CRISPR-Cas9** can be used to target specific genes that regulate the development of male or female flowers in crops like cucumbers, melons, or tomatoes. By editing these genes, breeders can create plants that produce a higher proportion of female flowers or develop male-sterile plants that are ideal for hybrid seed production. This level of precision will allow breeders to more effectively manipulate sex forms to maximize crop performance and ensure that hybrid varieties meet the needs of commercial growers and consumers alike. Moreover, as climate change continues to impact agricultural systems, the ability to manipulate sex forms in vegetable crops will become increasingly important for ensuring that plants can adapt to changing environmental conditions. Breeders may develop new sex forms that are more resilient to stress factors such as high temperatures, drought, or disease pressure, ensuring that hybrid varieties remain productive and profitable in the face of these challenges.

Chapter - 4

Mutation and Polyploidy Breeding for Improvement of Vegetable Crops

A. Mutation Breeding: Improvement of Asexually Propagated Vegetable Crops and Vegetables Suitable for a Protected Environment

Mutation breeding is an essential tool for improving asexually propagated vegetable crops and those grown in protected environments like greenhouses or high tunnels. Asexual propagation refers to the reproduction of plants from vegetative parts such as tubers, bulbs, rhizomes, and cuttings. This method ensures genetic uniformity, preserving desirable traits, but it also limits the natural introduction of genetic variability. Mutation breeding, through the use of mutagens like radiation or chemicals, introduces variability that can lead to the development of new and improved vegetable varieties.

Improvement of Asexually Propagated Vegetable Crops Through Mutation Breeding

Mutation breeding has a profound impact on the improvement of asexually propagated crops. These plants do not undergo sexual recombination, which limits genetic diversity. The lack of genetic variation in asexually propagated crops means that, while advantageous traits are preserved, there is limited potential for natural improvement or adaptation. Therefore, inducing mutations becomes a critical pathway for introducing beneficial traits like disease resistance, higher yields, better nutritional profiles, and tolerance to environmental stresses.

Methods for Inducing Mutations

1. Physical Mutagens

Physical mutagens are among the most widely used tools in mutation breeding, especially for inducing genetic variations in asexually propagated vegetable crops. These mutagens consist of ionizing radiation such as X-rays, gamma rays, and neutron rays. The radiation interacts with the DNA within the plant cells, causing breaks or alterations at various points in the genetic material. These alterations can lead to mutations that may result in new traits

DOI: 10.1201/9781003790976-4

or characteristics. Mutation breeding through physical mutagens has been an essential technique in enhancing crops by creating novel genetic variability, especially when other breeding methods are limited.

Ionizing radiation primarily affects the DNA by causing direct damage to the molecular structure, leading to breaks in the DNA strands or alterations in the nucleotides that form the genetic code. These changes can be either beneficial or harmful. The beneficial mutations can be selected for further breeding, leading to the development of crops with improved characteristics such as increased disease resistance, better yields, or enhanced nutritional quality. In contrast, harmful mutations are discarded during the selection process.

One of the most significant advantages of using physical mutagens is that they can create a wide range of mutations within a single generation, which accelerates the breeding process compared to traditional methods. For instance, when a crop is exposed to gamma radiation, the radiation randomly interacts with the DNA, causing a diverse set of mutations. These mutations can affect various genes responsible for traits such as growth rate, pest resistance, or environmental adaptability. Once these mutations are introduced, breeders can screen the plants for desirable traits, selecting the mutants that display the most beneficial characteristics.

Gamma Radiation and Its Impact on Asexual Crops

Gamma radiation, in particular, is one of the most commonly used forms of ionizing radiation in mutation breeding. It has proven effective in creating beneficial mutations in various asexually propagated vegetable crops such as potatoes, garlic, and onions. These crops are propagated through vegetative means (e.g., tubers, bulbs, and cuttings), meaning that the plants reproduce without undergoing sexual recombination. This lack of sexual recombination limits genetic diversity, making mutation breeding through physical mutagens an ideal approach to introducing new genetic variability.

For instance, in potato breeding, gamma radiation has been successfully used to induce mutations that improve the crop's resistance to diseases like late blight. Late blight, caused by the pathogen *Phytophthora infestans*, is a devastating disease that can lead to significant yield losses in potato crops. By exposing potato plants to gamma radiation, breeders have been able to generate new varieties that are more resistant to late blight, enhancing the crop's overall productivity and resilience. Similarly, gamma radiation has been used to develop potato varieties with improved storage capabilities, reducing post-harvest losses due to spoilage. In garlic and onion breeding, where vegetative propagation is the primary method of reproduction, gamma radiation has also been employed to introduce genetic diversity. These crops are often prone to viral diseases, and traditional breeding methods are limited in their ability to address these challenges. By inducing mutations through gamma radiation, breeders have been able to develop garlic and onion varieties with enhanced resistance to viral diseases, improving their overall health and productivity.

X-rays and Neutron Radiation in Mutation Breeding

Apart from gamma radiation, other types of ionizing radiation, such as X-rays and neutron rays, are also used in mutation breeding. X-rays, like gamma rays, are a form of electromagnetic radiation that can cause breaks or alterations in DNA. They have been used in various crops to induce mutations that lead to improved traits, such as increased disease resistance or enhanced growth characteristics. While X-rays are effective in inducing mutations, they are less commonly used than gamma rays due to their limited penetration depth, which can restrict their effectiveness in treating larger plants or seeds. Neutron radiation, on the other hand, is a type of particle radiation that interacts with atomic nuclei within the plant cells. Neutrons have a higher energy level than X-rays or gamma rays, making them more potent in causing DNA damage. As a result, neutron radiation is highly effective in inducing a wide range of mutations, although it requires specialized equipment and expertise to apply. In some cases, neutron radiation has been used to create new vegetable crop varieties with improved tolerance to environmental stressors such as drought or high temperatures, making it a valuable tool in the context of climate change and agricultural sustainability.

Advantages and Challenges of Using Physical Mutagens

One of the primary advantages of using physical mutagens in mutation breeding is their ability to generate a large number of mutations within a relatively short period. This allows breeders to quickly screen for beneficial traits and develop new varieties of crops. Additionally, physical mutagens can be applied to a wide range of crops, including those that are asexually propagated, making them a versatile tool in plant breeding.

However, there are also challenges associated with using physical mutagens. One of the main challenges is the random nature of the mutations induced by radiation. Since the mutations occur at random points in the DNA, not all of them will result in beneficial traits. In some cases, the mutations may lead to undesirable characteristics, such as reduced growth or increased susceptibility to diseases. This means that breeders must carefully screen the mutated plants to ensure that only those with positive traits are selected for further breeding. Another challenge is the potential for off-target effects. While physical mutagens can induce beneficial mutations, they can also cause unintended damage to other parts of the genome. This can result in the loss of important traits or the introduction of new vulnerabilities in the plants. To mitigate these risks, breeders must carefully control the dosage of radiation and use advanced screening techniques to identify and eliminate any unwanted mutations.

In recent years, advances in genomics and molecular biology have improved the precision and efficiency of mutation breeding with physical mutagens. Techniques such as genome sequencing and marker-assisted selection have made it easier for breeders to identify beneficial mutations and track their inheritance in subsequent generations. These tools have greatly enhanced the success of mutation breeding, allowing for the development of crops with improved traits while minimizing the risks associated with random mutations.

Applications in Modern Agriculture

Mutation breeding through physical mutagens continues to play a vital role in modern agriculture, particularly in the context of climate change and food security. As global temperatures rise and extreme weather events become more frequent, there is an increasing need for crop varieties that can withstand these challenges. Physical mutagens provide a way to introduce new genetic variability that can enhance the resilience of crops to environmental stressors such as drought, heat, and salinity.

In addition to improving crop resilience, physical mutagens have also been used to enhance the nutritional quality of vegetable crops. For example, in tomato breeding, gamma radiation has been used to develop varieties with higher levels of vitamins and antioxidants, improving the health benefits of the crop. Similarly, in leafy greens such as spinach and lettuce, mutation breeding has been used to enhance the nutritional content by increasing the levels of essential minerals such as iron and calcium. Physical mutagens have also been applied to improve the shelf life and storage capabilities of vegetable crops. In crops like potatoes, onions, and garlic, gamma radiation has been used to develop varieties that are more resistant to spoilage and degradation, reducing post-harvest losses and improving food security. These improvements are particularly important in regions where access to refrigeration and storage facilities is limited, as they help to ensure a stable supply of food throughout the year.

2. Chemical Mutagens in Vegetable Crops

Chemical mutagens play a crucial role in plant breeding, especially for inducing point mutations that alter the DNA sequence at a molecular level. These mutagens, such as ethyl methanesulfonate (EMS) and sodium azide, have been widely applied in breeding programs to introduce genetic variability in vegetable crops. By creating specific changes in the DNA sequence, chemical mutagens offer the potential to develop new traits in crops, such as enhanced disease resistance, improved yield, better storage qualities, and adaptability to various environmental stresses. One of the most successful applications of chemical mutagens has been seen in crops like sweet potatoes, where their use has led to the development of improved varieties with advantageous traits.

Mechanism of Action of Chemical Mutagens

Chemical mutagens work by interacting with the DNA in plant cells, inducing changes in the nucleotide sequence that can lead to point mutations. These mutations occur when a single base pair in the DNA sequence is altered, leading to the substitution, deletion, or insertion of nucleotides. Point mutations are particularly valuable in plant breeding because they allow for the precise alteration of specific genes responsible for desired traits, such as disease resistance or yield improvement. Unlike physical mutagens, which may cause more widespread damage to the DNA, chemical mutagens often result in targeted and specific changes, making them a powerful tool for crop improvement.

The process of inducing mutations using chemical agents typically involves treating plant seeds, embryos, or tissue cultures with the mutagen. The exposure time and concentration of the mutagen are carefully controlled to ensure that the desired level of mutation is achieved without causing excessive damage to the plant's genetic material. After treatment, the plants are grown and screened for desirable traits, and those that exhibit beneficial mutations are selected for further propagation.

Ethyl Methanesulfonate (EMS) as a Chemical Mutagen

Ethyl methanesulfonate (EMS) is one of the most commonly used chemical mutagens in plant breeding. EMS is a potent alkylating agent that works by adding alkyl groups to the DNA, specifically to the guanine base. This alkylation process causes the guanine to pair incorrectly with thymine instead of cytosine, resulting in point mutations in the DNA. These point mutations can lead to changes in the function of specific genes, which may result in new and advantageous traits in the plant. One of the key advantages of EMS is its ability to induce a high frequency of point mutations while causing relatively little chromosomal damage. This allows breeders to introduce genetic variability into crops without causing widespread genetic instability, making it an effective tool for developing new crop varieties. EMS has been widely applied in vegetable crops such as tomatoes, carrots, and sweet potatoes, where it has led to the development of improved traits such as disease resistance, enhanced yield, and better quality.

EMS-Induced Mutations in Sweet Potatoes

Sweet potatoes (*Ipomoea batatas*) are a globally important root crop, particularly in tropical and subtropical regions. Sweet potatoes are vegetatively propagated through cuttings rather than seeds, which limits the natural genetic diversity of the crop. This lack of genetic variability can make it challenging to develop new varieties with improved traits using traditional breeding methods. However, mutation breeding using chemical mutagens like EMS has been successful in overcoming this limitation and introducing beneficial mutations into sweet potatoes. One of the major challenges in sweet potato production is disease susceptibility, particularly to viral and fungal pathogens. Sweet potatoes are vulnerable to diseases such as sweet potato feathery mottle virus (SPFMV), sweet potato leaf curl virus (SPLCV), and fungal infections like Fusarium wilt. These diseases can significantly reduce yield and crop quality, making it essential to develop disease-resistant varieties. EMS has been used to induce point mutations in sweet potato cells, leading to the development of new varieties with enhanced resistance to these diseases.

For example, EMS-induced mutations have been successful in developing sweet potato varieties that exhibit improved resistance to sweet potato feathery mottle virus (SPFMV). This viral disease can cause stunted growth and reduced yield in infected plants, and the introduction of genetic resistance through mutation breeding has had a significant impact on improving sweet potato production. The mutations introduced by EMS alter specific genes involved in the plant's immune response, enabling the sweet potato to better defend against viral infections. In addition to disease resistance, EMS has been used to improve the storage qualities of sweet potatoes. Post-harvest deterioration is a significant issue in sweet potato production, particularly in regions with limited storage facilities. Mutations that enhance the crop's ability to resist spoilage during storage have been introduced through EMS treatments. These mutations result in improved resistance to sprouting, rotting, and microbial infections during storage, ensuring that the sweet potatoes remain in good condition for extended periods after harvest. This has significant implications for food security and reducing post-harvest losses, especially in developing regions.

Applications of EMS in Other Vegetable Crops

Beyond sweet potatoes, EMS has been applied to various other vegetable crops, with similarly successful outcomes. In tomatoes, for example, EMS-induced mutations have led to the development of new varieties with improved fruit quality, better resistance to diseases such as tomato mosaic virus (ToMV) and Fusarium wilt, and increased yield. The ability to induce specific point mutations with EMS has allowed breeders to target genes involved in fruit ripening, disease resistance, and plant growth, resulting in superior tomato varieties.

In carrots, EMS has been used to develop varieties with improved root quality and disease resistance. Carrots are susceptible to diseases such as Alternaria leaf blight and root-knot nematodes, which can cause significant yield losses. EMS-induced mutations have introduced resistance to these diseases, improving the overall productivity and profitability of carrot farming.

Sodium Azide as a Chemical Mutagen

Sodium azide is another widely used chemical mutagen in plant breeding. Like EMS, sodium azide induces point mutations by altering the DNA sequence at the molecular level. Sodium azide works by inhibiting certain enzymes involved in DNA replication and repair, leading to the introduction of mutations in the DNA. These mutations can result in the development of new traits in the plant, such as improved yield, disease resistance, or stress tolerance. One of the advantages of sodium azide is its ability to induce a high frequency of point mutations with relatively few chromosomal aberrations. This makes it an effective tool for creating genetic diversity in crops without causing widespread genetic damage. Sodium azide has been successfully applied in vegetable crops such as lettuce, radishes, and spinach to develop new varieties with enhanced traits.

Sodium Azide-Induced Mutations in Lettuce

Lettuce (*Lactuca sativa*) is a major leafy vegetable crop grown worldwide. However, lettuce is susceptible to a range of diseases, including downy mildew, a fungal disease that can devastate crops. Sodium azide has been used in lettuce breeding programs to induce mutations that confer resistance to downy mildew. By exposing lettuce seeds or tissue cultures to sodium azide, breeders have been able to develop new varieties that exhibit improved resistance to this disease, reducing the need for chemical fungicides and improving crop productivity. In addition to disease resistance, sodium azide has been used to improve the nutritional quality of lettuce. Mutations that increase the levels of vitamins, such as vitamin C and folate, have been introduced through sodium azide treatments. These improvements have made lettuce a more valuable crop from a nutritional standpoint, contributing to better health outcomes for consumers.

Sodium Azide in Other Crops

Sodium azide has also been applied in crops such as radishes and spinach to improve disease resistance, yield, and adaptability to environmental stresses. In radishes, sodium azide-induced mutations have led to the development of varieties with improved resistance to fungal diseases, such as black root rot, which can significantly reduce crop yields. Similarly, in spinach, sodium azide has been used to introduce mutations that enhance the crop's tolerance to abiotic stressors such as drought and salinity, making it more resilient to changing environmental conditions.

Advantages of Using Chemical Mutagens in Vegetable Crops

Chemical mutagens offer several advantages in plant breeding, particularly in inducing specific point mutations that lead to desirable traits. Some of the key benefits include:

1. **Targeted Mutations**: Chemical mutagens such as EMS and sodium azide induce point mutations that cause specific changes in the DNA sequence. This allows breeders to target particular genes responsible for traits like disease resistance, yield improvement, or stress tolerance.
2. **Increased Genetic Variability**: One of the primary goals of mutation breeding is to introduce genetic variability into crops, particularly those that reproduce asexually or have limited genetic diversity. Chemical mutagens are highly effective at generating a wide range of mutations, providing breeders with a larger pool of genetic material from which to select.
3. **Cost-Effectiveness**: Compared to more advanced genetic technologies, such as gene editing, the use of chemical mutagens is relatively inexpensive. This makes it an attractive option for breeding programs, particularly in regions with limited resources.
4. **Application to Asexually Propagated Crops**: Chemical mutagens are particularly useful in crops that propagate asexually, such as sweet potatoes, garlic, and bananas, where traditional breeding methods involving sexual recombination are limited. By introducing mutations through chemical mutagens, breeders can overcome the challenges associated with asexual propagation and develop new varieties with improved traits.

Challenges and Limitations of Chemical Mutagens in Plant Breeding

While chemical mutagens offer significant advantages, there are also challenges and limitations associated with their use. Some of the main challenges include:

1. **Random Nature of Mutations**: Although chemical mutagens can induce targeted point mutations, the mutations occur randomly throughout the genome. This means that not all mutations will result in beneficial traits, and breeders must screen large populations of plants to identify those with desirable characteristics.
2. **Potential for Harmful Mutations**: In some cases, chemical mutagens can induce mutations that negatively affect the plant's growth, development, or fertility. These harmful mutations must be identified and eliminated during the breeding process, which can be time-consuming and labor-intensive.
3. **Safety Concerns**: The use of chemical mutagens requires careful handling and safety precautions, as many of these agents are toxic to humans and the environment. Proper safety protocols must be followed to ensure that the mutagens are used safely and effectively.
4. **Limited Precision**: While chemical mutagens are effective at inducing point mutations, they do not offer the same level of precision as modern genome-editing technologies like CRISPR-Cas9. Breeders must rely on trial and error to identify beneficial mutations, which can slow down the breeding process.

3.Somatic Mutagenesis in Vegetable Crops

Somatic mutagenesis is a powerful tool in the field of plant breeding, particularly for asexually propagated crops. This technique involves applying mutagens to somatic cells, which are non-reproductive cells of the plant. The altered somatic cells are then regenerated into whole plants using tissue culture techniques. The regenerated plants can exhibit new traits, such as enhanced resistance to diseases or pests, improved yield, or better adaptability to environmental conditions. Somatic mutagenesis has been particularly useful in breeding crops like cassava and taro, where traditional breeding methods are limited due to the nature of vegetative propagation.

Asexual Propagation and the Challenges of Genetic Diversity

In crops that propagate asexually, such as cassava, taro, garlic, and potatoes, the reproduction process involves vegetative parts of the plant—tubers, bulbs, or cuttings—rather than seeds. This method ensures that the offspring are genetically identical to the parent plant, which is advantageous for preserving desirable traits. However, a major challenge of asexual propagation is the lack of genetic diversity. Without sexual reproduction, there is no genetic recombination to introduce new variations. This uniformity can be a disadvantage when it comes to developing resistance to diseases, pests, or environmental stressors.Mutation breeding offers a solution to this problem by inducing genetic variation through mutagenesis. In the case of somatic mutagenesis, the process is applied directly to somatic cells, which can come from any part of the plant, including leaves, stems, roots, or other vegetative tissues. These cells are exposed to physical or chemical mutagens, and the resulting mutations can lead to new traits that were not previously present in the plant population.

Somatic Cells and Their Role in Mutagenesis

Somatic cells are any cells of the plant body that are not involved in sexual reproduction. They include cells from the leaves, stems, roots, and other vegetative organs. In somatic mutagenesis, these cells are treated with mutagens—either physical (such as radiation) or chemical (such as EMS or sodium azide)—to induce mutations in the DNA. Once the cells have been altered, they can be cultured in a laboratory setting through a process known as tissue culture, which involves growing the cells in a controlled environment with specific nutrients, hormones, and growth factors that allow them to regenerate into whole plants. Tissue culture is a critical component of somatic mutagenesis because it enables the propagation of the mutated cells into entire plants. The new plants that arise from these cells are then screened for desirable traits, such as disease resistance, improved yield, or stress tolerance. If the mutations are beneficial, the new plants can be propagated further to create a new variety with enhanced traits.

Somatic Mutagenesis in Cassava

Cassava (*Manihot esculenta*) is a major staple crop in tropical regions, particularly in Africa, Asia, and Latin America. It is propagated vegetatively through stem cuttings, which means that traditional breeding methods involving sexual reproduction are not widely used. Cassava is also vulnerable to several pests and diseases, including cassava mosaic virus (CMV), cassava brown streak virus (CBSV), and cassava mealybug. These challenges make cassava a prime candidate for somatic mutagenesis, which can introduce new genetic diversity and help develop varieties with enhanced resistance to these pests and diseases.

Somatic Mutagenesis for Disease Resistance in Cassava

Somatic mutagenesis has been applied successfully to cassava in efforts to develop varieties resistant to cassava mosaic virus (CMV). CMV is a significant threat to cassava production, particularly in sub-Saharan Africa, where it can cause yield losses of up to 80%. Traditional breeding methods have had limited success in combating CMV due to the vegetative nature of cassava propagation, which limits genetic diversity. In somatic mutagenesis, cassava stem cells are exposed to mutagens, such as gamma rays or EMS, to induce mutations. These mutated cells are then cultured in vitro to regenerate whole plants, which are subsequently screened for resistance to CMV. Through this process, researchers have been able to identify cassava plants with mutations that confer resistance to the virus, allowing for the development of new CMV-resistant varieties.

Improving Cassava Yield and Adaptability Through Somatic Mutagenesis

In addition to disease resistance, somatic mutagenesis has been used to improve the yield and adaptability of cassava. Cassava is often grown in poor soils and under harsh environmental conditions, which can limit its productivity. By applying somatic mutagenesis, researchers have been able to develop cassava varieties that are better adapted to low-nutrient soils, drought, and other abiotic stressors. For example, somatic mutagenesis has been used to introduce mutations that improve cassava's tolerance to drought by altering its water-use efficiency. These mutations enable the plant to conserve water more effectively, allowing it to thrive in arid environments. Similarly, somatic mutagenesis has been applied to improve the nutrient-use efficiency of cassava, enabling it to grow in soils with low levels of essential nutrients such as nitrogen or phosphorus. These improvements have significant implications for food security in regions where cassava is a staple crop and where soil quality is poor.

Somatic Mutagenesis in Taro

Taro (*Colocasia esculenta*) is another asexually propagated crop that has benefited from somatic mutagenesis. Taro is an important root crop in tropical and subtropical regions, particularly in the Pacific Islands, Southeast Asia, and parts of Africa. Like cassava, taro is propagated vegetatively, which limits genetic diversity and makes it susceptible to various pests and diseases. Somatic mutagenesis has been used to introduce genetic variability into taro, leading to the development of new varieties with improved traits.

Somatic Mutagenesis for Pest Resistance in Taro

One of the major challenges in taro cultivation is the damage caused by taro leaf blight, a disease caused by the fungal pathogen *Phytophthora colocasiae*. Taro leaf blight can devastate taro crops, particularly in humid tropical regions where the fungus thrives. Traditional breeding methods have been ineffective in combating taro leaf blight due to the vegetative nature of taro propagation. Somatic mutagenesis has been used to address this challenge by introducing mutations that confer resistance to taro leaf blight. In this process, taro plantlets or corm cells are exposed to mutagens such as gamma rays or chemical agents. The mutated cells are then cultured in vitro to regenerate whole plants, which are screened for resistance to taro leaf blight. Through this approach, researchers have been able to develop taro varieties that are more resistant to the disease, reducing the need for chemical fungicides and improving crop productivity.

Improving Taro Yield and Nutritional Quality Through Somatic Mutagenesis

In addition to pest resistance, somatic mutagenesis has been used to improve taro yield and nutritional quality. Taro is an important source of carbohydrates in many regions, but its yield can be limited by various factors, including poor soil conditions and environmental stressors. By applying somatic mutagenesis, researchers have been able to develop taro varieties that are better adapted to challenging growing conditions, such as low-nutrient soils or waterlogged environments. Somatic mutagenesis has also been used to improve the nutritional content of taro. For example, mutations that increase the levels of essential nutrients, such as vitamins or minerals, have been introduced through somatic mutagenesis. These nutritional improvements make taro a more valuable food source for populations that rely on the crop as a staple, contributing to improved food security and public health.

Tissue Culture in Somatic Mutagenesis

Tissue culture is a critical component of somatic mutagenesis, as it allows for the regeneration of whole plants from mutated somatic cells. In tissue culture, small pieces of plant tissue, such as leaf segments, stem cuttings, or root sections, are placed in a nutrient-rich medium that contains essential growth hormones and nutrients. The tissue is then exposed to a controlled environment with specific light, temperature, and humidity conditions to promote cell division and growth. In the case of somatic mutagenesis, the plant tissue is first exposed to mutagens to induce mutations in the DNA. Once the mutations are introduced, the tissue is cultured in vitro to regenerate into whole plants. This process is known as somatic embryogenesis, where the mutated cells develop into embryos and eventually grow into fully formed plants. The regenerated plants are then screened for desirable traits, such as disease resistance or improved yield, and those with beneficial mutations are selected for further breeding. Tissue culture techniques are particularly useful in somatic mutagenesis because they allow for the rapid multiplication of mutated plants. Once a beneficial mutation is identified, tissue culture can be used to propagate the new variety on a large scale. This accelerates the breeding process and enables the rapid dissemination of new varieties to farmers.

Applications of Somatic Mutagenesis in Other Vegetable Crops

While somatic mutagenesis has been particularly successful in crops like cassava and taro, it has also been applied to other asexually propagated vegetable crops. For example, in potatoes, somatic mutagenesis has been used to develop varieties with improved resistance to pests and diseases. Potatoes are typically propagated through tubers, and traditional breeding methods are limited in their ability to introduce new genetic diversity. Somatic mutagenesis offers a way to overcome this limitation by inducing mutations in potato cells and regenerating whole plants through tissue culture.In garlic, somatic mutagenesis has been used to develop varieties with enhanced resistance to viral diseases. Garlic is propagated through cloves, which means that genetic diversity is limited, and viral diseases can spread rapidly through clonal propagation. By inducing mutations in garlic cells, researchers have been able to develop new varieties that are more resistant to viral infections, improving the health and productivity of garlic crops.

Advantages and Challenges of Somatic Mutagenesis

Advantages of Somatic Mutagenesis

One of the primary advantages of somatic mutagenesis is its ability to introduce genetic diversity in asexually propagated crops. This is particularly important in crops that have limited genetic variability due to vegetative propagation. By inducing mutations at the somatic cell level, breeders can develop new traits that would be difficult or impossible to achieve through traditional breeding methods. Another advantage of somatic mutagenesis is its ability to target specific traits. By applying mutagens to somatic cells and screening the resulting plants, breeders can select for traits such as disease resistance, improved yield, or enhanced stress tolerance. This targeted approach allows for the development of new varieties with specific improvements, making it a valuable tool in plant breeding programs.

Challenges of Somatic Mutagenesis

Despite its advantages, somatic mutagenesis also presents challenges. One of the main challenges is the randomness of the mutations induced by mutagens. Not all mutations will result in beneficial traits, and some may even be harmful to the plant. This means that breeders must carefully screen large populations of mutated plants to identify those with desirable characteristics.Another challenge is the complexity of the tissue culture process. While tissue culture is essential for regenerating plants from mutated cells, it requires specialized equipment, expertise, and facilities. This can be a limiting factor for breeding programs in regions with limited resources or technical capacity.

Screening and Selection of Mutants

Once mutations are induced, the next crucial step in mutation breeding is the screening and selection of plants with desirable traits. This process involves growing the mutated plants and evaluating them for characteristics like disease resistance, yield, and environmental tolerance. For asexually propagated crops, this process may be time-consuming, as the vegetative parts must first grow and then undergo evaluation. However, tissue culture techniques have accelerated the process by enabling faster multiplication of mutated cells into whole plants.

Vegetables Suitable for Protected Environments and Mutation Breeding

Protected environments like greenhouses, high tunnels, and controlled environment agriculture systems are increasingly being used for vegetable production. In these systems, environmental variables such as temperature, humidity, and light can be precisely controlled, allowing year-round cultivation and protecting crops from adverse weather conditions. However, these environments also present specific challenges, such as the need for varieties adapted to controlled conditions, pest and disease management, and efficient resource use. Mutation breeding provides a solution to some of these challenges by allowing the development of vegetable varieties better suited to growth in protected environments. By introducing beneficial mutations, breeders can create crops that are more efficient in their use of resources like water and nutrients, have enhanced resistance to pests and diseases, and are better suited to the microclimates of controlled environments.

Examples of Vegetables for Protected Environments Improved Through Mutation Breeding

- **Tomatoes**: Greenhouse tomato production requires varieties that can thrive under controlled environmental conditions. Mutation breeding has been used to develop tomato varieties with increased resistance to common greenhouse pests and diseases, such as whiteflies and powdery mildew. In addition, mutation breeding has been used to improve the flavor, texture, and shelf life of tomatoes, making them more suitable for long-term storage and transport.
- **Cucumbers**: Cucumbers are another crop that is commonly grown in protected environments. Mutation breeding has contributed to the development of varieties that are resistant to diseases like downy mildew and powdery mildew, which can be prevalent in greenhouse conditions. These mutations have also improved the yield and quality of cucumbers, making them more adaptable to the specific requirements of controlled environments.
- **Lettuce**: Lettuce is highly sensitive to environmental conditions, particularly temperature and light. In protected environments, where conditions are carefully controlled, mutation breeding has led to the development of lettuce varieties that are more heat-tolerant and resistant to bolting (premature flowering), ensuring consistent production throughout the year.
- **Peppers**: Peppers are another vegetable commonly cultivated in greenhouses. Mutation breeding has enabled the development of pepper varieties with improved resistance to pests like aphids and diseases such as Phytophthora blight. Additionally, mutations have led to better fruit quality and extended shelf life, which are important traits for commercial production in protected environments.

Advantages of Mutation Breeding for Vegetables in Protected Environments

1. **Resource Efficiency**: Mutation breeding can lead to the development of vegetable varieties that are more efficient in their use of resources, such as water and nutrients. This is particularly important in controlled environments, where optimizing resource use is key to reducing production costs and environmental impact.
2. **Pest and Disease Resistance**: In protected environments, pest and disease pressure can be higher due to the relatively confined space and stable conditions. Mutation breeding can be used to develop varieties with enhanced resistance to pests and diseases, reducing the need for chemical inputs and improving the sustainability of production systems.

3. **Adaptability to Controlled Conditions**: Vegetables grown in greenhouses or other protected environments must be adapted to the specific microclimates of these systems. Mutation breeding can introduce traits that make plants more tolerant to the temperature, humidity, and light levels found in controlled environments, improving yield and quality.
4. **Extended Growing Seasons**: Protected environments allow for year-round vegetable production, but this requires varieties that can perform well across different seasons. Mutation breeding can develop varieties with enhanced resilience to varying light and temperature conditions, ensuring stable production throughout the year.

B. Polyploidy Breeding in Vegetable Crops: Improvement of Asexually Propagated Vegetable Crops and Vegetables Suitable for Protected Environments

Polyploidy breeding is a fundamental approach in plant breeding that involves altering the number of chromosome sets in a plant. In asexual crops and those grown in controlled or protected environments, polyploidy offers the potential to enhance traits such as yield, disease resistance, and environmental adaptability. Polyploidy is achieved through chromosome doubling or introducing multiple sets of chromosomes beyond the normal diploid number, often resulting in significant changes in plant characteristics.

Mitotic and meiotic divisions in plants are highly precise processes that maintain chromosome stability. However, irregularities during these divisions can lead to changes in chromosome number, giving rise to plants with chromosome counts different from their species' norm. Such changes, particularly in the context of polyploidy, have had a significant impact on crop evolution and are highly valuable in plant breeding.

Types of Chromosome Changes in Polyploidy Breeding

In the study of plant breeding, the chromosome number of a species is critical. The normal somatic chromosome number is referred to as 2n, while the gametic chromosome number is denoted as n. Plants with a chromosome number equal to their gametic number are known as haploids. The basic chromosome number, x, refers to the number of chromosomes that make up a genome. Diploid species, for example, have two sets of chromosomes (2n), while polyploid species have more than two sets of chromosomes.In polyploidy breeding, changes in chromosome number can occur through two primary processes: aneuploidy and euploidy. Aneuploidy refers to the gain or loss of one or a few chromosomes, while euploidy refers to changes involving whole genomes. Euploid plants have a chromosome number that is a multiple of the basic chromosome number, whereas aneuploid plants do not. Polyploids are a type of euploid, characterized by having multiple copies of entire genomes.

Polyploidy can be categorized into two main types:

» **Autopolyploidy**: Occurs when multiple sets of identical genomes are present in a species. Autopolyploids often exhibit enhanced size and vigor due to their increased genetic content.

» **Allopolyploidy**: Involves the combination of distinct genomes from different species, leading to hybrids that possess traits from both parent species. Allopolyploidy is common in hybrid vegetable breeding, where breeders seek to combine desirable traits from two species into a single plant.

1. Polyploidy Breeding for Asexually Propagated Crops

Asexually propagated vegetable crops, such as garlic, potatoes, and sweet potatoes, pose unique challenges for traditional plant breeding due to their method of reproduction. These crops reproduce vegetatively, meaning they use parts of the plant such as tubers, bulbs, or cuttings instead of seeds. This method ensures that the offspring are genetically identical to the parent plant, preserving desirable traits across generations. However, this genetic uniformity limits genetic diversity, which is critical for developing resistance to diseases, pests, and environmental stressors. Traditional breeding techniques, which rely on sexual reproduction to introduce new genetic traits, are not applicable to these crops. Polyploidy breeding offers a viable solution by inducing genetic changes that enhance the traits of asexually propagated crops without requiring sexual recombination.

Polyploidy, the condition of having more than two sets of chromosomes, introduces genetic variability into crops by altering their chromosomal structure. This method allows breeders to enhance the characteristics of asexually propagated crops such as garlic and potatoes by doubling or combining chromosome sets. Through polyploidy breeding, new vegetable varieties can be developed with improved disease resistance, higher yields, and better adaptability to environmental conditions, addressing the limitations posed by asexual propagation. This approach has proven effective in improving several key vegetable crops, including garlic, potatoes, bananas, and cassava.

Challenges in Breeding Asexually Propagated Crops

Asexual propagation, while advantageous in maintaining desirable traits, presents several challenges for plant breeders. The primary challenge is the lack of genetic diversity, which makes asexually propagated crops more vulnerable to diseases, pests, and environmental changes. In sexually reproducing crops, the recombination of genes during meiosis introduces variability that can be selected for traits like disease resistance or improved yield. However, in asexually reproducing crops, this natural process of genetic recombination does not occur. As a result, the genetic makeup of the crop remains unchanged across generations, limiting its ability to adapt to new challenges.

Another challenge is the difficulty in introducing new traits through traditional breeding methods. In most plant breeding programs, hybridization between two parent plants introduces genetic variability, which can be exploited to select for improved traits. However, because asexually propagated crops do not undergo sexual reproduction, hybridization is not possible. This restricts the options for introducing new traits into these crops and slows the progress of breeding programs aimed at improving asexually propagated vegetables. Additionally, asexually propagated crops are more susceptible to disease and pest outbreaks because of their genetic uniformity. When a population of plants is genetically identical, it lacks the diversity needed to fight off pathogens and pests effectively. A disease that affects one plant is likely to affect the entire crop, leading to significant losses. For instance, the Irish potato famine of the 1840s was caused by the rapid spread of late blight in genetically uniform potato crops. A more genetically diverse population might have included plants with natural resistance to the disease, helping to mitigate the disaster. Given these challenges, polyploidy breeding emerges as a critical tool for improving asexually propagated crops. By altering the chromosome number, breeders can introduce new genetic variation into these crops, helping them to overcome their inherent limitations and making them more resilient and productive.

Polyploidy Breeding: Enhancing Genetic Diversity

Polyploidy breeding involves manipulating the number of chromosome sets in a plant to increase genetic diversity and improve plant traits. This process can involve either autopolyploidy, where the chromosome sets come from a single species, or allopolyploidy, where the chromosome sets come from different species. Both types of polyploidy offer unique advantages for breeding asexually propagated crops, allowing breeders to introduce genetic variation without the need for sexual reproduction. Polyploidy is particularly effective in asexually propagated crops because it allows breeders to create genetic diversity in crops that otherwise lack it. By doubling or combining chromosome sets, polyploidy can introduce new traits such as disease resistance, improved yield, and enhanced environmental adaptability. For example, in crops like potatoes and garlic, which are propagated through vegetative means, polyploidy breeding has been used to develop varieties with enhanced disease resistance and greater productivity.

Polyploidy breeding also offers the advantage of increased vigor, known as hybrid vigor or heterosis, in polyploid plants. Polyploid plants often exhibit enhanced growth, larger fruit size, and greater resistance to environmental stressors compared to their diploid counterparts. This is because the increased number of chromosome sets leads to a greater variety of gene combinations, which can result in improved plant performance. These benefits make polyploidy breeding an attractive option for improving asexually propagated vegetable crops.

Autopolyploidy in Asexually Propagated Crops

Autopolyploidy is a form of polyploidy in which an organism has more than two sets of chromosomes, all derived from the same species. In plant breeding, autopolyploidy involves doubling the chromosome number within a single species, leading to the development of plants with enhanced traits. This method is particularly useful in asexually propagated crops because it increases genetic diversity and introduces new traits without requiring sexual reproduction. In autopolyploidy, the additional chromosome sets can result in larger, more vigorous plants with improved characteristics. This is especially important in crops like potatoes, garlic, and sweet potatoes, where traditional breeding methods are not feasible due to the lack of sexual reproduction. By doubling the chromosome number, autopolyploidy allows breeders to develop new varieties of these crops with improved traits, such as greater resistance to diseases and pests, higher yields, and better adaptability to environmental conditions.

Autopolyploidy in Potatoes

Potatoes (*Solanum tuberosum*) are a key global staple food, and improving their disease resistance and yield is a major focus of breeding programs. Potatoes are propagated vegetatively through tubers, which limits their genetic diversity and makes them susceptible to diseases like late blight, a fungal infection caused by *Phytophthora infestans*. Traditional breeding methods have had limited success in improving potato varieties due to the challenges posed by asexual propagation.

Autopolyploidy has been successfully used in potato breeding to develop tetraploid varieties with enhanced disease resistance and higher yields. Tetraploid potatoes, which have four sets of chromosomes instead of the usual two, exhibit greater resistance to late blight and other diseases. The increased genetic content in tetraploid potatoes allows for a more robust immune response, enabling them to better resist infections and recover from disease. Additionally, tetraploid potatoes often produce larger tubers and have higher overall yields compared to their diploid counterparts. One of the key benefits of autopolyploidy in potatoes is the ability to combine favorable traits from different genetic backgrounds. By doubling the chromosome number, breeders can introduce greater genetic diversity into potato populations, allowing for the selection of traits that improve disease resistance, yield, and storage qualities. For example, tetraploid potato varieties have been developed that are more resistant to bacterial wilt, a devastating soil-borne disease that affects potato crops in tropical regions. These varieties not only resist infection but also produce larger tubers, making them more profitable for farmers.

Autopolyploidy in Garlic

Garlic (*Allium sativum*) is another asexually propagated crop that has benefited from autopolyploidy breeding. Garlic is propagated through cloves, which are vegetative parts of the plant. This method of reproduction ensures that the offspring are genetically identical to the parent plant, preserving desirable traits such as flavor and yield. However, it also limits genetic diversity, making garlic more susceptible to diseases like garlic mosaic virus, a viral infection that can spread rapidly through vegetative propagation.

Autopolyploidy has been used in garlic breeding to create varieties with larger bulbs, more robust plants, and improved resistance to viral diseases. By doubling the chromosome number in garlic, breeders have been able to introduce new genetic variability that helps the crop withstand viral infections and produce higher yields. For example, autopolyploid garlic varieties have been developed that are more resistant to garlic mosaic virus, reducing the need for chemical treatments and improving crop productivity. In addition to disease resistance, autopolyploid garlic varieties often exhibit larger bulbs and higher yields compared to their diploid counterparts. This is due to the increased genetic content in autopolyploid plants, which allows for greater cell size and more efficient nutrient uptake. As a result, autopolyploid garlic varieties are more vigorous and productive, making them an attractive option for commercial garlic production.

Allopolyploidy in Asexually Propagated Crops

Allopolyploidy is a form of polyploidy that involves combining chromosome sets from two or more different species. This method is particularly useful in hybrid vegetable breeding, where breeders seek to combine desirable traits from different species into a single plant. Allopolyploidy allows for the creation of hybrids that possess traits from both parent species, such as improved disease resistance, higher yields, and better environmental adaptability.In asexually propagated crops, allopolyploidy offers the advantage of introducing new genetic material without the need for sexual reproduction. This is especially important in crops like bananas, plantains, and cassava, which are vegetatively propagated and have limited genetic diversity. By combining chromosome sets from different species, allopolyploidy allows breeders to develop hybrids that are more resilient and productive.

Allopolyploidy in Bananas and Plantains

Bananas (*Musa spp.*) and plantains are important staple crops in tropical regions, where they serve as a primary source of food and income for millions of people. These crops are propagated vegetatively through suckers, which limits their genetic diversity and makes them susceptible to diseases like black Sigatoka, a fungal infection that can devastate banana plantations. Traditional breeding methods have had limited success in improving banana and plantain varieties due to the challenges posed by improved crop characteristics. However, for asexually propagated crops like garlic, potatoes, and sweet potatoes, this process is not feasible because these crops reproduce from vegetative parts rather than seeds. This limitation necessitates alternative methods, such as polyploidy breeding, to introduce genetic diversity and enhance traits.

Autopolyploidy in Asexual Crops

Autopolyploidy refers to the condition where a plant has multiple sets of chromosomes all derived from a single species. This is achieved by chromosome doubling, which results in plants with more than two sets of chromosomes. Autopolyploidy is particularly beneficial for crops that are vegetatively propagated, as it can lead to enhanced plant vigor, increased size, and improved resistance to diseases.

Potatoes are a prime example of how autopolyploidy can be applied to asexually propagated crops. Traditionally, potatoes are propagated from tubers, which results in limited genetic diversity. By inducing autopolyploidy, breeders have developed tetraploid potato varieties, which have four sets of chromosomes instead of the usual two. These tetraploid potatoes exhibit several advantages over their diploid counterparts. They tend to have higher yields, as the increased chromosome number contributes to greater plant growth and tuber production. Additionally, tetraploid potatoes have shown improved resistance to diseases such as late blight (Phytophthora infestans), a major threat to potato crops worldwide. The increased genetic content in tetraploid potatoes enhances their ability to resist infection and recover from disease, thereby improving their overall resilience and productivity.

Garlic, another asexually propagated crop, has also benefited from autopolyploidy. Garlic is traditionally propagated through cloves, which limits genetic variability. By inducing chromosome doubling in garlic, breeders have created larger bulbs and more robust plants. These autopolyploid garlic varieties are often more resistant to viral diseases, such as garlic mosaic virus, which can spread rapidly through vegetative propagation. The increased chromosome number introduces new genetic variability that helps the crop withstand viral infections and produce higher yields. The enhanced traits of autopolyploid garlic make it more competitive in the market and more resilient to common agricultural challenges.

Sweet potatoes are another example where autopolyploidy has been utilized to improve crop traits. By doubling the chromosome number in sweet potatoes, breeders have developed varieties with increased tuber size, improved disease resistance, and better storage qualities. The increased genetic content in autopolyploid sweet potatoes contributes to their enhanced adaptability to different environmental conditions and their ability to recover from stress. This results in more productive and resilient sweet potato varieties that can thrive in a wider range of growing conditions.

Allopolyploidy in Asexual Crops

Allopolyploidy involves the combination of chromosome sets from two different species, resulting in a hybrid plant with multiple sets of chromosomes from distinct parental species. This process introduces new genetic material into the plant, which can lead to the development of new traits that are beneficial for crop improvement. Allopolyploidy is particularly useful in hybrid vegetable breeding, where breeders aim to combine desirable traits from different species.

Bananas and plantains are examples of crops that have benefited from allopolyploidy. These crops are propagated asexually through suckers or shoots, which limits genetic diversity. By creating allopolyploid varieties, breeders have been able to combine the genetic material of different banana and plantain species to develop new varieties with improved traits. For instance, allopolyploid bananas have been developed with increased resistance to diseases such as Black Sigatoka (*Mycosphaerella fijiensis*), a major fungal disease affecting banana plantations globally. The introduction of additional chromosome sets from related species enhances the plant's ability to resist infection and recover from disease, resulting in more resilient and productive banana and plantain varieties.

Cassava, another important asexually propagated crop, has also benefited from allopolyploidy. Cassava is propagated through cuttings, which restricts genetic variability. By creating allopolyploid cassava varieties, breeders have introduced new genetic material from related species, resulting in plants with improved traits such as increased tuber yield, better resistance to pests and diseases, and enhanced adaptability to different growing conditions. The increased genetic diversity in allopolyploid cassava contributes to its improved performance and resilience in diverse environments.

Applications and Benefits of Polyploidy Breeding

The application of polyploidy breeding in asexually propagated crops offers several benefits. By introducing additional chromosome sets, breeders can create new plant varieties with enhanced traits such as increased yield, improved disease resistance, and better adaptability to environmental stressors. Polyploidy breeding provides a means to overcome the limitations of asexual propagation, which typically restricts genetic diversity and limits the ability to introduce new traits through traditional breeding methods.

Enhanced Yield: One of the primary benefits of polyploidy breeding is the potential for increased yield. Polyploid plants often exhibit greater plant vigor and larger sizes, leading to higher production of tubers, bulbs, or other vegetative parts. For example, tetraploid potatoes and autopolyploid sweet potatoes have been shown to produce larger tubers and higher yields compared to their diploid counterparts. The increased genetic content in polyploid plants contributes to their enhanced growth and productivity.

Improved Disease Resistance: Polyploidy breeding can also lead to improved disease resistance. The introduction of additional chromosome sets can enhance the plant's ability to resist diseases and recover from infections. This is particularly important for asexually propagated crops, which may lack genetic diversity and be more susceptible to diseases. For instance, tetraploid potatoes and autopolyploid garlic varieties have shown increased resistance to diseases such as late blight and garlic mosaic virus, respectively. The enhanced disease resistance contributes to the overall health and productivity of the crops.

Better Adaptability: Polyploidy breeding can also improve the adaptability of crops to different environmental conditions. Polyploid plants often exhibit greater tolerance to environmental stressors such as drought, temperature fluctuations, and soil conditions. This enhanced adaptability allows the crops to thrive in a wider range of growing conditions and improves their overall performance. For example, allopolyploid cassava and sweet potatoes have demonstrated better adaptability to diverse environmental conditions, resulting in more resilient and productive varieties.

Increased Genetic Diversity: By introducing additional chromosome sets, polyploidy breeding increases the genetic diversity of asexually propagated crops. This increased genetic variability allows breeders to develop new plant varieties with novel traits and improved performance. The ability to introduce new genetic material through polyploidy provides a valuable tool for enhancing the traits of asexually propagated crops and addressing the challenges associated with limited genetic diversity.

Challenges and Considerations

While polyploidy breeding offers significant benefits, it also presents challenges and considerations. The process of inducing polyploidy and developing new varieties requires careful management and expertise. Breeders must consider factors such as the stability of the polyploid plants, their potential for producing viable offspring, and the overall impact on crop performance.

Stability and Viability: Polyploid plants may exhibit varying degrees of stability and viability, depending on the type of polyploidy and the specific crop. Some polyploid plants may have difficulties in producing viable seeds or may exhibit reduced fertility. Breeders must ensure that the polyploid varieties they develop are stable and capable of producing viable offspring to ensure the success of the breeding program.

Performance and Adaptability: Polyploid plants may also exhibit changes in performance and adaptability compared to their diploid counterparts. Breeders must carefully evaluate the performance of polyploid varieties in different growing conditions to ensure that they meet the desired traits and perform well in the target environments. The potential benefits of polyploidy must be balanced with considerations of plant performance and adaptability.

Economic and Practical Considerations: The development of polyploid varieties requires investment in research, technology, and resources. Breeders must consider the economic and practical aspects of polyploidy breeding, including the costs associated with inducing polyploidy, managing tissue cultures, and evaluating plant performance. The benefits of polyploidy must be weighed against the costs and practical considerations to ensure the success of the breeding program.

2. Polyploidy Breeding for Protected Environments

Vegetables cultivated in protected environments such as greenhouses, high tunnels, or indoor farms are exposed to a unique set of challenges and opportunities. These controlled settings offer numerous benefits, including the ability to grow crops year-round and safeguard them from harsh weather conditions. However, the specific demands of these environments necessitate that vegetable crops be specially adapted to thrive under controlled conditions. Polyploidy breeding has emerged as a powerful tool to develop vegetable varieties that are better suited to the microclimates and resource constraints typical of protected environments. By manipulating the number of chromosome sets, polyploidy breeding introduces genetic variations that enhance essential traits for optimal performance in these settings.

Challenges and Adaptations in Protected Environments

Protected environments create a microclimate that differs significantly from open-field conditions. This artificial climate—characterized by controlled temperature, humidity, and light—imposes distinct requirements on vegetable crops. To optimize growth and productivity, crops must be specifically adapted to these conditions. Polyploidy breeding addresses these needs by enhancing traits that improve plant performance in protected environments.

1. Temperature Regulation:

Maintaining consistent temperatures is crucial in protected environments. Fluctuations in temperature can impact plant physiology, affecting processes such as flowering, fruit set, and overall growth. In greenhouses and high tunnels, temperatures are regulated to create optimal growing conditions, but these conditions may not always align with the natural preferences of the plants. Polyploidy breeding can help by developing vegetable varieties with improved temperature tolerance. For instance, certain polyploid tomato varieties exhibit enhanced adaptability to the temperature ranges typically found in greenhouses. These varieties are capable of maintaining stable growth and fruit production even when temperatures deviate from the ideal range. Similarly, polyploid peppers can be bred to withstand temperature variations, ensuring consistent yield and quality.

2. Humidity Management:

High humidity levels prevalent in protected environments create conditions that are conducive to fungal diseases and mold. Limited air circulation in greenhouses and high tunnels exacerbates these issues, increasing the risk of pathogens such as powdery mildew and downy mildew. Polyploidy breeding offers a solution by developing vegetable varieties with improved disease resistance and thicker foliage. For example, polyploid cucumbers with increased leaf thickness have demonstrated greater resilience to powdery mildew, a common problem in high-humidity environments. By enhancing the plant's natural defenses, polyploidy breeding reduces the need for chemical fungicides, promoting more sustainable cultivation practices.

3. Nutrient and Space Efficiency:

Protected environments often involve high-density planting and restricted space, which can limit root growth and nutrient availability. Efficient use of resources is essential for maximizing plant productivity. Polyploidy breeding can enhance the efficiency with which plants absorb and utilize nutrients. For instance, polyploid vegetable varieties may exhibit improved root development and nutrient uptake, allowing them to thrive even in confined spaces. This is particularly beneficial in high-density planting scenarios where competition for resources is intense. By improving resource-use efficiency, polyploidy breeding helps ensure that plants receive adequate nutrition and space to support optimal growth and yield.

4. Improved Stress Tolerance:

Vegetable crops in protected environments are exposed to different types of stress compared to those grown in open fields. These stresses may include variations in light intensity, humidity, and temperature. Polyploidy breeding can enhance stress tolerance by developing varieties that are better equipped to handle these controlled conditions. For example, polyploid varieties of lettuce and spinach have shown improved resistance to light stress, allowing them to grow more effectively under artificial lighting conditions in greenhouses. Similarly, polyploid varieties of herbs such as basil and cilantro can be bred to withstand variations in humidity and temperature, ensuring consistent quality and yield throughout the growing season.

Enhancing Yield and Disease Resistance

One of the primary benefits of polyploidy breeding in protected environments is the potential for increased yield. Polyploid vegetables often exhibit larger fruit size, greater biomass, and improved overall productivity due to their increased genetic content. This is particularly important in commercial greenhouse production, where consistency in quality and size is critical for marketability.

1. Increased Fruit Size and Biomass: Polyploidy breeding can lead to the development of vegetable varieties with larger fruit size and greater biomass. For instance, tetraploid tomatoes grown in greenhouses have been found to produce larger and more uniform fruit compared to diploid varieties. This increase in fruit size and uniformity is crucial for meeting market standards and maximizing profitability. Similarly, polyploid varieties of cucumbers and peppers can exhibit enhanced growth and fruit production, making them more competitive in the commercial market.

2. Enhanced Disease Resistance: Protected environments are susceptible to specific diseases and pests that can impact crop health and yield. Polyploidy breeding can enhance disease resistance by developing varieties with thicker leaves, improved immune responses, or other traits that reduce susceptibility to pathogens. For example, polyploid cucumbers with enhanced resistance to powdery mildew have demonstrated better performance in high-humidity environments, reducing the need for chemical treatments and promoting more sustainable cultivation practices. Likewise, polyploid peppers with improved resistance to fungal diseases can help ensure a more reliable and high-quality yield.

Resource Efficiency and Stress Tolerance

Polyploidy breeding can also address the challenges of resource efficiency and stress tolerance in protected environments. Efficient use of resources such as water, nutrients, and space is essential for optimizing plant growth and productivity.

1. Improved Resource Utilization: Protected environments often involve high-density planting and limited space, which can strain resources such as water and nutrients. Polyploidy breeding can enhance the efficiency with which plants absorb and utilize these resources. For example, polyploid vegetable varieties may exhibit improved root development and nutrient uptake, allowing them to make better use of available resources. This improved resource utilization is particularly beneficial in high-density planting scenarios, where competition for resources is intense.

2. Enhanced Stress Tolerance: Vegetable crops in protected environments are exposed to various types of stress, including fluctuations in light intensity, humidity, and temperature. Polyploidy breeding can enhance stress tolerance by developing varieties that are better equipped to handle these controlled conditions. For instance, polyploid lettuce and spinach varieties have shown improved resistance to light stress, allowing them to grow more effectively under artificial lighting conditions. Similarly, polyploid herbs such as basil and cilantro can be bred to withstand variations in humidity and temperature, ensuring consistent quality and yield throughout the growing season.

Case Studies and Examples

1. Polyploid Tomatoes: Polyploidy breeding has been successfully applied to tomatoes grown in greenhouses. Tetraploid tomatoes, with four sets of chromosomes, have been developed to produce larger and more uniform fruit. These varieties exhibit improved yield and quality, making them highly desirable for commercial greenhouse production. The enhanced disease resistance and better adaptability to greenhouse conditions contribute to their success in the market.

2. Polyploid Cucumbers: Polyploid cucumbers have demonstrated improved resistance to powdery mildew, a common problem in high-humidity environments. By increasing the chromosome number, breeders have developed cucumber varieties with thicker leaves and enhanced disease resistance. These polyploid cucumbers are better suited to greenhouse conditions, reducing the need for chemical fungicides and promoting more sustainable cultivation practices.

3. Polyploid Peppers: Polyploid peppers have been bred to exhibit enhanced resistance to fungal diseases and improved fruit production. The increased genetic content in polyploid peppers contributes to better stress tolerance and resource efficiency, making them well-suited to the controlled conditions of protected environments. These varieties have shown promise in greenhouse production, where consistent quality and disease resistance are crucial.

Applications and Advantages of Polyploidy Breeding in Vegetable Crops

Polyploidy breeding has emerged as a significant advancement in the field of plant breeding, offering various benefits for the improvement of both asexually propagated vegetable crops and those cultivated in protected environments. By manipulating the chromosome number in plants, polyploidy breeding introduces genetic variations that can lead to enhanced traits and performance. Here, we explore the key applications and advantages of polyploidy breeding in vegetable crops, emphasizing its impact on yield, disease resistance, resource efficiency, adaptability, and genetic diversity.

1. Increased Yield and Vigor

Polyploidy often results in plants that are larger, more vigorous, and capable of producing higher yields. The increase in chromosome number can lead to several beneficial physiological changes in the plant. For instance, polyploid plants often exhibit increased cell size, which can translate into larger fruit or tubers. This trait is particularly valuable in commercial vegetable production, where higher yields directly contribute to increased profitability.

Increased vigor is another notable advantage of polyploidy. Polyploid plants tend to have a more robust growth habit, enabling them to better compete with weeds and withstand adverse conditions. For example, in potatoes, polyploid varieties have been observed to produce more substantial tubers and higher overall biomass compared to their diploid counterparts. This increase in biomass and yield is beneficial for commercial growers seeking to maximize productivity and profitability. Additionally, polyploidy can enhance the overall resilience of the plant, making it better equipped to handle stress factors such as drought or nutrient deficiencies. The improved vigor associated with polyploid plants contributes to their ability to maintain high levels of productivity even under challenging growing conditions. This trait is particularly important in protected environments, where controlled conditions can sometimes introduce stressors that affect plant performance.

2. Improved Disease Resistance

Polyploidy breeding can significantly enhance a plant's ability to resist diseases. By increasing the genetic content of a plant, polyploidy introduces additional copies of genes, which can contribute to a more robust immune system. This is especially important in asexually propagated crops, which are often susceptible to disease outbreaks due to their genetic uniformity. The lack of genetic diversity in these crops means that a single disease or pest can potentially wipe out an entire crop.

For instance, in crops like garlic, which are propagated through cloves and thus have limited genetic diversity, polyploidy has been used to develop varieties with improved resistance to viral and fungal diseases. The increased genetic content in polyploid garlic plants enhances their ability to mount a defense against pathogens, reducing the risk of disease and improving overall plant health. In protected environments, where the risk of fungal and bacterial diseases can be heightened due to high humidity and limited airflow, polyploidy breeding can be particularly beneficial. Polyploid cucumbers and peppers, for example, have demonstrated improved resistance to common greenhouse pathogens such as powdery mildew and downy mildew. By developing varieties with enhanced disease resistance, polyploidy breeding helps reduce the need for chemical treatments and supports more sustainable cultivation practices.

3. Enhanced Resource Efficiency

Efficient use of resources such as water, nutrients, and space is crucial in protected environments where resources can be limited. Polyploidy breeding can improve a vegetable crop's ability to absorb and utilize nutrients more effectively. This enhanced resource efficiency is particularly valuable in high-density planting systems common in greenhouses and high tunnels.

Polyploid vegetables often exhibit improved nutrient uptake and utilization, allowing them to thrive in confined spaces and with limited resources. For example, polyploid varieties of tomatoes and peppers have been shown to utilize nutrients more efficiently, leading to better growth and yield despite the restricted availability of nutrients in protected environments. This improved efficiency not only supports higher productivity but also reduces the need for excessive fertilization, contributing to more sustainable cultivation practices. In addition to nutrient efficiency, polyploidy breeding can enhance water-use efficiency. Polyploid plants may exhibit improved water retention and reduced transpiration rates, which helps them cope with limited water availability. This trait is especially important in protected environments where water resources may be constrained, and efficient water use is essential for maintaining plant health and productivity.

4. Adaptability to Controlled Environments

Protected environments impose specific conditions that can differ significantly from open-field cultivation. These controlled settings, including greenhouses and high tunnels, require vegetable crops to be adapted to factors such as temperature, humidity, and light intensity. Polyploidy breeding offers a way to develop varieties that are better suited to these conditions, improving their adaptability and performance in protected environments.

Polyploidy breeding can produce vegetable varieties with traits that enhance their ability to thrive in controlled microclimates. For example, polyploid tomatoes have been bred to exhibit better performance under the specific temperature and light conditions found in greenhouses. These varieties may have improved heat tolerance, allowing them to grow and produce fruit consistently even when temperatures fluctuate. Similarly, polyploidy breeding can enhance a crop's ability to adapt to high humidity levels typical of protected environments. Varieties with thicker foliage or more robust immune responses can better resist fungal diseases and maintain healthy growth. This adaptability is crucial for maintaining consistent production and quality in controlled environments where external conditions can be tightly regulated.

5. Increased Genetic Diversity

Asexually propagated crops, by their nature, tend to have low genetic diversity. This lack of diversity can make them more susceptible to diseases, pests, and environmental stresses. Polyploidy breeding addresses this issue by introducing new genetic variability into these crops. The increased chromosome number resulting from polyploidy creates additional genetic variation, which can enhance the plant's resilience and adaptability. For example, in crops like potatoes and garlic, which are propagated vegetatively and thus have limited genetic diversity, polyploidy breeding introduces new genetic material that can improve disease resistance and overall performance. By creating polyploid varieties, breeders can develop crops that are better able to withstand diseases, adapt to changing environmental conditions, and perform more consistently under various stressors. In addition to enhancing resilience, increased genetic diversity from polyploidy breeding can also facilitate the development of new varieties with desirable traits. Breeders can exploit this genetic variability to introduce novel characteristics such as improved flavor, texture, or yield. The ability to create a broader range of genetic combinations enhances the potential for developing innovative and high-performing vegetable varieties.

Chapter - 5

Ideotype Breeding and Molecular Techniques in Vegetable Improvement

1. Ideotype Breeding

1.1 Introduction

Ideotype breeding is an innovative and highly targeted approach in plant breeding, aimed at developing crops that possess a predefined set of ideal traits. These traits are carefully selected based on their potential to maximize crop productivity, resource use efficiency, resistance to biotic and abiotic stresses, and overall performance in specific environments. The word "ideotype" originates from the concept of an ideal plant model, which serves as a blueprint for breeding new plant varieties. Unlike traditional breeding, which focuses on selecting plants based on their yield performance across multiple environments, ideotype breeding narrows its focus to designing plants with optimal traits for specific environments and agricultural practices.

This approach emphasizes the need for a holistic understanding of plant morphology, physiology, and the interaction of plants with their environment. By crafting plants with an array of desired traits, ideotype breeding aims to revolutionize agricultural productivity, address the challenges of climate change, and enhance food security. This methodology has gained significant traction as plant breeders recognize that merely selecting plants based on yield without understanding the underlying factors contributing to that yield may not be enough to achieve sustainable agriculture. Instead, the design of an "ideal" plant structure can lead to more predictable and efficient results.

The concept of ideotype breeding was first introduced by **C.M. Donald** in 1968. His seminal work laid the foundation for understanding plant productivity in terms of structure-function relationships. Donald's ideotype theory proposed that certain structural traits, such as reduced leaf area or plant height, could significantly enhance the efficiency of resource use in crops like cereals. Over time, the concept evolved to include a broader range

DOI: 10.1201/9781003790976-5

of traits such as root architecture, canopy structure, photosynthetic efficiency, and resistance to pests and diseases. Today, ideotype breeding is considered a powerful tool for addressing the multifaceted challenges of modern agriculture, from increasing crop yield to enhancing resilience against climate variability. The increasing demand for food, the need for climate-resilient crops, and the growing understanding of plant-environment interactions have fueled the application of ideotype breeding. Plant breeders across the globe are working to integrate this approach into their programs, aiming to develop superior vegetable and cereal crops that cater to diverse environmental conditions. As the world faces challenges in sustaining agricultural productivity, ideotype breeding holds the promise of developing crops that are more efficient, sustainable, and adaptive.

1.2 Evolution and Concept of Ideotype Breeding

The idea of ideotype breeding was conceived by C.M. Donald in 1968 as part of his efforts to enhance cereal crop yields. Donald's vision for ideotype breeding was to move away from conventional breeding strategies that relied heavily on the natural selection of high-yielding plants and instead focus on designing plants with a precise set of traits that contribute to increased productivity. In his work with wheat, Donald identified several traits that were considered ideal for maximizing resource use efficiency, including reduced plant height, limited tillering, and a high harvest index.

Donald argued that selecting for individual plant traits based on their contribution to the whole plant's performance would be more effective than simply selecting for high yield in diverse environments. This approach, he believed, would enable breeders to develop crops that were more predictable in their performance and less influenced by environmental variability. The early work of Donald paved the way for further research and application of the ideotype concept in a variety of crops, including cereals, legumes, and vegetables.

One of the significant contributions of ideotype breeding is its ability to integrate knowledge of plant physiology and environmental interactions into the breeding process. By identifying and selecting for specific traits such as photosynthetic efficiency, canopy structure, and root architecture, breeders can create plants that are more efficient in their use of water, nutrients, and sunlight. This has proven to be particularly useful in developing crops for regions prone to drought, heat, and poor soil fertility. The application of ideotype breeding has expanded over the years to include a broader range of traits beyond those related to yield alone. Modern ideotype breeding now considers traits related to biotic and abiotic stress resistance, such as resistance to pests and diseases, tolerance to drought and salinity, and the ability to thrive under low-input farming systems. As a result, ideotype breeding has become a versatile tool for improving the performance of crops in diverse agro-ecological zones, addressing both the need for higher productivity and the challenges of environmental sustainability.

1.3 Principles of Ideotype Breeding

The principles of ideotype breeding are centered around the idea of designing plants that conform to an "ideal" set of characteristics. These characteristics are typically determined based on the crop's intended use, the environment in which it will be grown, and the specific challenges the crop is likely to face. The principles of ideotype breeding can be broken down into several key elements:

1. **Trait Selection Based on Crop Functionality**: In ideotype breeding, the first step is to define the ideal set of traits that a crop should possess to achieve optimal performance. These traits are selected based on the crop's intended use (e.g., grain production, vegetable yield) and the environment in which it will be grown. Traits that enhance resource use efficiency, such as improved photosynthetic capacity or better nutrient uptake, are often prioritized.
2. **Environmental Considerations**: The environmental conditions in which the crop will be grown play a critical role in determining the ideal set of traits for the ideotype. For example, crops grown in drought-prone areas may require traits that enhance water use efficiency, while those grown in regions with high pest pressure may need strong resistance to specific diseases or insects. This targeted approach allows breeders to develop crops that are better suited to their environments, resulting in improved performance and resilience.
3. **Integration of Biotic and Abiotic Stress Resistance**: Modern ideotype breeding incorporates resistance to biotic (pests, diseases) and abiotic (drought, salinity, temperature extremes) stresses into the ideal plant model. By breeding plants with inherent resistance to these stresses, ideotype breeding reduces the need for chemical inputs, making the crops more environmentally sustainable and economically viable for farmers.
4. **Focus on Morphological and Physiological Traits**: Ideotype breeding places significant emphasis on the morphological and physiological traits of plants. For example, traits such as reduced plant height, improved root architecture, and optimized leaf area can all contribute to a crop's overall performance. These traits are selected not in isolation but in combination with other functional traits to ensure that the plant's overall architecture supports enhanced yield and stress tolerance.
5. **Harvest Index Improvement**: One of the primary goals of ideotype breeding is to improve the harvest index, which is the ratio of economic yield (e.g., grain or vegetable) to the total above-ground biomass. Crops with a high harvest index allocate more of their energy to producing the harvestable parts (e.g., grains, fruits) rather than vegetative parts like stems and leaves. This improves yield efficiency and maximizes the economic return from the crop.

1.4 Applications in Vegetable Breeding

In vegetable breeding, the application of ideotype breeding principles has been particularly successful in developing high-yielding, resource-efficient, and stress-tolerant varieties. By identifying the ideal traits for vegetable crops, such as improved root systems, better water-use efficiency, or enhanced resistance to common pests, breeders have been able to create vegetable varieties that are better adapted to modern agricultural practices. For instance, in tomatoes, ideotype breeding has been used to develop varieties with traits like compact plant structure, disease resistance, and better fruit setting under high-temperature conditions. Similarly, in root vegetables like carrots and beets, ideotype breeding has focused on optimizing root structure and enhancing nutrient uptake efficiency. This not only improves the yield of the crop but also enhances its nutritional value, which is a significant advantage for both producers and consumers.

Ideotype breeding has also been applied to develop vegetable varieties that can thrive under low-input or organic farming systems. In these systems, plants must be able to compete effectively with weeds, resist pests without chemical interventions, and efficiently utilize limited nutrients. By designing plants with these characteristics, ideotype breeding supports sustainable agricultural practices while ensuring high productivity and profitability for farmers. Ideotype breeding represents a promising avenue for the future of vegetable crop improvement. As global agriculture faces increasing challenges related to climate change, population growth, and resource scarcity, ideotype breeding offers a strategic approach to developing crops that are resilient, resource-efficient, and highly productive.

1.5 Mechanism of Ideotype Breeding

The mechanism of ideotype breeding involves a systematic approach to the design, selection, and development of plants with ideal traits. This process is rooted in the understanding of plant physiology, genetics, and their interaction with the environment. The goal of ideotype breeding is to create a plant model that maximizes performance under specific conditions. The following steps outline the detailed mechanism of ideotype breeding:

1. **Defining the Ideotype**:

The first step in ideotype breeding is the definition of an ideal plant model. This involves determining the specific traits or characteristics that will make the crop perform optimally in the target environment. This process is highly specialized, as different environments (e.g., arid regions, high-rainfall areas, or pest-prone zones) require different plant attributes. In defining the ideotype, breeders consider a wide range of traits, including morphological traits (plant height, leaf area), physiological traits (photosynthesis efficiency, water use efficiency), and biochemical traits (nutrient uptake, resistance to pests or diseases).

» **Morphological traits**: This includes physical attributes such as plant height, leaf structure, root architecture, and tillering ability. For example, a crop ideotype may include a short plant height to minimize lodging (falling over) in areas with high winds or heavy rains.

» **Physiological traits**: Traits like photosynthetic efficiency, water-use efficiency, and nutrient assimilation rates are critical for determining how well a plant can perform in a given environment. Breeders may focus on improving these traits to develop crops that can thrive in conditions of low water availability or poor soil fertility.

» **Biochemical traits**: Resistance to pathogens or pests, the ability to tolerate salinity or other environmental stresses, and efficient nutrient cycling are important biochemical traits considered in ideotype breeding.

2. Genetic Analysis and Trait Selection:

Once the ideotype is defined, the next step involves identifying and selecting the genetic traits that will enable the plant to conform to the ideal model. This is where modern genetic tools and techniques, such as marker-assisted selection (MAS) and genomic selection, come into play. By understanding the genetic basis of the desired traits, breeders can more efficiently select parent plants that carry the desired genes and combine them in offspring through controlled breeding.

- **Marker-assisted selection (MAS):** This technique uses molecular markers associated with specific genes of interest to track the presence of these genes in offspring. MAS allows breeders to identify plants that carry the ideal genetic traits early in the breeding process, reducing the time and effort required to develop the ideotype.
- **Genomic selection**: This involves selecting plants based on the overall genetic makeup (genome-wide selection) rather than focusing on individual genes. Genomic selection is particularly useful when breeding for complex traits that are controlled by multiple genes, such as drought tolerance or pest resistance.

3. Crossbreeding and Hybridization:

After identifying the parent plants with the desired traits, crossbreeding or hybridization is performed to combine these traits in the offspring. The selection of parent plants is based on the genetic traits that align with the defined ideotype. Crossbreeding involves controlled pollination between two plants, allowing breeders to combine the strengths of both parent plants into a new generation. The goal is to ensure that the offspring inherit the ideal set of traits from both parents.

» **Crossbreeding**: Crossbreeding is one of the most common methods in ideotype breeding, where two genetically different plants are crossed to combine their desirable traits. This process can take several generations, as breeders work to stabilize the desired traits in the new plant variety.

» **Hybridization**: Hybridization is another powerful technique, where two parent plants from different varieties or species are crossed to produce hybrid offspring. Hybrid plants often exhibit hybrid vigor, or heterosis, where the offspring perform better than either parent in terms of yield, disease resistance, or other traits.

4. **Phenotypic Selection**:

Following crossbreeding or hybridization, the resulting offspring are evaluated for their phenotypic traits – the observable characteristics of the plants that reflect the expression of their genetic makeup. Plants that closely match the predefined ideotype are selected for further breeding. This step involves rigorous field testing to assess how well the plants perform under various environmental conditions.

» **Field trials**: These trials are conducted in different environments to assess the performance of the offspring. Breeders examine the plants for their ability to express the desired traits in the target environment. For example, a drought-resistant ideotype might be tested in arid regions to evaluate its water use efficiency and overall productivity under water-limited conditions.

» **Multienvironmental testing**: To ensure that the developed ideotype can perform well across a range of environments, breeders often conduct multienvironmental testing. This allows them to identify plants that are stable and consistently perform well under different climatic or soil conditions.

5. **Backcrossing and Refinement:**

The next step is to refine the ideotype by backcrossing the selected plants with parent plants or other genotypes that carry additional desirable traits. Backcrossing allows breeders to stabilize the genetic traits of the ideotype and remove any unwanted characteristics. This step may be repeated several times over multiple generations until the ideal combination of traits is achieved.

» **Backcrossing**: This is a method used to introduce specific traits from one parent into an established variety. Breeders will cross the selected offspring back to one of the original parent plants to "fix" the desired traits in the progeny. This process can continue for several generations until the offspring closely resemble the ideal plant ideotype.

» **Stabilization**: Stabilizing the ideotype ensures that the desirable traits are consistently passed down to future generations. This involves multiple cycles of breeding and selection to produce a stable variety that reliably expresses the ideal characteristics.

6. **Evaluation and Validation**:

Once the ideotype has been refined through several breeding cycles, the new plant variety undergoes final evaluation to ensure it meets the predefined criteria. This includes assessing the plant's performance in terms of yield, resource use efficiency, resistance to stresses, and other traits that were selected for in the breeding process. The goal is to validate that the new variety performs as expected in real-world agricultural conditions.

» **Yield performance**: One of the key metrics for evaluating an ideotype is its yield performance. Breeders will assess whether the new variety produces higher yields than existing varieties under similar conditions.

» **Stress tolerance**: The ideotype is also evaluated for its ability to withstand biotic and abiotic stresses. This includes testing the plant's resistance to pests and diseases, its tolerance to drought or salinity, and its adaptability to different soil types.

7. Commercialization and Release:

Once the new variety has been thoroughly tested and validated, it can be released for commercial cultivation. This process involves obtaining regulatory approval, registering the variety, and preparing it for large-scale production. The ultimate goal is to provide farmers with a new crop variety that offers superior performance, higher yields, and better resilience to environmental stresses.

» **Regulatory approval**: Before a new ideotype variety can be released, it must undergo regulatory approval to ensure it meets national and international standards for crop safety, performance, and quality.

» **Market adoption**: Once approved, the new variety is introduced to farmers, who can adopt the ideotype in their agricultural practices. The success of an ideotype variety depends on its ability to deliver tangible benefits to farmers, such as higher yields, lower input costs, or improved resilience to climate variability.

2. Varietal Release Procedure

2.1 Introduction

The varietal release procedure represents a critical stage in plant breeding, as it transitions newly developed plant varieties from research settings to practical, large-scale agricultural use. It is a regulated process, overseen by governmental or regional agricultural agencies, to ensure that the variety meets specific performance and safety criteria. These criteria may include crop yield, disease resistance, adaptability to various climatic conditions, and consumer acceptability, especially in the case of vegetable crops. In modern agriculture, especially in the context of vegetable breeding, the varietal release process is vital due to the increasing global demand for food security, sustainability, and adaptability to changing environmental conditions. As vegetable crops vary significantly in terms of their environmental requirements, disease susceptibility, and market preferences, the release process involves rigorous testing across multiple environments. It ensures that the variety meets the criteria not just for yield, but also for qualities such as taste, appearance, shelf life, and nutrition.

The global agricultural landscape is increasingly challenged by factors such as climate change, pest and disease pressures, and the need for more sustainable farming practices. As such, the varietal release process is designed to validate that new varieties can withstand these challenges while delivering consistent productivity and quality. It also provides a safety net for farmers, reducing the risk of adopting varieties that may not perform well under their specific conditions. Farmers rely on the integrity of the varietal release process to ensure that the seeds they plant will yield economically viable crops, thus supporting their livelihoods. The introduction of new varieties is a culmination of years, often decades, of breeding efforts, and the varietal release process ensures that these efforts are thoroughly vetted. It also formalizes the variety's place within agricultural systems, where only approved varieties can be legally cultivated, marketed, and sold.

2.2 Step-by-Step Process of Varietal Release

The release of a new variety is conducted through a systematic process that involves several key steps: breeder's seed development, field trials, evaluation of resistance to pests and diseases, marketability assessments, and finally, official release and seed multiplication. Below is a detailed breakdown of these steps:

2.2.1 Breeder's Seed Development

The process begins with the development of the breeder's seed, which represents the initial and pure genetic material of the new variety. This is the most fundamental stage in plant breeding, where the selected plants have been genetically stabilized after many breeding cycles to ensure that the desired traits are consistently passed on to future generations. Breeder's seed serves as the cornerstone for the subsequent steps in the varietal release procedure. It is produced under highly controlled conditions to prevent any contamination or genetic drift that could compromise the integrity of the variety. The production of a breeder's seed is an exacting process, as it requires the maintenance of the variety's genetic purity. Typically, breeder's seed is produced in isolated fields or controlled greenhouse environments to prevent cross-pollination from other varieties or species.

In this stage, breeders focus on ensuring that the plants are true to type, meaning that the offspring exhibit the same characteristics as the parent plants. These characteristics might include traits such as plant height, resistance to specific diseases, or the quality of the edible parts of the plant, such as fruit size or flavor. The integrity of the breeder's seed is paramount, as it forms the genetic foundation for future generations of the variety. Breeders also focus on producing enough breeder's seed to allow for the next step in the process—field testing—while still maintaining strict quality control. Maintaining genetic purity at this stage is crucial because any variation or genetic drift could lead to a loss of the desirable traits that the breeding program worked so hard to develop. Therefore, specialized techniques are used to ensure that the breeder's seed remains genetically identical to the original selected plants. This often involves using molecular markers or other genetic tools to verify the purity of the seed.

2.2.2 Preliminary and Advanced Field Testing

Once the breeder's seed has been successfully produced, the next step is to test the new variety in real-world conditions. This phase, known as preliminary and advanced field testing, involves growing the variety in controlled research fields and then in diverse environments to assess its agronomic performance, yield potential, and adaptability to various conditions. Preliminary testing is conducted under controlled conditions to evaluate the basic performance of the variety. During these trials, breeders look at key agronomic traits such as plant vigor, time to maturity, and yield. They also assess how the plant responds to common agricultural inputs such as water and fertilizers. Preliminary tests provide an initial indication of the variety's performance and help identify any potential issues that need to be addressed before proceeding to more advanced testing.

After preliminary testing, the variety undergoes advanced field testing, which involves evaluating the plant under a wider range of conditions. These tests are conducted in real-world agricultural environments, where the variety is exposed to the same conditions that farmers will face, such as fluctuating weather, pests, and diseases. The goal of advanced testing is to confirm that the variety performs consistently across different environments and to identify any weaknesses that may not have been apparent in the controlled preliminary trials. The advanced trials are crucial for determining the variety's adaptability. Vegetable crops, in particular, may be subject to various environmental stresses, and it's essential to determine how the new variety responds to these stresses. This might include testing for drought tolerance, resistance to heat stress, or the ability to thrive in soils with low fertility. These tests help ensure that the variety will be successful in a range of growing environments, thereby increasing its commercial viability.

2.2.3 Multilocation Trials

The multilocation trial stage is one of the most critical aspects of the varietal release procedure, especially in vegetable breeding, where crops are grown in diverse environments. During this stage, the new variety is tested in multiple locations that represent different agro-climatic zones. These zones may vary in terms of temperature, rainfall, altitude, and soil types, all of which can influence the plant's performance.

Multilocation trials are designed to assess the adaptability of the variety across different conditions. This is important because a variety that performs well in one region may not necessarily perform well in another. For example, a tomato variety that produces high yields in a temperate climate might struggle in a tropical region due to differences in temperature, humidity, or disease pressure. Multilocation trials help breeders identify the environments where the variety will perform best and ensure that it can be recommended to farmers in those regions.

The trials also include stress testing, where the variety is exposed to adverse conditions such as drought, high salinity, or nutrient-poor soils. These tests are critical for determining the resilience of the variety and its ability to thrive in suboptimal conditions. In vegetable breeding, where environmental conditions can vary greatly between growing regions, this stage helps ensure that the variety is robust enough to succeed in a wide range of environments. Additionally, multilocation trials provide breeders with valuable data on the stability of the variety. Stability refers to the consistency of the plant's performance across different environments. A stable variety will produce reliable yields and quality regardless of variations in environmental conditions. This is an important trait for farmers, as it reduces the risk of crop failure and ensures a more predictable return on investment.

2.2.4 Evaluation for Pests and Disease Resistance

Pest and disease resistance is a critical factor in the success of any new variety, particularly in vegetable crops that are vulnerable to a wide range of pathogens and pests. The evaluation for pest and disease resistance is a key component of the varietal release procedure, as it helps breeders ensure that the new variety can withstand the most common threats it will face in the field. In this phase, the variety is tested for its resistance to specific pests and diseases that are prevalent in the regions where it will be grown. For example, a tomato variety might be evaluated for resistance to blight, while a lettuce variety might be tested for resistance to aphids or mildew. These tests are conducted both in controlled environments, where the pests or diseases are deliberately introduced, and in real-world agricultural settings, where the variety is exposed to natural infestations or infections.

Pest resistance is especially important in vegetable crops, where chemical control options may be limited or undesirable due to concerns about food safety and environmental sustainability. Breeding for natural resistance to pests and diseases can reduce the need for pesticides, thereby lowering production costs for farmers and minimizing the environmental impact of the crop. Moreover, varieties with strong resistance to pests and diseases are more likely to succeed in low-input or organic farming systems, where chemical control options are limited. The tests also assess the durability of the resistance, meaning how long the variety can maintain its resistance under continuous pest or disease pressure. In some cases, pests and pathogens can evolve to overcome the plant's defenses, so breeders need to ensure that the variety's resistance is durable and long-lasting.

2.2.5 Quality and Marketability Assessment

For vegetable crops, quality and marketability are just as important as yield when it comes to determining the success of a new variety. Consumers have specific preferences when it comes to the appearance, taste, and texture of vegetables, and these preferences can vary between markets. Therefore, the quality and marketability assessment is a key step in the varietal release process.

This stage involves evaluating the variety's sensory qualities, such as taste, texture, and appearance. For example, a cucumber variety might be assessed for its crispness, while a tomato variety might be evaluated for its sweetness and juiciness. In addition to sensory qualities, the nutritional content of the vegetable is also assessed. For many consumers, the nutritional value of vegetables is an important factor in their purchasing decisions, so breeders must ensure that the new variety meets or exceeds the nutritional standards for the crop.

Shelf life and post-harvest quality are also critical factors in the marketability of a vegetable variety. Vegetables that have a longer shelf life and better transportability are more likely to succeed in the market, as they can be shipped longer distances without losing quality. Breeders assess how well the variety holds up after harvest, including how it responds to storage conditions and whether it is prone to spoilage or bruising. Finally, market trials may be conducted to gauge consumer acceptance of the new variety. These trials involve introducing the variety to consumers in specific markets and gathering feedback on its quality and desirability. Based on this feedback, breeders can make final adjustments to the variety before it is officially released.

2.2.6 Official Release and Registration

The official release and registration stage marks the formal approval of the new variety for commercial cultivation. This stage involves submitting the variety to a regulatory body or agricultural authority for approval. The authority will review the data collected during the previous stages, including field trial results, pest and disease resistance evaluations, and marketability assessments, to ensure that the variety meets all the necessary criteria for release.

In most cases, the submission process includes a detailed report of the variety's performance, as well as supporting data from the trials. The regulatory body will evaluate this information to determine whether the variety is suitable for release. If the variety meets the standards, it is officially registered and assigned a unique variety name. Variety registration is an important step in the process, as it ensures that the variety is legally protected and can be marketed under its registered name. This protects the breeder's intellectual property and ensures that farmers and consumers know they are receiving the genuine variety.

2.2.7 Seed Multiplication and Distribution

After the variety is officially released, the process of seed multiplication begins. This involves producing large quantities of seeds for commercial sale. Seed multiplication is a highly regulated process, as the genetic purity of the seeds must be maintained throughout the multiplication process. The seeds are produced in stages, starting with foundation seed and moving on to certified seed, which is sold to farmers. Foundation seed is produced from the breeder's seed and is used to produce larger quantities of certified seed. Certified seeds must meet strict quality standards, including germination rates, purity, and freedom from disease. Once the certified seed is produced, it is distributed to farmers through commercial channels.

The success of the new variety ultimately depends on its adoption by farmers. Therefore, breeders often work closely with seed companies and agricultural extension services to promote the new variety and ensure that farmers understand its benefits and how to grow it successfully. The varietal release procedure is a critical component of the plant breeding process, ensuring that new varieties are thoroughly tested, validated, and protected before they are made available for commercial cultivation. This process helps safeguard the interests of farmers, consumers, and the environment, while also promoting innovation in agricultural production.

3. DUS Testing in Vegetable Crops

3.1 Introduction

DUS testing, which stands for Distinctness, Uniformity, and Stability testing, is an essential part of the plant variety protection (PVP) and varietal release process, especially in vegetable breeding. It is a legal and technical requirement that ensures that a new plant variety is distinct from existing varieties, uniform in its traits, and stable over successive generations. DUS testing is a critical step in the evaluation and registration of new plant varieties, as it determines whether the variety qualifies for plant breeders' rights (PBR) or can be registered for commercial use. The test aims to guarantee the identity and performance of new vegetable varieties, ensuring that they meet specific standards of quality and consistency.

Vegetable crops, given their high genetic diversity and the critical role they play in human diets, require precise and rigorous DUS testing. With global agriculture becoming increasingly competitive and regulated, ensuring that new varieties conform to international standards of distinctness, uniformity, and stability is essential. Without proper DUS testing, breeders cannot secure legal protection for their new varieties, and farmers may not have access to reliable, high-quality seeds. The concept of DUS testing is governed by international organizations such as the International Union for the Protection of New Varieties of Plants (UPOV), which provides standardized guidelines for conducting these tests across various crop types, including vegetables. These guidelines help breeders, researchers, and regulatory authorities establish whether a new variety can be legally registered and commercially distributed.

3.2 Distinctness in DUS Testing

The first pillar of DUS testing is distinctness. For a new vegetable variety to be considered distinct, it must exhibit clear differences in one or more important traits when compared to other existing, commonly known varieties. These differences must be easily observable and measurable, allowing the variety to be identified as unique. The distinct traits could be related to the plant's physical structure, such as leaf shape or color, or functional traits, such as resistance to a specific disease or pest.

» **Criteria for Distinctness**: In vegetable crops, distinctness is assessed based on morphological, physiological, and agronomic traits. For example, in tomatoes, distinctness may be determined by characteristics like fruit shape, color, or plant growth habit. In carrots, traits like root shape, color intensity, and leaf structure may be examined. The variety must demonstrate that it has at least one significant trait that sets it apart from other registered varieties, making it unique in terms of its identity and utility.

» **Distinctness Testing Process**: During DUS testing, the candidate variety is grown alongside several reference varieties (commonly known varieties) to ensure a direct comparison. Over the growing season, traits are carefully measured and documented, and any distinct differences are noted. The distinct traits must be consistent across multiple environments and over different growing seasons to ensure that the variety can be reliably distinguished from others.

» **Importance of Distinctness in Vegetable Breeding**: For vegetable breeders, establishing distinctness is critical, as it not only facilitates legal protection under plant breeders' rights but also adds value to the variety in the market. Consumers and farmers alike benefit from having access to new, distinct varieties that offer better productivity, disease resistance, or market appeal. Breeding programs often focus on creating varieties with distinct traits such as improved taste, extended shelf life, or better resistance to environmental stresses, all of which must be verified through DUS testing.

Distinctness helps breeders protect their intellectual property and prevents the unauthorized propagation or sale of their varieties. It also provides clarity and transparency in the seed market, ensuring that new varieties are identifiable and meet the specific needs of farmers and consumers.

3.3 Uniformity in DUS Testing

Uniformity is the second critical aspect of DUS testing. A vegetable variety is considered uniform if the individual plants within that variety exhibit consistent traits across all growing conditions. This means that the plants must display little variation in characteristics like plant height, leaf size, or fruit shape within a given population. For farmers, uniformity is essential because it ensures predictable outcomes, such as consistent yields and quality across their fields.

» **Criteria for Uniformity**: Uniformity in vegetable crops is measured by the consistency of key traits across all plants of the variety. For example, in a cucumber variety, traits like fruit length, skin color, and texture must be uniform across all plants in the field. Any significant deviations from these traits could disqualify the variety from passing the DUS test. In some cases, minor variations are allowed, but these must be within a narrowly defined range that does not affect the overall appearance or performance of the crop.

» **Uniformity Testing Process**: During the uniformity test, a large number of plants from the candidate variety are grown in various environments to assess how consistently they express the desired traits. The plants are measured for key attributes, and any variations are documented. For the variety to pass the uniformity test, the level of variation must be within acceptable limits, as defined by the testing guidelines. The consistency of traits is measured using statistical methods to ensure that any observed variations are not due to genetic differences within the variety but are instead the result of environmental factors.

- **Importance of Uniformity in Vegetable Breeding**: In vegetable crops, uniformity is particularly important for commercial production, as it ensures that farmers can achieve predictable results from planting to harvest. For instance, uniformity in fruit size and shape is critical for vegetables like peppers, cucumbers, and tomatoes, where consistent quality is essential for marketability. Moreover, uniformity helps streamline agricultural operations, such as mechanized harvesting, which relies on plants being of similar size and maturity.

Uniformity is not only vital for commercial agriculture but also for seed production. Seed companies need to ensure that the seeds they distribute will produce uniform plants, which is key to maintaining the trust of farmers and ensuring the variety's long-term success. In the absence of uniformity, farmers might face challenges such as uneven growth, inconsistent yields, or poor marketability, all of which can affect their profitability.

3.4 Stability in DUS Testing

The final component of DUS testing is stability. Stability refers to the ability of a variety to maintain its distinct and uniform traits over successive generations. In other words, the variety must be stable in its genetic expression, meaning that when propagated through seeds or vegetative means, it consistently produces offspring with the same traits as the parent plants. Stability ensures that the variety will continue to perform as expected in different environments and over time.

- **Criteria for Stability**: Stability is evaluated by growing the variety over several generations and observing whether its traits remain consistent. The variety must show that it can reliably produce plants with the same characteristics as the original generation. For instance, if a lettuce variety is stable, the plants in the third or fourth generation should have the same leaf shape, color, and growth habit as the first generation. If any traits begin to change significantly over time, the variety may fail the stability test.
- **Stability Testing Process**: Stability is assessed by propagating the variety through multiple generations and observing the offspring for any changes in traits. For seed-propagated crops, this might involve planting seeds from several generations of the variety and comparing the resulting plants for consistency in traits. For vegetatively propagated crops like potatoes or some herbs, the stability of traits is evaluated by observing the clones or cuttings over time. Stability testing ensures that the variety will continue to meet the DUS standards throughout its lifespan.
- **Importance of Stability in Vegetable Breeding**: Stability is crucial for both breeders and farmers. For breeders, demonstrating the stability of a variety ensures that the desirable traits they worked hard to develop will be preserved over time. This is important for maintaining the variety's value and ensuring that it continues to perform well in the field. For farmers, stability assures that the seeds or plants they

purchase will reliably produce the same results year after year. Without stability, a variety may begin to show undesirable traits over time, leading to inconsistent yields or reduced quality.

In vegetable breeding, where traits like disease resistance, yield potential, and marketable quality are essential, stability guarantees that these traits will remain intact through multiple growing cycles. It also helps seed companies maintain the integrity of their products, ensuring that farmers can rely on the seeds they purchase to produce uniform and stable plants.

3.5 DUS Testing in International and Regional Contexts

DUS testing is a standardized process that is conducted globally according to the guidelines set by the International Union for the Protection of New Varieties of Plants (UPOV). UPOV provides a uniform framework for DUS testing across various countries, ensuring that new plant varieties are assessed in a consistent and scientifically sound manner. The UPOV guidelines define the specific traits that must be examined for each crop and provide detailed instructions on how to conduct the tests.

» **UPOV Guidelines**: For vegetable crops, UPOV guidelines outline the specific characteristics that must be evaluated during DUS testing. These characteristics vary depending on the crop but generally include traits such as plant height, leaf shape, fruit size, and color. The guidelines ensure that DUS tests are conducted consistently across different regions, allowing breeders to register their varieties in multiple countries without having to undergo entirely new testing procedures.

» **Regional and National DUS Testing Frameworks**: While UPOV provides an international framework, many countries have their own national or regional systems for DUS testing. These systems may be tailored to specific local conditions or crop types but generally follow the same principles of distinctness, uniformity, and stability. For example, in India, the Protection of Plant Varieties and Farmers' Rights Authority (PPV&FRA) conducts DUS testing according to guidelines that are aligned with UPOV but adapted to the Indian agricultural context.

» **Global Impact of DUS Testing**: The global implementation of DUS testing has had a significant impact on the plant breeding industry. It has enabled breeders to protect their new varieties internationally, facilitating the exchange of plant genetic resources and the development of improved crop varieties worldwide. For vegetable breeders, this means that varieties developed in one region can be tested and released in other regions, providing farmers with access to a wider range of high-quality seeds.

3.6 Challenges and Future Prospects of DUS Testing in Vegetable Crops

While DUS testing is a critical component of vegetable breeding, it is not without challenges. One of the main difficulties in DUS testing is the need for extensive, long-term trials to assess uniformity and stability. These trials can be time-consuming and costly, especially for small-scale breeders or companies with limited resources. Moreover, environmental factors can sometimes complicate the testing process, as traits like plant height or fruit size may be influenced by growing conditions, making it difficult to distinguish between genetic and environmental variation.

- » **Technological Advancements**: Recent advancements in genetic analysis and molecular breeding techniques offer promising solutions to some of these challenges. For example, DNA markers and genomic tools can help breeders more accurately assess distinctness, uniformity, and stability, reducing the reliance on lengthy field trials. These technologies can also improve the precision of DUS testing by providing genetic confirmation of traits that are difficult to observe or measure in the field.
- » **Sustainability and Climate Resilience**: As the effects of climate change become more pronounced, DUS testing will play an increasingly important role in ensuring that new vegetable varieties are resilient to environmental stresses. In the future, DUS testing may need to incorporate additional traits related to climate adaptation, such as drought tolerance or heat stress resistance, to ensure that new varieties can meet the demands of a changing agricultural landscape.

4. Application of In-vitro and Molecular Techniques in Vegetable Improvement

4.1 Introduction

The application of in-vitro and molecular techniques has revolutionized vegetable breeding and improvement, offering advanced methods to accelerate the development of new, high-performing, and resilient varieties. These techniques provide breeders with tools to modify and select plant traits at the cellular and molecular levels, allowing for precise improvements that traditional breeding methods often cannot achieve. In-vitro techniques involve manipulating plant cells or tissues in controlled laboratory environments, while molecular techniques refer to the use of genetic tools to study, modify, or manipulate the plant genome. Together, these technologies have led to breakthroughs in areas such as disease resistance, yield enhancement, stress tolerance, and nutritional improvement in vegetable crops.

The growing global demand for high-quality vegetables, coupled with the challenges posed by climate change, pests, diseases, and diminishing agricultural resources, underscores the importance of these advanced techniques. In-vitro and molecular approaches allow breeders to rapidly introduce desirable traits into new vegetable varieties without the lengthy process associated with traditional breeding. They also provide solutions to specific challenges, such as enhancing resistance to biotic and abiotic stresses, improving nutritional content, and ensuring the sustainability of vegetable production systems. *In-vitro* and molecular techniques are now widely adopted in vegetable breeding programs worldwide. These techniques range from tissue culture methods like micropropagation and somatic embryogenesis to molecular approaches such as marker-assisted selection (MAS), transgenic technologies, and gene editing. The application of these tools has led to the development of new vegetable varieties with improved traits, contributing to the overall goal of food security and agricultural sustainability.

4.2 *In-vitro* Techniques in Vegetable Improvement

In-vitro techniques refer to a range of methods used to grow and manipulate plant cells, tissues, or organs in a controlled, sterile environment. These techniques provide a platform for regenerating plants from a small piece of tissue, offering significant advantages for vegetable improvement. Some key in-vitro techniques include micropropagation, somatic embryogenesis, and the development of somaclonal variants.

4.2.1 Micropropagation

Micropropagation, also known as plant tissue culture, is a widely used in-vitro technique for the rapid multiplication of plant material. It involves the growth of plant cells or tissues in a sterile, nutrient-rich medium under controlled conditions, leading to the generation of multiple clones of the original plant. This technique is particularly useful for the mass production of disease-free and genetically uniform plants, making it highly valuable in vegetable breeding.

Schematic Representation of Micropropagation

» **Advantages of Micropropagation**: Micropropagation allows breeders to produce large numbers of identical plants in a relatively short period. This is especially important for vegetables that are propagated vegetatively, such as potatoes, garlic, and sweet potatoes. The ability to produce disease-free planting material is one of the key benefits, as it reduces the need for chemical inputs and improves the overall health and vigor of the crop.

» **Process**: The process of micropropagation begins with the selection of a healthy plant, from which small tissue samples (explants) are taken. These explants are placed in a sterile growth medium containing the necessary nutrients and hormones to promote cell division and plant regeneration. Over time, the explants develop into new plants, which can be transferred to soil and grown to maturity. This process can be repeated to produce large quantities of uniform, high-quality plants.

» **Application in Vegetables**: Micropropagation has been successfully applied in the improvement of several vegetable crops, including potatoes, tomatoes, and eggplants. In potatoes, for example, micropropagation is used to produce virus-free seed tubers, which leads to higher yields and better crop quality. In tomatoes, micropropagation is used to rapidly multiply elite lines with desirable traits such as disease resistance and improved fruit quality.

4.2.2 Somatic Embryogenesis

Somatic embryogenesis is another in-vitro technique that involves the development of embryos from somatic (non-reproductive) cells. This process mimics the natural development of seeds but takes place in a laboratory setting. Somatic embryogenesis is widely used in the regeneration of plants from single cells, making it a powerful tool for genetic transformation and the production of new varieties.

» **Advantages of Somatic Embryogenesis**: One of the main advantages of somatic embryogenesis is that it allows for the regeneration of whole plants from a single cell. This makes it an ideal technique for the genetic transformation of vegetables, where specific genes can be introduced into the plant cells, and the resulting plants can be grown to maturity. Somatic embryogenesis also provides a platform for the production of synthetic seeds, which can be stored and distributed like traditional seeds but without the need for sexual reproduction.

» **Process**: The process of somatic embryogenesis involves inducing somatic cells to form embryos, which can then be developed into complete plants. This is done by culturing the somatic cells in a medium containing specific hormones and nutrients that trigger the formation of embryogenic tissue. Once the embryos are formed, they are transferred to a different medium to promote plant growth, and eventually, the regenerated plants can be grown in the soil.

Somatic embryogenesis in Banana

» **Application in Vegetables**: Somatic embryogenesis has been successfully applied in the improvement of various vegetable crops, such as carrots, celery, and peppers. For example, in carrot breeding programs, somatic embryogenesis is used to produce genetically modified plants with enhanced nutritional content or improved resistance to pests. In pepper breeding, this technique is used to introduce genes that confer resistance to diseases like bacterial wilt.

4.2.3 Somaclonal Variation

Somaclonal variation refers to the genetic variation that occurs in plants regenerated from tissue cultures. This variation can be harnessed by breeders to select plants with desirable traits that may not have been present in the original plant population. Somaclonal variation provides an additional source of genetic diversity, which is essential for the improvement of vegetable crops.

» **Advantages of Somaclonal Variation**: One of the key advantages of somaclonal variation is that it generates new genetic variation, which can be used to develop improved vegetable varieties. This variation may include traits such as enhanced disease resistance, improved yield, or better nutritional quality. Somaclonal variation also allows breeders to explore new traits that may not have been achievable through traditional breeding methods.

- **Process**: Somaclonal variation occurs naturally during the process of tissue culture, as cells undergo genetic changes while being cultured. These changes can result in plants with new and potentially useful traits. Breeders can screen the regenerated plants for desirable characteristics and select the best individuals for further breeding or commercial production.
- **Application in Vegetables**: Somaclonal variation has been applied in the improvement of several vegetable crops, including tomatoes, potatoes, and lettuce. In potato breeding, somaclonal variation has been used to develop new varieties with improved resistance to late blight, one of the most devastating diseases in potato production. In tomatoes, somaclonal variation has been harnessed to develop lines with enhanced flavor and better shelf life.

4.3 Molecular Techniques in Vegetable Improvement

Molecular techniques have brought a new dimension to vegetable breeding by allowing breeders to manipulate the plant genome directly. These techniques enable precise modifications, making it possible to introduce or delete specific genes, thereby improving the efficiency of breeding programs. Molecular tools such as marker-assisted selection (MAS), genetic engineering, and gene editing are now widely used to enhance the speed and accuracy of vegetable improvement.

4.3.1 Marker-Assisted Selection (MAS)

Marker-assisted selection (MAS) is a molecular breeding technique that involves the use of DNA markers to track the presence of desirable genes in plant breeding populations. These markers are closely linked to specific traits, allowing breeders to identify and select plants with the desired genetic makeup without the need for extensive field trials.

- **Advantages of MAS**: MAS accelerates the breeding process by enabling breeders to select plants at an early stage of development based on their genetic profile. This reduces the time and resources required for field testing, as breeders can focus on plants that are more likely to express the desired traits. MAS is particularly useful for selecting traits that are difficult to measure in the field, such as disease resistance or drought tolerance.
- **Process**: The process of MAS begins with the identification of DNA markers linked to the genes responsible for specific traits. Once these markers are identified, breeders can use molecular techniques to screen plant populations for the presence of these markers. Plants that carry the desired markers are selected for further breeding, while those that do not are discarded.
- **Application in Vegetables**: MAS has been widely used in the improvement of various vegetable crops, such as tomatoes, peppers, and cucumbers. For example, in tomato breeding, MAS is used to select plants with resistance to diseases like tomato yellow leaf curl virus (TYLCV) or late blight. In cucumber breeding, MAS is used to identify plants with improved resistance to powdery mildew and better fruit quality.

4.3.2 Genetic Engineering

Genetic engineering involves the direct manipulation of a plant's DNA to introduce new traits or enhance existing ones. This technique allows breeders to introduce genes from other species or organisms into vegetable crops, conferring traits such as pest resistance, herbicide tolerance, or improved nutritional content.

» **Advantages of Genetic Engineering**: Genetic engineering offers unparalleled precision in plant breeding, as it allows breeders to introduce specific genes directly into the plant genome. This bypasses the need for traditional cross-breeding and reduces the time required to develop new varieties. Genetic engineering also makes it possible to introduce traits that are not naturally found in the plant's gene pool, such as resistance to insect pests or tolerance to herbicides.

» **Process**: The process of genetic engineering begins with the identification of the gene or genes responsible for the desired trait. These genes are then inserted into the plant's genome using a variety of techniques, such as Agrobacterium-mediated transformation or gene gun technology. Once the gene is successfully integrated into the plant's DNA, the plant is grown and tested to ensure that the new trait is expressed correctly.

» **Application in Vegetables**: Genetic engineering has been successfully applied in the improvement of several vegetable crops, including tomatoes, eggplants, and sweet corn. For example, genetically modified tomatoes have been developed with delayed ripening traits, allowing for a longer shelf life and reduced post-harvest losses. In eggplants, genetic engineering has been used to introduce resistance to pests like the fruit and shoot borer, reducing the need for chemical pesticides.

4.3.3 Gene Editing (CRISPR-Cas9)

Gene editing is a revolutionary molecular technique that allows for precise modifications of the plant genome by directly targeting and altering specific DNA sequences. CRISPR-Cas9 is one of the most widely used gene-editing tools, enabling breeders to make targeted changes to the plant's genetic code with high accuracy and efficiency.

» **Advantages of Gene Editing**: One of the main advantages of gene editing is its precision. Unlike traditional genetic engineering, which often involves inserting foreign genes, gene editing allows breeders to modify the plant's own genes to enhance or suppress specific traits. This reduces the risk of unintended genetic changes and makes the process faster and more efficient. Gene editing is also more acceptable to consumers and regulators, as it does not involve the introduction of foreign DNA.

» **Process**: The CRISPR-Cas9 system works by using a guide RNA to target a specific DNA sequence in the plant genome. Once the target sequence is identified, the Cas9 enzyme makes a cut in the DNA, allowing for the deletion, insertion, or modification of genetic material. The plant's natural repair mechanisms then fix the DNA, resulting in the desired genetic change.

» **Application in Vegetables**: Gene editing has been applied to improve various traits in vegetable crops, such as disease resistance, yield potential, and nutritional content. For example, CRISPR-Cas9 has been used to develop tomatoes with enhanced lycopene content, improving their nutritional value. In lettuce, gene editing has been used to increase resistance to downy mildew, one of the most common diseases affecting leafy vegetables.

4.4 Challenges and Future Prospects of In-vitro and Molecular Techniques in Vegetable Improvement

While in-vitro and molecular techniques have greatly advanced vegetable improvement, they are not without challenges. One of the main issues is the high cost and technical expertise required to implement these techniques, which may limit their accessibility to small-scale breeders or developing countries. Additionally, public perception and regulatory hurdles surrounding genetically modified organisms (GMOs) can pose challenges to the widespread adoption of some molecular techniques. However, the prospects of these technologies are promising. Advances in gene editing, molecular markers, and tissue culture techniques are likely to make these methods more accessible, efficient, and cost-effective. As global agriculture faces increasing challenges from climate change, pests, and diseases, the continued development and application of in-vitro and molecular techniques will be crucial for ensuring the sustainability and productivity of vegetable crops.

Chapter - 6

Breeding of Crucifers, Cucurbitaceous, Solanaceous and Tuberous / Bulbs Vegetables

1. Cruciferous vegetable Breeding

1.1. Introduction

Cruciferous vegetables, which belong to the Brassicaceae family, are a significant group of crops that have played a pivotal role in global agriculture. This group encompasses a diverse range of vegetables, including cabbage (*Brassica oleracea*), broccoli (*Brassica oleracea* var. *italica*), cauliflower (*Brassica oleracea* var. *botrytis*), kale (*Brassica oleracea* var. *sabellica*), Brussels sprouts (*Brassica oleracea* var. *gemmifera*), turnip (*Brassica rapa*), and radish (*Raphanus sativus*). These vegetables have long been valued for their high nutritional content, particularly their rich levels of vitamins, minerals, and health-promoting phytochemicals. The significance of cruciferous vegetables in human diets and their potential health benefits have spurred a growing interest in the improvement and diversification of these crops through breeding.

Cruciferous vegetables are particularly noted for their glucosinolate content, which is responsible for many of their health-promoting properties, including potential anti-carcinogenic effects. Additionally, these vegetables are valued for their versatility in various cuisines worldwide. Given their importance, there has been a continued focus on enhancing their productivity, quality, disease resistance, and tolerance to abiotic stress. As a result, breeding programs for cruciferous vegetables have become increasingly sophisticated, employing both traditional methods and modern biotechnological approaches to address the challenges faced by producers and to meet the evolving demands of consumers.

1.2. Historical Background of Cruciferous Vegetable Breeding

The cultivation and breeding of cruciferous vegetables date back to ancient times. Some of the earliest evidence of Brassica cultivation comes from the Mediterranean region, where wild species of *Brassica oleracea* were domesticated thousands of years ago. From these wild progenitors, farmers selected for desirable traits, such as larger leaves or more compact heads, giving rise to the diverse varieties that we recognize today, such as cabbage and

DOI: 10.1201/9781003790976-6

cauliflower. Traditional breeding methods relied on mass selection and open pollination, where farmers selected the best-performing plants for further cultivation, allowing for the gradual improvement of the crop.

In the 19th and early 20th centuries, the development of formal plant breeding programs helped accelerate the improvement of cruciferous vegetables. Mendel's discoveries in genetics provided a scientific foundation for breeding, and plant breeders began to apply principles of heredity to improve crop traits systematically. Early breeding efforts focused on enhancing yield, disease resistance, and shelf life. This period also saw the introduction of hybrid breeding techniques, which allowed for the production of hybrid varieties with increased vigor and productivity.

The advent of modern genetics and biotechnology has further transformed cruciferous vegetable breeding. Advances in molecular biology, genomics, and tissue culture have opened new avenues for improving these crops. For instance, marker-assisted selection (MAS) allows breeders to select plants with desirable traits based on genetic markers, speeding up the breeding process and increasing precision. Furthermore, the development of transgenic technologies has enabled the introduction of novel traits, such as insect resistance, into cruciferous crops.

1.3. Importance of Cruciferous Vegetables in Human Nutrition

Cruciferous vegetables are renowned for their exceptional nutritional value. They are rich in vitamins, such as vitamin C, vitamin K, and several B vitamins. In addition, they are an important source of dietary fiber, which promotes digestive health, and essential minerals, such as calcium, magnesium, and potassium. However, it is the presence of bioactive compounds like glucosinolates and their breakdown products, including isothiocyanates and indoles, that has drawn significant attention to the health benefits of these vegetables. Several studies have shown that the consumption of cruciferous vegetables is associated with a reduced risk of various cancers, particularly cancers of the lung, colon, and breast. The bioactive compounds in these vegetables have been found to inhibit the growth of cancer cells, induce apoptosis (programmed cell death), and modulate detoxification enzymes. Additionally, cruciferous vegetables have anti-inflammatory properties and are linked to improved cardiovascular health. As such, breeding programs often focus not only on improving yield and resistance to pests but also on enhancing the levels of these health-promoting compounds.

1.4. Challenges in Cruciferous Vegetable Breeding

The breeding of cruciferous vegetables presents a range of challenges. One of the most significant issues is the need for resistance to diseases and pests. Cruciferous crops are susceptible to a variety of diseases, including black rot (caused by *Xanthomonas campestris*), clubroot (caused by *Plasmodiophorabrassicae*), and downy mildew (caused by *Peronospora parasitica*). These diseases can cause significant yield losses and reduce the quality of the produce.

Breeding for disease resistance is a major focus of cruciferous vegetable improvement programs. Traditional breeding methods have been effective in developing varieties with resistance to some diseases, but the emergence of new pathogen strains poses an ongoing challenge.In addition to disease resistance, pest management is a critical concern. Cruciferous vegetables are often targeted by pests such as the cabbage looper (*Trichoplusiani*), diamondback moth (*Plutellaxylostella*), and aphids. Breeding for insect resistance, whether through traditional or transgenic methods, is an important aspect of cruciferous vegetable improvement.

Abiotic stress tolerance is another key area of focus. Climate change has exacerbated the challenges faced by cruciferous vegetable producers, as these crops are sensitive to extremes in temperature and water availability. Breeding for drought tolerance, heat tolerance, and adaptability to different soil conditions is increasingly important for maintaining and improving productivity in the face of changing environmental conditions. Moreover, the breeding of cruciferous vegetables must also take into account consumer preferences. Factors such as taste, texture, color, and appearance play a significant role in the marketability of these crops. For example, broccoli with a deep green color and tight, uniform heads is preferred by consumers, while cabbages with fewer outer leaves and a compact shape are often more desirable in the market. Balancing these aesthetic and sensory traits with agronomic performance is a constant challenge for breeders.

1.5. Modern Approaches to Cruciferous Vegetable Breeding

In recent decades, the development of advanced breeding techniques has revolutionized the breeding of cruciferous vegetables. One of the most notable advances is the use of molecular markers to facilitate marker-assisted selection (MAS). By identifying genetic markers linked to specific traits, breeders can more efficiently select plants that possess desirable characteristics, such as disease resistance or improved nutritional content. This approach reduces the time and resources required for traditional breeding, which often involves multiple generations of crossing and selection.

Genomic selection is another powerful tool in cruciferous vegetable breeding. With the advent of high-throughput DNA sequencing technologies, it has become possible to analyze the entire genome of a plant and predict its breeding value based on its genetic composition. This method allows for the selection of plants with optimal combinations of traits, even if those traits are controlled by multiple genes.In addition to molecular approaches, tissue culture techniques have become indispensable in cruciferous vegetable breeding. Micropropagation, for example, allows for the rapid multiplication of disease-free plants, while somaclonal variation can be used to introduce new genetic diversity into breeding programs. Moreover, techniques such as embryo rescue and protoplast fusion have enabled the development of interspecific hybrids that would not be possible through traditional breeding methods.

Genetic engineering has also played a role in cruciferous vegetable breeding. For example, the development of transgenic crops with insect resistance has reduced the need for chemical pesticides in the cultivation of cruciferous vegetables. The incorporation of the *Bacillus thuringiensis* (Bt) gene into crops such as cabbage has provided effective protection against lepidopteran pests. However, the use of genetically modified organisms (GMOs) in food production remains controversial in some regions, and breeders must navigate the regulatory and consumer acceptance challenges associated with these technologies. The future of cruciferous vegetable breeding is likely to be shaped by the continued integration of traditional breeding techniques with cutting-edge biotechnologies. As the global population grows and the demand for nutritious, high-quality food increases, there will be an ongoing need to develop cruciferous vegetable varieties that are more productive, resilient, and nutritious. Advances in genome editing technologies, such as CRISPR-Cas9, offer exciting possibilities for precisely targeting and modifying genes associated with desirable traits, potentially accelerating the development of improved varieties. Furthermore, the growing awareness of the importance of sustainable agriculture is likely to influence breeding priorities. Breeding programs will need to focus on developing varieties that require fewer inputs, such as water, fertilizers, and pesticides while maintaining or improving yield and quality. This will involve a greater emphasis on breeding for traits such as nutrient-use efficiency, water-use efficiency, and pest and disease resistance.

A. Cauliflower (*Brassica oleracea* var. *botrytis*)

Family: Brassicaceae

Chromosome number: 2n = 2x = 18

Cauliflower is a cool-season vegetable and is one of the most important members of the *Brassica oleracea* family. Its edible part, the curd, is comprised of tightly clustered, immature flower buds. Cauliflower is widely consumed for its high nutritional value and its versatility in culinary applications, from salads and soups to complex dishes. The curd is typically white, although some cultivars produce curds that are purple, orange, or green, each offering unique nutritional advantages. This vegetable is an excellent source of vitamins, including Vitamin C, Vitamin K, and folate. It is also rich in dietary fiber and contains a wide range of antioxidants, glucosinolates, and sulforaphane, compounds that are associated with health benefits such as cancer prevention and anti-inflammatory properties. Globally, cauliflower is grown in a range of climatic conditions, although it prefers cool temperatures and well-drained soils. It is especially sensitive to heat, which can cause premature bolting or reduced curd quality. India is one of the largest producers of cauliflower, with a significant percentage of the crop being cultivated in the northern states, including Uttar Pradesh and West Bengal. The versatility of cauliflower in both fresh and processed forms makes it a staple in many cuisines, contributing to its steady global demand.

Origin and Domestication

The origin of cauliflower can be traced to the Mediterranean region, particularly Cyprus, where it was first cultivated. Its domestication involved selecting traits that favored curd formation and desirable curd characteristics, such as size, density, and texture. The primary center of origin is believed to be in the Mediterranean Basin, and from there, cauliflower spread to different parts of the world, especially Europe and Asia. The development of the modern cauliflower, as we know it today, involved numerous genetic modifications over centuries of cultivation.

Cauliflower, being a part of the *Brassica oleracea* species, shares a common ancestry with other cruciferous vegetables, such as cabbage, broccoli, kale, and Brussels sprouts. These crops have a similar genetic makeup and belong to the same species complex. Over time, the selection of specific traits in cauliflower, such as large, dense curds and increased tolerance to environmental stresses, has been pivotal in its evolution as a global crop. The most important aspect of domestication was the development of varieties that could adapt to diverse climatic conditions, as cauliflower is sensitive to temperature changes. The domestication process also involved the development of varieties that could produce curds in various colors. Purple, orange, and green varieties of cauliflower have been selectively bred for their nutritional qualities, such as higher levels of anthocyanins (in purple cauliflower) or beta-carotene (in orange cauliflower). These varieties provide additional health benefits and appeal to consumers looking for novelty and increased nutrition.

Related Genera

Cauliflower belongs to the genus *Brassica*, which includes several other important agricultural crops, many of which are closely related and have been used in breeding programs to enhance cauliflower's resilience and yield. The related genera that are significant in cauliflower breeding include:

- » **Brassica:** This genus is the primary one that cauliflower belongs to, and it includes many related crops, such as cabbage (*Brassica oleracea* var. *capitata*), broccoli (*Brassica oleracea* var. *italica*), and kale (*Brassica oleracea* var. *sabellica*).
- » **Raphanus:** Commonly known as radish, this genus has been used in breeding programs to introduce resistance to certain pests and diseases.
- » **Sinapis:** This genus includes mustard species, which are used in crop rotation to improve soil health and as green manure.

These genera are particularly important in hybridization and genetic engineering programs, where traits from related species can be introduced into cauliflower to enhance characteristics such as disease resistance, abiotic stress tolerance, and nutrient content.

Related Species

Cauliflower is part of the *Brassica oleracea* species complex, which is known for its genetic diversity. This group includes a variety of subspecies and cultivars, each with unique traits that make them valuable in breeding programs. The most closely related species include:

- » *Brassica oleracea var. capitata* (Cabbage): This species is known for its compact head of leaves, and it shares many genetic traits with cauliflower. Cabbage has been used in breeding programs to improve cauliflower's resistance to pests and its adaptability to different climates.
- » *Brassica oleracea var. italica* (Broccoli): Broccoli and cauliflower are genetically very similar, and they have been hybridized to create "broccoflower," a variety of cauliflower with green curds.
- » *Brassica oleracea var. gongylodes* (Kohlrabi): This species has a swollen stem and is grown for its edible stem, which can be eaten raw or cooked. It has been used in breeding programs to introduce traits related to abiotic stress tolerance in cauliflower.
- » **Brassica napus** (Rapeseed): Rapeseed is primarily grown for its oil, but it is also used in genetic studies to introduce traits such as disease resistance and enhanced nutrient content in cauliflower.

In addition to these species, wild relatives of cauliflower have been explored for their potential to introduce traits such as heat tolerance and improved curd quality. The genetic diversity found within the *Brassica* complex makes it a valuable resource for cauliflower improvement.

Types of Cauliflower on Basis of Colour of Curds:

Cauliflower has several important botanical varieties, which are classified based on curd color, shape, and growing season. These varieties are essential for breeding programs, as they offer a wide range of genetic diversity that can be harnessed to improve curd quality, yield, and adaptability. The primary botanical varieties include:

1. **White Cauliflower (*Brassica oleracea* var. *botrytis*):** This is the most commonly grown variety worldwide. It is characterized by its white, dense curds, which are formed by tightly packed immature flower buds. White cauliflower is preferred for its mild flavor and versatility in culinary applications.
2. **Purple Cauliflower (*Brassica oleracea* var. *botrytis*):** Purple cauliflower is known for its distinctive purple color, which is due to the presence of anthocyanins, powerful antioxidants that provide health benefits, including cancer-preventive properties.
3. **Orange Cauliflower (*Brassica oleracea* var. *botrytis*):** This variety contains high levels of beta-carotene, a precursor to Vitamin A, which gives it its orange hue. Orange cauliflower is valued for its enhanced nutritional content and is often marketed as a superfood.
4. **Green Cauliflower (Broccoflower) (*Brassica oleracea* var. *botrytis*):** A hybrid between cauliflower and broccoli, green cauliflower, or broccoflower, is noted for its unique green color and its combination of the flavor profiles of both broccoli and cauliflower.

Table 1: Major Types of Cauliflower on basis of Curd Color and their Special Features

S No	Botanical Variety	Description	Curd Color	Special Features
1	*Brassica oleracea* var. *botrytis*	White Cauliflower	White	Standard type, widely cultivated for fresh markets
2	*Brassica oleracea* var. *botrytis*	Purple Cauliflower	Purple	High in anthocyanins, offering antioxidant benefits
3	*Brassica oleracea* var. *botrytis*	Orange Cauliflower	Orange	Rich in beta-carotene, enhanced nutritional value
4	*Brassica oleracea* var. *botrytis*	Green Cauliflower (Broccoflower)	Light Green	Hybrid of cauliflower and broccoli, unique flavour

Growth Habit

Cauliflower is generally grown as an annual crop, but in its native Mediterranean habitat, it behaves as a biennial. The plant's vegetative growth consists of large, broad leaves that are arranged in a rosette around the central curd. These leaves serve to protect the developing curd from sunlight, which is crucial for maintaining the white color in most varieties. The curd itself is a dense cluster of meristematic tissue that forms before the plant enters its reproductive phase. The ideal temperature range for cauliflower growth is between 15°C and 20°C, making it highly sensitive to heat stress. Excessive heat can lead to bolting or curd discoloration, while low temperatures can delay curd development. Cauliflower requires well-drained soils with a pH of 6.0 to 7.5 for optimal growth. Irrigation is critical, especially during dry periods, to ensure that the curd develops fully without becoming loose or "ricey." Riceyness occurs when the curd separates into individual flower buds, reducing market value. In commercial production, cauliflower is often grown under controlled conditions to ensure uniform curd development and high yield.

Floral Morphology

Cauliflower's floral morphology is a significant focus of both botanical studies and breeding programs due to the importance of the curd, which is an arrested floral meristem. The inflorescence of cauliflower is essentially a cluster of immature flower buds that do not fully develop, resulting in the compact, edible curd. These buds are densely packed and arranged in a spiral pattern. Unlike other members of the *Brassica oleracea* species complex, cauliflower does not allow the development of full flowers under normal growth conditions. The floral parts remain undeveloped until bolting, which occurs when the plant is exposed to unfavorable environmental conditions such as high temperatures.

The structure of the cauliflower flower, when it does develop, is typical of cruciferous plants. The flowers are composed of four sepals and four petals, arranged in a cross shape. The petals are usually white or pale yellow, and the flowers are small and borne in dense clusters. The reproductive organs include six stamens, four of which are long and two are short (a characteristic feature of the Brassicaceae family), and a single pistil. The ovary is superior, with a single locule, and it develops into a siliqua (a long, slender pod) if fertilized. However, in cauliflower cultivation, the goal is to prevent flowering and maintain the curd in its immature state. The floral biology of cauliflower is crucial for successful breeding and seed production. The development of the curd, and the prevention of flowering, is controlled by both genetic and environmental factors. Breeders select for traits that promote curd formation under a variety of conditions and prevent premature bolting, which can significantly reduce the market value of the crop. Additionally, knowledge of floral biology is essential for hybrid seed production, where controlled pollination techniques are employed to ensure the purity of hybrid varieties.

Breeding Objectives

The breeding objectives for cauliflower focus on several key areas, including curd quality, yield, disease resistance, and abiotic stress tolerance. Each of these objectives is crucial for developing high-performing cauliflower varieties that meet the demands of both producers and consumers.

1. **Curd Quality:** One of the primary objectives in cauliflower breeding is to improve

curd quality. A high-quality curd is large, dense, and uniform, with a smooth surface and a bright white color. Varieties are selected for their ability to produce curds that are free from physiological defects such as riceyness (where the curd separates into individual flower buds), hollow stem, or discoloration. The texture of the curd is also important, with breeders aiming for varieties that produce firm, compact heads that are suitable for both fresh market and processing.

2. **Yield Improvement:** Increasing the yield potential of cauliflower is another major breeding objective. Yield is determined by several factors, including the size of the curd, the plant's ability to produce consistently under various environmental conditions, and resistance to pests and diseases. Breeders select for traits that enhance the overall productivity of the plant, such as vigorous vegetative growth, early maturity, and the ability to produce large curds even in suboptimal conditions.

3. **Disease Resistance:** Cauliflower is susceptible to a range of diseases, including clubroot (*Plasmodiophorabrassicae*), black rot (*Xanthomonas campestris*), and downy mildew (*Peronospora parasitica*). Developing varieties with resistance to these diseases is critical for ensuring stable production and reducing the need for chemical inputs. Breeders utilize both conventional breeding techniques, such as backcrossing with resistant varieties, and modern molecular breeding methods, such as marker-assisted selection, to incorporate disease resistance into elite cauliflower lines.

4. **Abiotic Stress Tolerance:** Cauliflower is particularly sensitive to abiotic stresses, including high temperatures, drought, and salinity. Heat tolerance is one of the most important traits for breeders, especially in tropical and subtropical regions where cauliflower is increasingly grown. Varieties with improved heat tolerance are able to maintain curd formation and quality under higher temperatures, which extends the growing season and improves yield. Drought tolerance is another important trait, particularly in regions where water availability is limited. Breeding for salinity tolerance is also gaining importance as soil salinization becomes a growing concern in many cauliflower-growing areas.

5. **Nutritional Quality:** In recent years, there has been increasing interest in breeding cauliflower varieties with enhanced nutritional qualities. Purple cauliflower, for example, is rich in anthocyanins, which are powerful antioxidants that provide health benefits. Orange cauliflower contains high levels of beta-carotene, a precursor to Vitamin A. Breeding for increased glucosinolate content is another objective, as these compounds are associated with cancer prevention and other health benefits. Consumers are increasingly looking for cauliflower varieties that not only taste good but also offer additional nutritional advantages.

Table 2: Major Breeding Objectives for Cauliflower and Released Varieties

S No	Objective	Description	Example Varieties
1	Curd Quality	Large, dense, uniform curds with a smooth surface	White Cloud, Snowball, Early Snowball
2	Yield Improvement	High yield potential under diverse environmental conditions	Pacifica, Fremont, Candid Charm
3	Disease Resistance	Resistance to clubroot, black rot, and downy mildew	Amazing, Markanta, Clarify
4	Abiotic Stress Tolerance	Tolerance to heat, drought, and salinity	Candid Charm, Fremont
5	Nutritional Quality	Enhanced levels of antioxidants and glucosinolates	Purple Cape, Cheddar, Orange Bouquet

Breeding Methods

The development of new cauliflower varieties involves a combination of traditional and modern breeding methods. These methods are aimed at improving specific traits, such as yield, curd quality, and disease resistance, while also introducing new traits that make cauliflower more adaptable to changing climatic conditions and consumer preferences. Below is an expansion of the major breeding methods used in cauliflower improvement.

Conventional Breeding Methods

1. **Introduction:** This method involves introducing cauliflower germplasm from different regions into breeding programs. The goal is to increase genetic diversity and introduce new traits that may not be present in local populations. For example, varieties from Europe may be introduced into tropical breeding programs to improve curd quality under high-temperature conditions.
2. **Selection:** Selection is one of the most fundamental breeding methods. Mass selection involves choosing the best plants from a population based on specific traits, such as curd size or disease resistance. Pure line selection involves selecting the best individuals from a self-pollinating crop and propagating them over successive generations to develop a uniform variety.
3. **Hybridization:** Hybridization is the process of crossing two genetically distinct cauliflower varieties to combine desirable traits from both parents. Hybrid varieties are often more vigorous and productive than traditional open-pollinated varieties. They are also more uniform, which is important for commercial production. Hybridization is commonly used to develop varieties with improved yield, curd quality, and disease resistance.
4. **Pedigree Method:** The pedigree method involves selecting superior plants from the

progeny of a cross between two cauliflower varieties. This method allows breeders to track the inheritance of specific traits over multiple generations. It is particularly useful for selecting for complex traits, such as disease resistance or heat tolerance.

5. **Bulk Method:** In the bulk method, seeds from a large number of plants are harvested together and grown as a mixed population. Selection is deferred until later generations, allowing natural selection to play a role in the early stages of breeding. This method is useful for selecting for traits that are influenced by the environment, such as abiotic stress tolerance.
6. **Backcross Method:** The backcross method is used to transfer a specific gene or trait from one variety (the donor parent) into another variety (the recurrent parent). For example, if a variety with high yield potential is susceptible to clubroot, breeders may backcross it with a clubroot-resistant variety to introduce resistance while maintaining the high yield potential of the recurrent parent.
7. **Mutation Breeding:** Mutation breeding involves inducing mutations in the plant's genome using chemical agents or radiation. This method has been used to create novel traits, such as the orange and purple curd colors found in some modern cauliflower varieties.

Non-Conventional Breeding Methods

1. **Tissue Culture:** Tissue culture is a biotechnology technique that allows breeders to propagate cauliflower plants from small pieces of tissue in a controlled environment. This method is particularly useful for developing disease-free planting material and for multiplying plants with desirable traits rapidly. Tissue culture also enables the regeneration of plants from genetically modified tissues, making it an important tool in modern cauliflower breeding.
2. **Genetic Engineering (Recombinant DNA Technology):** Genetic engineering allows breeders to introduce specific genes from other species into cauliflower to enhance traits such as pest resistance, herbicide tolerance, or nutritional content. For example, genes for insect resistance from *Bacillus thuringiensis* (Bt) can be introduced into cauliflower to reduce the need for chemical insecticides. Genetic engineering also enables the development of varieties with enhanced nutritional qualities, such as increased levels of glucosinolates or carotenoids.

Breeding for Disease Resistance

Cauliflower is prone to several serious diseases, including clubroot, black rot, and downy mildew. These diseases can significantly reduce yield and quality, making disease resistance a top priority in cauliflower breeding programs. Breeders use a combination of conventional and molecular breeding techniques to develop varieties that can withstand these pathogens.

1. **Clubroot Resistance:** Clubroot is caused by the soil-borne pathogen *Plasmodiophora brassicae*, which infects the roots of cauliflower plants, leading to swollen, distorted roots and stunted growth. Clubroot-resistant varieties have been developed by introgressing resistance genes from wild relatives of cauliflower, such as *Brassica rapa*. Marker-assisted selection is also used to identify and select for clubroot resistance genes in breeding populations.
2. **Black Rot Resistance:** Black rot, caused by *Xanthomonas campestris*, is a bacterial disease that affects the vascular system of cauliflower, leading to leaf chlorosis, wilting, and curd rot. Breeding for black rot resistance involves selecting varieties that can limit the spread of the pathogen within the plant. Molecular markers linked to black rot resistance are being developed to speed up the selection process in breeding programs.
3. **Downy Mildew Resistance:** Downy mildew, caused by *Peronospora parasitica*, is a fungal disease that causes yellowing and necrosis of cauliflower leaves. It can also affect the curd, reducing its quality. Breeding for downy mildew resistance involves selecting varieties with genetic resistance to the pathogen. In addition to conventional breeding methods, researchers are exploring the use of genetic engineering to introduce resistance genes from other species.

Table 3: Major Diseases of Cauliflower and Resistant Varieties

S No	Disease	Causal Agent	Resistant Varieties
1	Clubroot	*Plasmodiophorabrassicae*	Amazing, Clarify, Gilboa
2	Black Rot	*Xanthomonas campestris*	Fremont, Clarify, Celio
3	Downy Mildew	*Peronospora parasitica*	Fremont, Early Snowball, Clarify

Breeding for Abiotic Stresses in Cauliflower

Abiotic stresses such as extreme temperatures, drought, flooding, moisture stress, soil salinity, and water logging can severely affect cauliflower production. Breeding for abiotic stress tolerance is crucial for maintaining yield and quality under adverse environmental conditions. Several cauliflower varieties and wild relatives have been developed or identified with inherent resistance to these stresses.

Key Environmental Conditions Affecting Cauliflower:

» **Low and High Temperatures:** Cauliflower is highly sensitive to temperature fluctuations. High temperatures can lead to poor curd formation or bolting, while low temperatures can delay curd development.

» **Drought Conditions:** Water availability is critical for curd formation. Varieties with enhanced drought tolerance are needed, particularly in arid regions.

» **Flooding and Waterlogging:** Excess water can lead to root rot and other fungal diseases, so breeding for waterlogging tolerance is important.

» **Moisture Stress:** Consistent water supply is essential, but varieties tolerant of moisture stress can maintain curd quality even in fluctuating water conditions.

» **Soil Salinity:** High salinity levels in soil can impede nutrient uptake and affect plant growth. Salt-tolerant varieties are particularly valuable in coastal and saline-prone areas.

Some Important Related Wild Species of Cauliflower and Features Tolerant to Abiotic Stresses

S No	Variety Name	Abiotic Stress Tolerance Features
1	*Brassica oleracea* var. *cheesamanii*	Salt tolerance, heat tolerance for curd development
2	*Brassica oleracea* var. *pimpinellifolium*	Heat tolerance for curd formation, drought tolerance
3	*Brassica oleracea* var. *chilense*	Drought tolerance, cold resistance
4	*Brassica oleracea* var. *hirsutum*	Cold tolerance, chilling tolerance, salt tolerance
5	*Brassica oleracea* var. *pennelii*	Drought tolerance, salt tolerance
6	*Brassica oleracea* var. *lycopersicoides*	Cold tolerance, resilience to low temperature during curd development

The above wild relatives have contributed to modern breeding programs for developing cauliflower varieties with enhanced tolerance to these stresses. Through gene introgression and molecular breeding techniques, these wild species have been crossed with cultivated cauliflower varieties to produce more resilient plants.

Abiotic Stress Resistant Cauliflower Varieties

Several varieties of cauliflower have been developed with resistance to various abiotic stresses, enhancing their adaptability to diverse environmental conditions. These varieties are particularly valuable for growers in regions with extreme weather patterns or limited water availability.

S No	Variety Name	Key Features
1	Pusa Hybrid-1	Tolerant to high night temperatures (up to 28°C), excellent fruit set even in heat
2	Pusa Sadabahar	Resistant to both low (6°C) and high (30°C) night temperatures
3	Sabour Suphala	Salt-tolerant at the seed germination stage
4	Arka Vikas	Tolerant to moisture stress, adapted to arid conditions

These varieties have been developed through both conventional breeding and molecular methods to ensure stable yield and high-quality curds in regions prone to abiotic stresses. Continuous improvement is being made to develop varieties that can withstand multiple environmental challenges simultaneously.

Varietal Achievements in Cauliflower Breeding

The development of cauliflower varieties with specific traits such as disease resistance, curd quality, and adaptability to environmental stresses is a key achievement of modern breeding programs. Below are some notable cauliflower varieties developed by various agricultural institutions:

S No	Variety Name	Institution	Parentage	Key Features
1	Pusa Ruby	IARI, New Delhi	Sioux x Improved Meeruti	Indeterminate, early, firm, medium-sized curds, uniform ripening
2	Pusa Early Dwarf	IARI, New Delhi	Improved Meeruti x Red Cloud	Dwarf, early maturing, medium-sized curds, compact growth
3	HS 101	HAU, Hisar	Sel. 2-3 x Exotic culture	Determinate, dwarf, medium-sized juicy curds, wide adaptability
4	HS 110	HAU, Hisar	Selection from an exotic line	Determinate, late, large curds, suitable for table purposes
5	Hisar Arun	HAU, Hisar	Pusa Early Dwarf x K1	Early, high-yielding, dwarf plant, medium to large curds
6	Hisar Lalima	HAU, Hisar	Pusa Early Dwarf x HS 101	Early, round, large curds, high yielding
7	Hisar Lalit	HAU, Hisar	HS 101 x Resistant Bangalore	Semi-determinate, medium to large curds, resistant to nematodes

These varieties represent significant improvements in curd size, quality, and plant architecture, making them ideal for both small-scale farmers and commercial growers. The success of these varieties in different growing regions highlights the importance of localized breeding efforts to meet the specific needs of cauliflower growers across various climates.

Breeding Methods and Hybrid Varieties in Cauliflower

To achieve the desired traits in cauliflower, both conventional and non-conventional breeding methods are utilized. Hybrid varieties have gained prominence due to their high yield potential, uniformity, and resilience. Hand emasculation and pollination are the most commonly used techniques for producing hybrid seeds, ensuring the purity and vigor of the offspring.

Key Breeding Methods:

1. **Introduction and Selection:** Introducing new germplasm and selecting for desired traits, such as curd quality and stress resistance.

2. **Hybridization:** Crossing different cauliflower lines to combine traits such as disease resistance, abiotic stress tolerance, and high yield.
3. **Mutation Breeding:** Inducing genetic mutations to create novel varieties with unique traits.
4. **Backcrossing:** Used to incorporate specific traits from wild relatives into cultivated varieties.
5. **Tissue Culture:** For mass propagation of disease-free plants and to accelerate breeding cycles.

Hybrid Varieties in Cauliflower:

S. No	Variety Name	Institution	Features
1	Pusa Hybrid-1	IARI, New Delhi	Determinate plant, prolific curd set, high-yielding
2	Pusa Hybrid-2	IARI, New Delhi	Semi-determinate, dwarf plant, medium-sized curds
3	Arka Samrat	IIHR, Bangalore	Resistant to major diseases, including bacterial wilt and early blight
4	Arka Rakshak	IIHR, Bangalore	Triple disease resistance to ToLCV, bacterial wilt, and early blight
5	COTH1	TNAU, Coimbatore	High curd yield, suitable for both fresh market and processing
6	COTH2	TNAU, Coimbatore	Resistant to leaf curl virus, adapted to high-density planting

These hybrid varieties represent the cutting-edge of cauliflower breeding, offering superior yield, disease resistance, and abiotic stress tolerance. Hybridization has allowed breeders to develop cauliflower varieties with greater uniformity and adaptability, making them highly desirable for commercial production.

Commercial Hybrid Varieties from the Private Sector

Several private-sector companies have also developed high-performing cauliflower hybrids, which are now widely available to growers. These hybrids are bred for specific traits, such as curd quality, stress tolerance, and disease resistance, making them popular choices for both domestic and export markets.

S. No	Company	Hybrid Varieties
1	Century	Century-12, Maitri, Rishi
2	Indo American	Karnataka, Mangla, Vaishali, Rupali
3	Mahyco	MTH-1, MTH-2, MTH-6, MTH-15, MTH-16
4	Namdhari	NS-386, NS-815, Summerset, Cross B, Gotya

5	Nath	NA-501, NA-601
6	Nijjar	NH-15, NH-25, NH-38
7	Pioneer	LIHB-230
8	Sandoz	Learika, Rasika, Avinash 11
9	Sungrow	Arjuna, Krishna, Bhim

These commercial hybrids offer growers the ability to produce consistent, high-quality curds across a variety of environmental conditions. The private sector plays a crucial role in expanding the genetic base of cauliflower through innovative breeding programs and the introduction of hybrids that meet the demands of both domestic and international markets.

Technology of Hybrid Seed Production

The production of hybrid cauliflower seeds involves careful selection of parent lines with strong combining ability. The most common method is hand emasculation followed by hand pollination. This ensures the purity of the hybrid seeds and maximizes the desired traits in the offspring.

1. **Hand Emasculation and Pollination:** Female flowers are emasculated manually using forceps or needles, typically 12-15 hours before anthesis. Pollination is carried out the following morning using freshly collected pollen.

2. **Emasculation Process:** Sepals are removed from pollinated flowers for easy identification, and the ovary begins swelling within 4-5 days, indicating successful pollination.
3. **Seed Production:** The seeds are produced under controlled conditions to maintain a 12:1 ratio of seed plants to pollen parents. Hybrid seed production typically results in more than 90% fruit set.

This method ensures that the hybrid seeds produced are vigorous and capable of yielding high-quality curds with strong resistance to both biotic and abiotic stresses.

B. Cabbage (*Brassica oleracea* var. *capitata*)

Family: Brassicaceae

Chromosome number: 2n = 2x = 18

Cabbage, a member of the *Brassica oleracea* species, is one of the most widely grown vegetables worldwide. It is highly valued for its dense, compact head of leaves, which can be consumed both fresh and cooked. The vegetable is highly nutritious, being a rich source of Vitamin C, Vitamin K, and dietary fiber, and is commonly included in diets for its low-calorie content. The leaves of cabbage contain a variety of health-promoting compounds such as glucosinolates, flavonoids, and polyphenols, which have been linked to cancer-preventive properties. Additionally, cabbage provides a range of essential minerals including calcium, potassium, and magnesium, which contribute to maintaining overall health.

Cabbage is an extremely versatile vegetable and is used in numerous culinary traditions around the world. It can be eaten raw in salads, shredded for coleslaw, cooked in soups and stews, or fermented into products like sauerkraut and kimchi. In recent years, the growing trend towards plant-based diets has further increased the popularity of cabbage as a nutritious, low-calorie food.Global production of cabbage is dominated by China, which produces more than half of the world's supply. India is the second-largest producer, with significant cultivation in the states of West Bengal, Bihar, and Odisha. The increasing demand for cabbage in both fresh and processed forms has led to the development of new varieties that are resistant to pests, diseases, and adverse environmental conditions, making it a profitable crop for farmers.

Origin

The domestication of cabbage dates back thousands of years to the Mediterranean region, where its wild ancestor, *Brassica oleracea* var. *oleracea*, grew along the coastal cliffs of southern Europe. This wild form was a leafy, non-heading plant that closely resembled kale or collard greens. Cabbage was first cultivated by early farmers who selected for larger leaves and more compact growth habits, traits that eventually led to the formation of the familiar cabbage head.

The Mediterranean Basin is widely considered the primary center of origin for cabbage, but its cultivation spread quickly throughout Europe. By the time of the Roman Empire, cabbage was a staple crop in the diets of both the wealthy and the poor, prized for its storage qualities and nutritional benefits. Cabbage could be harvested late in the season and stored for long periods, making it an essential winter food in northern climates where fresh vegetables are scarce.

Cabbage's spread from its Mediterranean origins to other parts of the world was facilitated by its adaptability to different climates. It was introduced to Northern Europe, where the cooler temperatures proved ideal for growing cabbage, and it became a key ingredient in traditional dishes such as sauerkraut in Germany and borscht in Russia. In Asia, cabbage varieties were developed independently, with Chinese cabbage (*Brassica rapa* var. *pekinensis*) and bok choy (*Brassica rapa* var. *chinensis*) emerging as important crops in China, Japan, and Korea.

The domestication process not only focused on increasing head size and leaf density but also on improving the flavor and reducing the bitterness often associated with wild brassicas. Over centuries, farmers selectively bred plants that produced sweeter, more tender leaves. This process led to the diversification of cabbage into the many varieties we see today, including green cabbage, red cabbage, and Savoy cabbage, each adapted to different climates and culinary uses.

Related Genera

Cabbage belongs to the *Brassica* genus, which includes several other economically and nutritionally important vegetables. The *Brassica* genus is part of the Brassicaceae family, which also includes plants like mustard, radishes, and horseradish. Species within the *Brassica* genus share many genetic similarities, making them valuable in breeding programs aimed at improving disease resistance, abiotic stress tolerance, and nutritional content in cabbage. Here are the key related genera:

» **Brassica:** The primary genus to which cabbage belongs, *Brassica* includes vegetables like broccoli (*Brassica oleracea* var. *italica*), cauliflower (*Brassica oleracea* var. *botrytis*), Brussels sprouts (*Brassica oleracea* var. *gemmifera*), and kale (*Brassica oleracea* var. *sabellica*). All of these crops have been domesticated from the same wild ancestor, *Brassica oleracea*, and they share similar growth habits, nutritional profiles, and genetic traits. This makes them highly useful for hybridization and genetic research.

» **Raphanus:** This genus includes radishes, which are commonly grown for their edible roots. Radishes share many of the same pest and disease resistance traits as cabbage, and they are often used in intercropping systems to reduce pest pressure. The genetic similarities between radishes and cabbage also allow for the transfer of desirable traits between the two crops.

» **Sinapis:** Mustard plants belong to the Sinapis genus, and they are closely related to cabbage. Mustard greens are commonly grown for their spicy leaves and seeds, and they are often used in crop rotations with cabbage to improve soil health. Mustard plants are also valuable in breeding programs for their resistance to certain pests and diseases that affect cabbage.

These related genera are important not only for understanding the evolutionary history of cabbage but also for their practical applications in modern breeding programs. Hybridization between species in the *Brassica* genus has led to the development of new varieties with improved traits, such as disease resistance, cold tolerance, and increased nutritional value.

Related Species

Within the *Brassica* genus, cabbage is closely related to several other species, many of which are also important crops. These species share similar genomes and can be crossed with cabbage to introduce desirable traits such as disease resistance, abiotic stress tolerance, and improved head formation. Some of the most closely related species include:

- » *Brassica oleracea var. capitata* (Cabbage): This is the primary variety of cabbage, known for its tight, dense head of leaves. Cabbage is grown for its edible leaves and is used in a wide range of culinary applications.
- » *Brassica oleracea var. italica* (Broccoli): Broccoli is closely related to cabbage and shares many of the same growth characteristics. It is valued for its edible flower heads, which are rich in vitamins and antioxidants. Broccoli and cabbage are often grown in rotation to reduce pest pressure and improve soil health.
- » *Brassica oleracea var. botrytis* (Cauliflower): Cauliflower is another close relative of cabbage, and it is grown for its large, white curd, which is composed of undeveloped flower buds. Cauliflower shares many of the same pest and disease challenges as cabbage, and breeding efforts often focus on developing varieties that are resistant to common threats like black rot and clubroot.
- » **Brassica rapa** (Turnip and Chinese Cabbage): While not as closely related as other *Brassica oleracea* varieties, turnip and Chinese cabbage share many genetic similarities with cabbage. These species are often used in breeding programs to introduce cold tolerance and improved head formation traits into cabbage varieties. Chinese cabbage (*Brassica rapa* var. *pekinensis* and *chinensis*) is a particularly valuable crop in Asia, where it is used in a variety of dishes, including stir-fries, soups, and pickled preparations.
- » **Brassica napus** (Rapeseed): Rapeseed, grown primarily for its oil-rich seeds, is closely related to cabbage. In addition to being an important oilseed crop, rapeseed is used in cabbage breeding programs to introduce traits such as disease resistance and drought tolerance. The genetic diversity found within *Brassica napus* makes it a valuable resource for improving cabbage varieties.

These related species provide important genetic diversity for breeding programs aimed at improving cabbage's resilience to pests, diseases, and environmental stressors. By incorporating genes from related species, breeders can develop cabbage varieties that are better suited to the challenges of modern agriculture.

Types of Cabbage on Basis of Morphological and Agronomical Characteristics

Cabbage is classified into several botanical varieties, each of which has distinct morphological and agronomic characteristics. These varieties are typically distinguished based on head shape, leaf texture, and color. The primary botanical varieties of cabbage include green cabbage, red cabbage, and Savoy cabbage. Each of these varieties has been selected for different growing conditions and culinary uses, making cabbage a versatile and adaptable crop.

1. **Green-Headed Cabbage (*Brassica oleracea* var. *capitata* f. *alba*):** This is the most common variety of cabbage, characterized by its green leaves and tightly packed, round heads. Green cabbage is widely grown in temperate regions and is used for both fresh consumption and processing. It is commonly found in salads, soups, and fermented dishes like sauerkraut. Green cabbage is valued for its high yield and adaptability to a wide range of climates.
2. **Red Cabbage (*Brassica oleracea* var. *capitata* f. *rubra*):** Red cabbage is known for its deep purple-red leaves, which are due to the presence of anthocyanins, a type of antioxidant. Red cabbage is often eaten raw in salads or pickled. It has a slightly more peppery taste compared to green cabbage and is rich in vitamins C and K. Red cabbage is also more cold-tolerant than green cabbage, making it a popular choice for cooler climates.
3. **Savoy Cabbage (*Brassica oleracea* var. *capitata* f. *sabauda*):** Savoy cabbage is easily recognized by its crinkled, textured leaves. It is considered the most tender of the cabbage varieties and has a mild, sweet flavor. Savoy cabbage is often used in salads, wraps, and cooked dishes. It is less dense than green or red cabbage, and its leaves have a more delicate texture. Savoy cabbage is grown primarily for the fresh market, as it does not store as well as other cabbage varieties.
4. **Chinese Cabbage (*Brassica rapa* var. *pekinensis* and *chinensis*):** Although technically a different species, Chinese cabbage is often grouped with European cabbage due to its similar use in culinary applications. Chinese cabbage forms a loose head of pale green leaves, and it is widely used in Asian dishes such as stir-fries, soups, and dumpling fillings. Bok choy (*Brassica rapa* var. *chinensis*) is another form of Chinese cabbage, which has dark green leaves and white stalks. Both varieties of Chinese cabbage are highly valued for their mild flavor and high water content.

Table 1: Major Botanical Varieties of Cabbage

S No	Botanical Variety	Description	Head Color	Special Features
1	*Brassica oleracea* var. *capitata* f. *alba*	Green Headed Cabbage	Green	Most common, dense heads, versatile culinary uses
2	*Brassica oleracea* var. *capitata* f. *rubra*	Red Cabbage	Red/Purple	High anthocyanin content, cold-hardy, used raw or cooked
3	*Brassica oleracea* var. *capitata* f. *sabauda*	Savoy Cabbage	Green	Crinkled leaves, tender texture, milder flavour
4	*Brassica rapa* var. *pekinensis*	Chinese Cabbage	Light Green	Loose head, widely used in Asian cuisines

5	*Brassica rapa* var. *chinensis*	Bok Choy	Dark Green	Open, leafy form, used in soups and stir-fries

Each of these varieties has been selected and bred to meet specific market demands and growing conditions. Green cabbage is by far the most widely cultivated variety, but red cabbage and Savoy cabbage are increasingly popular for their unique flavors, textures, and nutritional profiles.

Growth Habit

Cabbage is a biennial plant by nature, but it is typically cultivated as an annual for the production of its edible head. The plant forms a rosette of broad, thick leaves that overlap to form a dense head at the center. The growth habit of cabbage can vary depending on the variety and environmental conditions, but it is generally divided into two main types: compact-heading and loose-heading. In compact-heading varieties, the leaves grow tightly together, creating a firm, dense head. This is typical of most commercial cabbage varieties. Loose-heading types, such as Savoy cabbage, have more open heads with loosely arranged leaves, giving them a softer texture. The growth of cabbage is divided into distinct phases: the vegetative phase, during which the plant forms its leaves and head, and the reproductive phase, which occurs if the plant is left to grow beyond its normal harvest period. During the reproductive phase, cabbage bolts, meaning it produces a flowering stalk. The flowers are small and yellow, typical of the Brassicaceae family. Bolting is undesirable in cabbage cultivation as it marks the end of the head's marketability.

Cabbage is considered a cool-season crop and performs best in temperatures between 15°C and 20°C. Higher temperatures can lead to bolting or result in poor head formation, while frost can damage young plants. This makes cabbage a common crop in temperate climates, although it can be grown in subtropical regions during the cooler months. Cabbage is also photoperiod-sensitive, meaning its development can be influenced by the length of daylight hours. Breeding efforts have focused on developing varieties that can be grown in a wide range of environments, extending the growing season and ensuring consistent yields. To maximize yields, cabbage is often grown in well-drained, fertile soils with a pH between 6.0 and 6.8. Adequate irrigation is essential, especially during head formation, as moisture stress can lead to smaller, less dense heads. Cabbage is also a heavy feeder and benefits from regular applications of nitrogen, phosphorus, and potassium. Modern farming techniques, including precision agriculture and controlled-environment farming, have been used to optimize cabbage production, particularly in regions with challenging climates or soil conditions.

Floral Morphology

The floral morphology of cabbage is typical of plants in the Brassicaceae family. Cabbage produces small, bright yellow flowers arranged in clusters known as racemes. The flowers are complete, meaning they contain both male and female reproductive structures. Each flower has four sepals, four petals arranged in a cross shape (which gives the Brassicaceae family its common name, Cruciferae), six stamens (four long and two short), and a single pistil. The ovary is superior, meaning it sits above the other floral parts, and it develops into a siliqua, a long, narrow fruit that contains the seeds. Under normal cultivation, cabbage is harvested before it flowers, as the edible portion is the leafy head rather than the flowers. However, if left to grow through its full life cycle, cabbage will bolt and produce a tall flowering stalk. Bolting can be triggered by environmental stresses such as high temperatures or water stress, but it can also occur naturally as part of the plant's biennial growth cycle. In the second year of growth, cabbage naturally enters its reproductive phase and produces flowers.

The flowers of cabbage are self-incompatible, meaning that they require cross-pollination from other plants to produce viable seeds. This trait is important for maintaining genetic diversity in cabbage populations. Pollination is typically carried out by insects, particularly bees, which transfer pollen from one flower to another. Once pollinated, the ovary develops into a siliqua, which can contain dozens of seeds. These seeds are harvested for use in the next planting season or for breeding programs. The reproductive biology of cabbage plays a crucial role in hybrid seed production. In hybrid breeding programs, controlled pollination is used to ensure that only the desired crosses occur. This is achieved by isolating plants or using physical barriers to prevent unwanted pollination. The production of hybrid cabbage seeds is labor-intensive but results in varieties with improved traits such as disease resistance, abiotic stress tolerance, and higher yields.

Floral Biology

The floral biology of cabbage is closely related to its ability to reproduce and produce seeds for subsequent generations. Cabbage flowers are hermaphroditic, containing both male (stamens) and female (pistil) reproductive organs. However, due to self-incompatibility, cabbage requires cross-pollination for successful seed production. Self-incompatibility is a genetic mechanism that prevents a plant from fertilizing itself, thereby promoting outcrossing and increasing genetic diversity within the population. This trait is common among many species in the Brassicaceae family and plays a vital role in maintaining healthy, vigorous cabbage plants.

The timing of flowering, or anthesis, is influenced by several environmental factors, including temperature, day length, and water availability. In temperate regions, cabbage typically flowers in the second year of growth when exposed to longer days and warmer temperatures. The flowers open sequentially, starting from the bottom of the flowering stalk and moving upward. This ensures that pollination can occur over an extended period, increasing the chances of successful fertilization.

Pollination in cabbage is primarily carried out by insects, with bees being the most common pollinators. The bright yellow flowers of cabbage are highly attractive to pollinators, and the large amounts of pollen they produce make them an important food source for bees.

Cross-pollination occurs when pollen from one plant is transferred to the stigma of another, leading to fertilization. Once fertilized, the ovary develops into a siliqua, which matures over several weeks before the seeds are fully developed. Seed production is an important part of cabbage cultivation, particularly for hybrid varieties. Hybrid seeds are produced by carefully controlling the pollination process to ensure that only the desired crosses occur. This is achieved by isolating plants or using hand-pollination techniques to prevent unwanted pollen from contaminating the breeding program. Hybrid cabbage varieties are prized for their uniformity, vigor, and resistance to pests and diseases, making them popular among commercial growers.

Breeding Objectives

The primary breeding objectives for cabbage focus on improving yield, head quality, disease resistance, and abiotic stress tolerance. These traits are crucial for developing cabbage varieties that can thrive in a wide range of environmental conditions while meeting the demands of both farmers and consumers.

1. **Yield Improvement:** Increasing yield is a key goal in cabbage breeding, particularly for varieties grown for processing or fresh market sales. Yield is influenced by several factors, including head size, uniformity, and the plant's ability to resist pests and diseases. Breeders select varieties that produce large, dense heads with a high leaf-to-head ratio, ensuring maximum marketable yield. In addition to improving yield potential, breeders also focus on reducing the time to maturity, allowing growers to harvest multiple crops per year.
2. **Head Quality:** Head quality is one of the most important traits in cabbage breeding. High-quality cabbage heads are uniform in size and shape, with tightly packed leaves that are free from defects such as splitting or bolting. Breeders aim to develop varieties that produce heads with a smooth, consistent texture and a mild, sweet flavor. In addition to visual and taste characteristics, breeders also consider the nutritional quality of cabbage, selecting varieties with higher levels of vitamins and antioxidants.
3. **Disease Resistance:** Cabbage is susceptible to a wide range of diseases, including black rot (*Xanthomonas campestris*), clubroot (*Plasmodiophorabrassicae*), and downy mildew (*Peronospora parasitica*). Disease-resistant varieties are essential for reducing the need for chemical pesticides and ensuring stable yields. Breeding programs focus on developing varieties with resistance to multiple diseases, using both conventional breeding methods and modern molecular techniques such as marker-assisted selection.
4. **Abiotic Stress Tolerance:** Cabbage is sensitive to environmental stresses such as temperature fluctuations, drought, and salinity. Breeding for abiotic stress tolerance involves selecting for traits such as heat tolerance, cold hardiness, and drought resistance. These traits are especially important in regions where climate change is leading to more unpredictable weather patterns. Abiotic stress-tolerant varieties help ensure that cabbage can be grown successfully in a wider range of environments.

Breeding Methods

The breeding methods for cabbage are primarily focused on improving traits such as yield, head quality, disease resistance, and abiotic stress tolerance. To achieve these goals, both traditional and modern breeding techniques are employed. The key breeding methods used in cabbage improvement include introduction, selection, hybridization, mutation breeding, and advanced molecular techniques like marker-assisted selection and genetic engineering.

1. **Introduction:**

   » The introduction method involves introducing varieties or germplasm from one region into another to test their adaptability and performance under different environmental conditions. This method is particularly useful for introducing varieties with specific traits, such as disease resistance or drought tolerance, into regions where those traits are needed.

   » In the case of cabbage, many successful introductions have been made over the years, particularly from Europe to other parts of the world. For instance, many European cabbage varieties were introduced to Asia, where they were further selected and adapted to local growing conditions. These introductions have contributed to the global spread of cabbage cultivation.

2. **Selection**:

   » Selection is one of the oldest and most fundamental breeding methods. It involves choosing the best-performing plants from a population and using them to produce the next generation. This method is effective for improving simple traits, such as head size, shape, or disease resistance. There are two main types of selection:

   - **Mass Selection:** In mass selection, superior plants are selected based on their phenotype, or observable traits, such as head size or disease resistance. The seeds from these plants are then bulked together and grown in the next generation. This method is simple but can be slow, as it relies on natural variation within the population.
   - **Pure Line Selection:** In pure line selection, individual plants are selected and propagated over several generations to produce a uniform, homozygous line. This method is particularly useful for improving traits that are controlled by a few genes, such as disease resistance or head quality.

3. **Hybridization**:

   » Hybridization is one of the most powerful breeding methods used in cabbage improvement. It involves crossing two genetically distinct parents to produce offspring, or hybrids, that combine the desirable traits of both parents. Hybrid cabbage varieties are prized for their vigor, uniformity, and resistance to pests and diseases. Hybridization is particularly important for developing varieties that can perform well under a wide range of environmental conditions.

- » **F1 Hybrids:** The most common type of hybrid in cabbage breeding is the F1 hybrid. F1 hybrids are produced by crossing two inbred lines, each of which has been selected for specific traits. The resulting F1 hybrids often exhibit hybrid vigor, or heterosis, which leads to improved growth, yield, and stress tolerance compared to either parent. F1 hybrids are widely used in commercial cabbage production due to their uniformity and high performance.
- » **Backcross Breeding:** Backcross breeding is used to introduce specific traits, such as disease resistance, into an otherwise well-performing variety. In this method, a desired trait is introduced from a donor parent into a recurrent parent through repeated crossings. For example, a disease-resistant wild relative of cabbage might be crossed with a commercial variety to introduce resistance while maintaining the other desirable traits of the commercial variety.

4. **Mutation Breeding:**
   - » Mutation breeding involves exposing cabbage seeds or plants to chemicals or radiation to induce random mutations in their genetic material. These mutations can lead to new traits that may not occur naturally, such as improved disease resistance, abiotic stress tolerance, or novel head shapes and colors.
   - » One famous example of mutation breeding in cabbage is the development of a purple cabbage variety with increased levels of anthocyanins, a type of antioxidant that gives the cabbage its distinctive color. These kinds of traits can be useful in both enhancing the crop's appeal to consumers and improving its nutritional profile.
5. **Molecular Breeding:**
   - » Advances in molecular biology have allowed breeders to use DNA markers to select for specific traits in cabbage. Marker-assisted selection (MAS) is a technique in which genetic markers linked to desirable traits, such as disease resistance or abiotic stress tolerance, are used to select plants in the early stages of breeding. This reduces the time and resources required to develop new varieties.
   - » **Genetic Engineering:** Genetic engineering allows breeders to introduce specific genes from other species into cabbage. This can be particularly useful for introducing traits that are not present in the cabbage gene pool, such as resistance to certain pests or herbicides. However, the use of genetically modified organisms (GMOs) in cabbage is still limited, and much of the current research focuses on developing non-GMO varieties through traditional breeding and molecular techniques.

Breeding for Biotic Stresses

Biotic stresses, such as diseases and insect pests, are among the most significant challenges in cabbage production. Cabbage is susceptible to a wide range of diseases caused by bacteria, fungi, viruses, and nematodes, as well as damage from insect pests such as aphids, caterpillars, and flea beetles. Breeding for resistance to biotic stresses is a major focus of cabbage breeding programs, as it reduces the need for chemical pesticides and helps ensure stable yields.

1. **Disease Resistance:**
   - Several diseases can affect cabbage, including black rot (*Xanthomonas campestris*), clubroot (*Plasmodiophorabrassicae*), downy mildew (*Peronospora parasitica*), and Fusarium wilt. Breeding for disease resistance involves identifying resistant varieties or wild relatives and incorporating those resistance genes into commercial cabbage varieties.
   - **Black Rot:** Black rot is one of the most destructive diseases of cabbage, causing yellowing and wilting of the leaves, and leading to significant yield losses. Resistance to black rot has been identified in some wild relatives of cabbage, such as *Brassica carinata*, and these resistance genes have been introgressed into commercial cabbage varieties through backcrossing.
   - **Clubroot:** Clubroot is a soil-borne disease caused by the pathogen *Plasmodiophorabrassicae*. It causes swelling and deformation of the roots, which can lead to stunted growth and reduced yields. Clubroot-resistant varieties have been developed by crossing cabbage with resistant wild relatives, and marker-assisted selection has been used to accelerate the breeding process.
2. **Pest Resistance:**
   - Insect pests such as aphids, diamondback moths, cabbage worms, and flea beetles can cause significant damage to cabbage crops. Breeding for pest resistance involves selecting for traits such as leaf toughness, trichomes (plant hairs), and natural plant defenses that make the plant less attractive to pests.
   - **Aphid Resistance:** Aphids can transmit viral diseases and cause direct damage by sucking sap from the leaves. Some cabbage varieties have been bred to produce higher levels of glucosinolates, a natural defense compound that deters aphids.
   - **Caterpillar Resistance:** Caterpillars, such as those of the cabbage looper and the diamondback moth, can defoliate cabbage plants, leading to reduced yields and poor head formation. Breeding for resistance to these pests often involves selecting for varieties with tougher leaves or increased levels of natural insecticidal compounds.

Breeding for Abiotic Stresses

Abiotic stresses such as extreme temperatures, drought, salinity, and waterlogging can significantly affect cabbage growth and yield. Breeding for abiotic stress tolerance is becoming increasingly important as climate change leads to more unpredictable weather patterns and harsher growing conditions. The goal is to develop cabbage varieties that can withstand these stresses while maintaining high yields and quality.

1. **Temperature Tolerance:**
   - Cabbage is a cool-season crop that thrives in temperatures between 15°C and 20°C. However, many regions experience temperature fluctuations that can affect cabbage growth, particularly during head formation. Breeding for temperature tolerance focuses on developing varieties that can withstand both high and low temperatures.
   - **Heat Tolerance:** High temperatures during the growing season can cause bolting (premature flowering), reduced head formation, or poor head quality. Breeding for heat tolerance involves selecting varieties that can maintain head quality and yield even when exposed to higher temperatures. Some wild relatives of cabbage, such as *Brassica oleracea* var. *sabellica* (kale), have been used to introduce heat tolerance into commercial varieties.
   - **Cold Tolerance:** Cabbage is relatively cold-tolerant compared to other crops, but severe frost can damage young plants and reduce yield. Breeding for cold tolerance involves selecting for varieties that can survive freezing temperatures during the early stages of growth. Cold-tolerant varieties are especially important in regions with short growing seasons or unpredictable frosts.
2. **Drought and Salinity Tolerance:**
   - Water availability is critical for cabbage, particularly during head formation. Drought stress can lead to smaller heads and lower yields, while excess salinity in the soil can inhibit nutrient uptake and stunt plant growth. Breeding for drought and salinity tolerance involves selecting varieties with deep root systems, improved water-use efficiency, and the ability to maintain growth under suboptimal conditions.
   - **Drought Tolerance:** Some wild relatives of cabbage, such as *Brassica napus* (rapeseed), have been used to introduce drought tolerance into commercial cabbage varieties. These varieties are able to withstand periods of water stress without significant reductions in yield.
   - **Salinity Tolerance:** Salinity tolerance is important in coastal regions or areas with poor-quality irrigation water. Salinity-tolerant cabbage varieties have been developed by selecting for traits such as ion exclusion (the plant's ability to prevent the uptake of harmful salts) and osmotic adjustment (the plant's ability to maintain water uptake in salty conditions).

Important Genes in Cabbage Breeding

In cabbage breeding, identifying and utilizing genes that control desirable traits such as disease resistance, abiotic stress tolerance, head formation, and nutritional content is crucial. Advances in molecular biology have enabled breeders to locate and use specific genes more efficiently, leading to faster breeding cycles and more targeted improvements. Below are some of the important genes involved in cabbage breeding:

S. No	Trait	Gene	Function
1	Black Rot Resistance	**BRR1**	Confers resistance to *Xanthomonas campestris*
2	Clubroot Resistance	**CRa**	Provides resistance to *Plasmodiophorabrassicae*
3	Head Shape and Size	**HED1**	Controls head size and density
4	Cold Tolerance	**COLD1**	Increases cold hardiness in low-temperature conditions
5	Heat Tolerance	**HT1**	Enhances heat tolerance during head formation
6	Fusarium Wilt Resistance	**FWR1**	Confers resistance to *Fusarium oxysporum*
7	Glucosinolate Production	**GS1**	Increases glucosinolate levels for pest deterrence
8	Anthocyanin Production	**ANTH1**	Increases anthocyanin content in red cabbage
9	Drought Tolerance	**DRT1**	Improves water-use efficiency in drought conditions

The genes listed above are involved in various pathways that contribute to cabbage's adaptability, nutritional value, and resistance to pests and diseases. Molecular breeding techniques like marker-assisted selection (MAS) and genome editing tools like CRISPR are increasingly used to manipulate these genes, leading to the development of improved cabbage varieties.

Varietal Achievements

Over the years, cabbage breeding programs have developed a range of varieties tailored to different climatic conditions, pest pressures, and consumer preferences. These varietal achievements reflect advancements in traditional breeding techniques and modern molecular breeding technologies. Below are some notable cabbage varieties and their key features:

S. No	Variety Name	Parentage	Key Features	Organization
1	Pusa Drumhead	Selection from local germplasm	Large, firm heads, late maturing, suitable for tropical regions	IARI, New Delhi
2	Golden Acre	Cross between European varieties	Early-maturing, compact heads, good for fresh market	PAU, Ludhiana
3	Hisar Pragati	Cross between tropical and temperate lines	Heat-tolerant, early-maturing, moderate pest resistance	HAU, Hisar
4	Hybrid Cabbage 1	Hybrid of heat-tolerant and cold-tolerant parents	Tolerant to both high and low temperatures, uniform heads	TNAU, Coimbatore

5	Red Cabbage KS-2	Pure line selection	High anthocyanin content, cold-hardy, used for salads and pickling	Kalyanpur
6	Savoy Express	Cross between Savoy cabbage lines	Tender, crinkled leaves, cold-tolerant, high yield	IIHR, Bangalore
7	Copenhagen Market	European selection	Round, firm heads, ideal for fresh market and storage	European origin
8	Pant Bahar	Pure line selection	Heat-tolerant, ideal for late-season cultivation, good for tropical climates	GBPUAT, Pantnagar

These varietal achievements demonstrate how cabbage breeding has adapted to different environmental and market needs. Varieties like Pusa Drumhead and Hisar Pragati are specifically bred for tropical and subtropical regions, while Red Cabbage KS-2 and Savoy Express cater to colder climates and niche markets.

Hybrid Varieties in Cabbage

Hybrid varieties are becoming increasingly popular in cabbage cultivation due to their superior performance in terms of yield, uniformity, and resistance to biotic and abiotic stresses. Hybrid breeding combines the best traits from two parental lines, resulting in plants that exhibit hybrid vigor (heterosis). Below are some important hybrid cabbage varieties:

S No	Hybrid Name	Parentage	Key Features	Organization
1	Pusa Hybrid-1	Cross of local varieties	Heat tolerance, good head formation, early maturing	IARI, New Delhi
2	Green Express	Cross of high-yielding lines	Early-maturing, uniform heads, disease-resistant	PAU, Ludhiana
3	Hisar Hybrid-2	Cross of heat and cold-tolerant lines	High yield, wide adaptability to various climates	HAU, Hisar
4	TNAU Hybrid 3	Cross between temperate and tropical parents	Heat tolerance, good for processing and fresh market	TNAU, Coimbatore
5	COTH-2	Cross of cold-tolerant varieties	High cold tolerance, large heads, good for fresh market	IIHR, Bangalore
6	Savoy King	Hybrid of Savoy cabbage lines	Crinkled leaves, tender texture, cold-tolerant, high yield	European origin
7	Pant Hybrid 1	Cross of tropical varieties	High resistance to pests, heat-tolerant, early maturing	GBPUAT, Pantnagar
8	Arka Rakshak	Triple disease-resistant hybrid	Resistant to black rot, downy mildew, and Fusarium	IIHR, Bangalore

Hybrid cabbage varieties offer superior performance compared to open-pollinated varieties. They are especially useful in regions with extreme weather conditions or where disease pressure is high. In recent years, hybrid varieties have played a key role in increasing cabbage productivity and ensuring crop resilience.

Technology of Hybrid Seed Production

The production of hybrid seeds in cabbage involves a series of controlled pollination techniques to ensure genetic purity and the desired combination of traits from the parent plants. Below are the key steps involved in hybrid seed production:

1. **Parent Selection:**
   - » The first step in hybrid seed production is selecting the parental lines. These parents are chosen based on their ability to complement each other in terms of traits like head quality, disease resistance, and stress tolerance. Typically, one parent (the male) provides traits like vigor and disease resistance, while the other parent (the female) contributes traits like head size and uniformity.
2. **Isolation:**
   - » To prevent unwanted cross-pollination, hybrid seed production requires physical or spatial isolation of the parent plants. This can be achieved by planting the parents at a considerable distance from other cabbage crops or by using mechanical barriers like nets. The goal is to ensure that only the selected male and female plants cross-pollinate, thus preserving the genetic purity of the hybrid seed.
3. **Hand Emasculation and Pollination:**
   - » In cabbage, the female flowers need to be emasculated (removal of the male parts) to prevent self-pollination. This is typically done by hand, using forceps or a similar tool. After emasculation, pollen from the male parent is manually applied to the stigma of the female flower to achieve controlled pollination. This ensures that only the desired genetic traits from both parents are passed on to the hybrid offspring.
4. **Seed Harvesting:**
   - » Once pollination has occurred, the fertilized flowers develop into siliques, which contain the hybrid seeds. These siliques are allowed to mature on the plant, and when they are fully ripe, they are harvested. The seeds are then extracted from the siliques, cleaned, and processed for commercial use.
5. **Seed Testing**:
   - » Before hybrid seeds are released to farmers, they undergo rigorous testing for germination rates, seed purity, and genetic stability. This ensures that the hybrid seeds will produce the desired plants with consistent performance in the field. Seed companies also conduct field trials to assess the performance of hybrid varieties under different environmental conditions.

S No	Hybrid Seed Production Step	Key Description
1	Parent Selection	Selection of male and female plants with complementary traits
2	Isolation	Ensuring physical isolation to prevent unwanted pollination
3	Hand Emasculation and Pollination	Manual removal of male parts from female flowers, controlled pollination
4	Seed Harvesting	Collection of mature siliques containing hybrid seeds
5	Seed Testing	Testing for germination, purity, and genetic stability

Hybrid seed production is labor-intensive but critical for producing high-quality, uniform cabbage varieties that meet the demands of modern agriculture. The resulting hybrid seeds provide farmers with reliable, high-yielding crops that perform well under a variety of conditions.

C. Radish (*Raphanus sativus*)

Family: Brassicaceae

Chromosome number: 2n = 2x = 18

Radish (*Raphanus sativus*) is an annual or biennial root vegetable grown for its edible, fleshy root and tender leaves. It is one of the oldest known cultivated crops, consumed both as a fresh salad vegetable and in cooked dishes. Radishes come in many shapes, sizes, and colors, ranging from the small, round red radish to the long, white *daikon* type. Rich in nutrients like Vitamin C, folate, and potassium, radish also contains glucosinolates, which are sulfur-containing compounds with potential cancer-preventive properties.

Radish is valued for its fast-growing nature; some varieties can be harvested as early as three weeks after sowing, making it an ideal crop for quick rotation in vegetable gardens. The crop is also popular for its crunchy texture and mildly peppery flavor, with variations depending on the variety and growing conditions. Radish can be grown year-round in different regions, and it is widely cultivated in temperate and subtropical climates.

Origin and Domestication

Radish is believed to have originated from Southeast Asia, specifically in regions that include modern-day China and India. It has been cultivated for thousands of years, with evidence of radish consumption dating back to ancient Egypt and Greece. The wild ancestors of radish are believed to be native to the Mediterranean region, but through domestication and selection, modern radish varieties have evolved into numerous forms and adapted to various environments. The domestication of radish likely began with early farmers selecting plants that produced larger and tastier roots. Over time, different regions developed their distinct types of radishes. For example, the small, round red radish commonly found in Europe and North America differs significantly from the long, white *daikon* radish popular in East Asia. These selections led to the diversification of radish into various shapes, sizes, and colors.

Domestication also involved selecting fast-growing plants, as radish was often used as a catch crop or a crop grown between the main planting seasons. Today, radish is grown on a large scale in many parts of the world, and new varieties continue to be developed through breeding programs focused on improving yield, disease resistance, and flavor.

Related Genera

Radish belongs to the Brassicaceae family, which includes a wide range of important agricultural crops. Several related genera share genetic traits with radish, making them valuable for crossbreeding and genetic improvement efforts.

- **Brassica:** This genus includes crops such as cabbage, broccoli, cauliflower, and mustard. While radish and *Brassica* species are distinct, they share many genetic similarities, particularly in their responses to abiotic stresses and diseases. These similarities make *Brassica* species valuable in radish breeding programs, especially when selecting for disease resistance and improved nutritional content.

» **Raphanus:** Radish is part of the *Raphanus* genus, which also includes wild radish (*Raphanus raphanistrum*), a close relative of the cultivated radish. Wild radish is often considered a weed in agricultural fields but has been used in breeding programs to introduce traits such as pest resistance and tolerance to harsh growing conditions.

» **Sinapis:** Mustard plants, belonging to the *Sinapis* genus, are also closely related to radish. Mustard is often used in crop rotation with radish to improve soil health, as it can help reduce the buildup of soil-borne diseases and pests. In some breeding programs, mustard has been used to introduce resistance to pests and diseases into radish varieties.

The genetic diversity found within these related genera provides a wealth of resources for radish improvement, particularly in terms of disease resistance, abiotic stress tolerance, and nutritional enhancement.

Related Species

Several species within the *Raphanus* genus are closely related to the cultivated radish and have been used in breeding programs to improve various traits. Some of these species include:

» **Raphanus raphanistrum** (Wild Radish): Wild radish is one of the closest relatives of cultivated radish and shares many genetic traits. Although it is often considered a weed, wild radish has been used in breeding programs to introduce resistance to pests and diseases. It is particularly useful for improving drought tolerance in radish varieties grown in arid regions.

» **Raphanus sativus var. niger** (Black Radish): Black radish is a larger, tougher variety of radish with a black or dark brown skin and a pungent flavor. It is commonly grown in Europe and is valued for its medicinal properties. Black radish is more resistant to pests and diseases than other radish varieties, making it a valuable genetic resource in breeding programs aimed at improving pest resistance.

» **Raphanus sativus var. caudatus** (Rat-tail Radish): Rat-tail radish is a unique variety grown for its edible seed pods rather than its root. This variety is popular in Southeast Asia and is often used in salads and stir-fries. Rat-tail radish has been used in breeding programs to introduce heat tolerance and pest resistance into root radish varieties.

These related species provide a wide range of genetic diversity that can be harnessed in breeding programs to improve cultivated radish. Traits such as heat tolerance, disease resistance, and improved flavor can be introduced through crossbreeding with wild and related species.

Botanical Varieties of Radish

Radish is classified into several botanical varieties, each with distinct characteristics such as root shape, size, color, and flavor. These varieties are adapted to different climates and growing conditions, making radish a versatile crop that can be grown in various environments.

1. **Round Red Radish (*Raphanus sativus* var. *radicula*):** This is the most common type of radish, characterized by its small, round, red root. It has a crisp texture and a mildly peppery flavor, making it ideal for salads. This variety is fast-growing and can be harvested within 20-30 days of sowing. It is typically grown in temperate regions and is highly popular in Europe and North America.
2. **White Radish (*Raphanus sativus* var. *longipinnatus*):** Also known as *daikon* in Japan, this variety is characterized by its long, white root, which can grow up to 50 cm in length. It has a mild, slightly sweet flavor and is often used in Asian cuisine, both fresh and cooked. *Daikon* radish is slower-growing than round radish but produces much larger roots. It is commonly grown in East Asia and is gaining popularity in Western markets.
3. **Black Radish (*Raphanus sativus* var. *niger*):** This variety is known for its large, dark-skinned roots and strong, pungent flavor. Black radish is often used in traditional European cuisine, particularly in pickles and fermented dishes. It is also valued for its medicinal properties, particularly in promoting digestive health. Black radish is more cold-hardy than other varieties and can be grown in cooler climates.
4. **Rat-tail Radish (*Raphanus sativus* var. *caudatus*):** Unlike other radish varieties, rat-tail radish is grown for its edible seed pods rather than its root. The seed pods are long, thin, and crunchy, with a spicy flavor. This variety is popular in Southeast Asian cuisine, where the pods are used in salads and stir-fries. Rat-tail radish is heat-tolerant and grows well in tropical and subtropical regions.
5. **Watermelon Radish (*Raphanus sativus* var. *mantanghong*):** This variety is known for its striking appearance, with a green outer skin and bright pink or red flesh, resembling a watermelon. It has a mild, sweet flavor with a slight peppery aftertaste. Watermelon radish is typically grown in cooler climates and is often used in salads for its vibrant color and crunchy texture.

Table 1: Major Botanical Varieties of Radish

S No	Botanical Variety	Description	Root Color	Special Features
1	*Raphanus sativus* var. *radicula*	Round Red Radish	Red	Small, fast-growing, mildly peppery
2	*Raphanus sativus* var. *longipinnatus*	White Radish (*Daikon*)	White	Large, mild flavor, popular in Asian cuisine

3	*Raphanus sativus* var. *niger*	Black Radish	Black	Large, pungent, medicinal properties
4	*Raphanus sativus* var. *caudatus*	Rat-tail Radish	Green	Grown for edible seed pods, heat-tolerant
5	*Raphanus sativus* var. *mantanghong*	Watermelon Radish	Green/Pink	Sweet, colorful, striking appearance

Each of these varieties is adapted to different growing conditions and markets. Round red radish is the most common globally, while white radish (*daikon*) is popular in Asia. Other varieties, such as black radish and watermelon radish, serve niche markets and are valued for their unique flavors and uses.

Growth Habit

Radish is a fast-growing, herbaceous plant that is typically grown as an annual, though some varieties can be biennial. The plant forms a rosette of basal leaves, which vary in shape and size depending on the variety. The root, which is the primary edible part of the plant, develops underground, while the leaves emerge above the soil surface. Radish roots can vary greatly in size, shape, and color, ranging from small, round red roots to large, elongated white roots like the *daikon* radish.

The growth habit of radish is heavily influenced by environmental factors such as temperature, soil type, and moisture availability. Radish prefers cool growing conditions and performs best in temperatures between 10°C and 20°C. In warmer climates, radish tends to bolt, meaning it will prematurely produce a flowering stalk, which results in poor root development. To avoid this, radish is often grown during the cooler months of the year in subtropical and tropical regions.

Floral Morphology

Radish flowers exhibit typical characteristics of the Brassicaceae family. The flowers are arranged in racemes and are usually white, pink, or purple. Each flower consists of four petals arranged in a cross shape, which is a defining characteristic of the family, and gives them the alternative name "Cruciferae." The petals are often veined and can vary in color intensity depending on the variety. Below the petals, four sepals provide support to the flower's structure. Each flower also has six stamens, four long and two short, which is another trait commonly found in members of the Brassicaceae family.

The flowers are bisexual, containing both male (stamens) and female (pistil) reproductive structures. The ovary is superior, meaning it sits above the attachment point of the other floral parts. Once pollinated, the ovary develops into a pod-like structure known as a siliqua, which contains the seeds. These siliques are long and slender, characteristic of many Brassica and *Raphanus* species.

Radish flowers are generally insect-pollinated, with bees being the primary pollinators. However, some varieties of radish are self-compatible, meaning they can self-pollinate, although cross-pollination is more common in wild or non-domesticated species. In commercial seed production, controlled pollination techniques may be used to maintain genetic purity, particularly in hybrid varieties.

Floral Biology

The timing of flowering in radishes is influenced by environmental factors such as temperature, day length, and moisture availability. Radish typically flowers during its second year of growth if it is allowed to overwinter. However, some varieties will flower in the first year, especially if they experience stress conditions like high temperatures or water scarcity.The flowering process begins when the plant forms a tall, upright flowering stalk. This process, known as bolting, is undesirable in radish cultivation because it diverts energy from root development to flower and seed production, resulting in poor-quality roots. The plant's natural reproductive strategy is to flower and produce seeds after the vegetative phase, but modern radish varieties are bred to delay bolting to extend the harvestable period of the root.

Radish flowers are protandrous, meaning the male reproductive parts (stamens) mature before the female parts (pistils). This reduces the likelihood of self-pollination and encourages cross-pollination between different plants, thereby increasing genetic diversity. Once pollinated, the siliques develop, which contain the seeds that will be harvested for future planting.In commercial seed production, radish plants are often allowed to flower and set seed. After the flowers are pollinated, the siliques mature and dry on the plant before they are harvested. The seeds are then extracted, cleaned, and processed for commercial use.

Breeding Objectives

Breeding radish varieties with improved characteristics is a continuous goal for breeders worldwide. The main breeding objectives in radishes focus on improving yield, enhancing root quality, increasing disease resistance, and adapting to different environmental conditions.

1. **Yield Improvement:**
   - One of the primary goals in radish breeding is to increase yield per hectare. Yield is often influenced by factors such as root size, uniformity, and plant density. Breeders focus on developing varieties that produce large, uniform roots with minimal spacing between plants. Yield improvement is especially important for commercial growers, who require consistent, high-quality produce for the fresh market.
2. **Root Quality:**
   - Root quality is a critical breeding objective, as it directly affects consumer preference and marketability. Traits such as root size, shape, color, and texture are important considerations. Breeding programs aim to develop varieties with uniform, smooth roots that have a consistent shape and vibrant color. Flavor and texture are also

important, with breeders focusing on developing varieties that have a crisp, juicy texture and a mild, slightly peppery flavor.

3. **Disease Resistance:**

   » Radish is susceptible to several diseases, including clubroot (*Plasmodiophorabrassicae*), black rot (*Xanthomonas campestris*), and downy mildew (*Peronospora parasitica*). Breeding for disease resistance is essential to reduce crop losses and minimize the need for chemical pesticides. Resistance genes from wild relatives or related species are often used in breeding programs to introduce resistance to these pathogens.

4. **Abiotic Stress Tolerance:**

   » With the increasing impact of climate change, breeding for abiotic stress tolerance has become a key objective. Radish is sensitive to temperature fluctuations, particularly high temperatures, which can cause bolting. Breeding efforts are focused on developing varieties that can tolerate a wider range of temperatures, as well as drought and salinity. This is particularly important for regions with extreme weather conditions or poor soil quality.

5. **Improved Nutritional Content:**

   » Modern consumers are increasingly seeking vegetables with enhanced nutritional value. Breeding for improved nutritional content in radish focuses on increasing levels of beneficial compounds such as Vitamin C, antioxidants, and glucosinolates. These compounds are linked to various health benefits, including cancer prevention and improved immune function. Breeding programs aim to develop varieties that are not only visually appealing and flavorful but also nutritionally superior.

6. **Bolting Resistance:**

   » One of the main challenges in radish cultivation is bolting, especially in warmer climates. Breeding for bolting resistance involves selecting for plants that maintain root production without prematurely entering the flowering phase. Bolting resistance is particularly important for extending the growing season and ensuring high-quality root development.

7. **Adaptability to Different Climates:**

   » Radish is grown in a wide range of climates, from temperate to subtropical regions. Breeding programs focus on developing varieties that are adaptable to different environmental conditions, including temperature extremes, varying day lengths, and soil types. This adaptability is crucial for ensuring that radish can be grown year-round in diverse regions.

Breeding Methods

Radish breeding employs a combination of traditional and modern methods to improve various traits such as yield, disease resistance, and stress tolerance. The main breeding methods used in radish improvement include selection, hybridization, mutation breeding, and molecular techniques such as marker-assisted selection (MAS).

1. **Introduction:**

   » The introduction method involves introducing radish varieties from different regions or countries to assess their adaptability and performance in new environments. This method is especially useful for introducing varieties with specific traits, such as heat tolerance or disease resistance, into regions where those traits are needed.

2. **Selection:**

   » Selection is one of the most fundamental breeding methods in radish improvement. In this process, superior plants are chosen based on their phenotype, or observable traits, such as root size, shape, color, and disease resistance. The seeds from these selected plants are used to produce the next generation.

   » **Mass Selection:** This involves selecting the best-performing plants from a population and using their seeds to grow the next generation. Mass selection is effective for improving simple traits, such as root shape and color, but it can take several generations to achieve the desired improvements.

   » **Pure Line Selection:** In pure line selection, individual plants are selected and propagated over several generations to produce a uniform, homozygous line. This method is particularly useful for improving traits controlled by a few genes, such as disease resistance.

3. **Hybridization:**

   » Hybridization is one of the most widely used methods in radish breeding, especially for developing F1 hybrids. F1 hybrids are produced by crossing two genetically distinct parental lines, resulting in offspring that exhibit hybrid vigor, or heterosis. Hybrid radish varieties are prized for their uniformity, high yield, and resistance to biotic and abiotic stresses.

   » **Backcross Breeding:** This method involves crossing a desirable variety with a donor parent that carries a specific trait, such as disease resistance. The resulting offspring are then backcrossed with the original variety to introduce the trait while maintaining the desirable characteristics of the original variety.

4. **Mutation Breeding:**

   » Mutation breeding involves exposing radish seeds to chemicals or radiation to induce random mutations. These mutations can lead to the development of new traits, such as improved disease resistance, abiotic stress tolerance, or novel root shapes and colors.

5. Molecular Breeding:

» Advances in molecular biology have made it possible to use DNA markers to select for specific traits in radishes. Marker-assisted selection (MAS) allows breeders to identify plants that carry desirable genes at an early stage of growth, speeding up the breeding process. Molecular breeding is particularly useful for improving traits like disease resistance and nutritional content.

Breeding for Biotic Stresses

Biotic stresses, including diseases and pests, are a significant challenge in radish production. Breeding for biotic stress resistance is essential to reduce the impact of these stresses on crop yield and quality. Below are some of the key biotic stresses affecting radish and the breeding strategies used to combat them:

1. Clubroot Resistance:

» Clubroot, caused by the soil-borne pathogen *Plasmodiophorabrassicae*, is one of the most destructive diseases of radish. It causes swelling and deformation of the roots, leading to stunted growth and reduced yields. Breeding for clubroot resistance involves selecting varieties that can withstand infection and continue to produce healthy roots. Resistance genes from wild relatives, such as *Raphanus raphanistrum*, have been incorporated into commercial radish varieties through traditional breeding and marker-assisted selection.

2. Black Rot Resistance:

» Black rot, caused by the bacterium *Xanthomonas campestris*, leads to yellowing and wilting of the leaves, as well as the eventual death of the plant. Breeding for resistance to black rot involves identifying resistant varieties or wild relatives and incorporating those resistance genes into commercial radish varieties.

3. Downy Mildew Resistance:

» Downy mildew, caused by *Peronospora parasitica*, is a common fungal disease that affects radish leaves, leading to reduced photosynthesis and poor root development. Breeding for resistance to downy mildew is essential to reduce crop losses, especially in regions with high humidity. Resistance genes from related species have been introduced into commercial radish varieties through hybridization.

4. Insect Pest Resistance:

» Radish is vulnerable to several insect pests, including aphids, flea beetles, and root maggots. Breeding for pest resistance involves selecting varieties with natural defense mechanisms, such as high levels of glucosinolates, which deter pests. Additionally, some breeding programs focus on developing radish varieties with tougher leaves, making them less attractive to pests.

Breeding for Abiotic Stresses

Abiotic stresses such as extreme temperatures, drought, salinity, and waterlogging are major limiting factors in radish cultivation. Breeding for abiotic stress tolerance is essential for developing varieties that can thrive in challenging environments while maintaining high yield and quality. Modern radish breeding programs aim to develop varieties that can withstand these stresses, ensuring stable production even under suboptimal growing conditions.

1. **Temperature Tolerance:**
   - **Heat Tolerance:** Radish is primarily a cool-season crop, but in many regions, it is subjected to high temperatures during certain parts of the growing season. High temperatures can lead to bolting (premature flowering), which reduces root quality and yield. Breeding for heat tolerance involves selecting varieties that can continue to produce high-quality roots even when exposed to higher temperatures. Heat-tolerant varieties of radish are particularly important for tropical and subtropical regions, where temperature fluctuations are common.
   - **Cold Tolerance:** Although radish is generally tolerant of cooler temperatures, frost or prolonged cold periods can damage young plants and inhibit root development. Breeding for cold tolerance focuses on developing varieties that can withstand frost and low temperatures, particularly in regions with short growing seasons. Cold-tolerant varieties of radish are also important for winter production in temperate regions.
2. **Drought Tolerance:**
   - Drought stress is another major challenge in radish cultivation, particularly in arid or semi-arid regions. Drought stress during critical growth stages, such as root formation, can result in smaller, tougher roots and reduced yields. Breeding for drought tolerance involves selecting for traits such as deeper root systems, improved water-use efficiency, and the ability to maintain growth under limited water availability. Drought-tolerant radish varieties are increasingly important in areas affected by water scarcity, where irrigation is limited or unreliable.
3. **Salinity Tolerance:**
   - Soil salinity is a growing problem in many agricultural regions, particularly in areas with poor irrigation management or coastal farming zones. High salt concentrations in the soil can reduce radish growth, inhibit nutrient uptake, and lead to poor root development. Breeding for salinity tolerance involves selecting varieties that can maintain growth and yield in saline soils. Salinity-tolerant radish varieties are bred to exclude excess salts from the root zone or to tolerate higher internal salt concentrations without affecting growth.
4. **Waterlogging Tolerance:**
   - Waterlogging, or the saturation of the soil with excess water, is another abiotic stress

that can affect radish production, particularly in regions with heavy rainfall or poor drainage. Waterlogged soils restrict oxygen availability to plant roots, leading to root rot and reduced growth. Breeding for waterlogging tolerance focuses on selecting varieties that can maintain root function and growth under waterlogged conditions. Waterlogging-tolerant radish varieties are especially important for regions with high rainfall or areas prone to flooding.

Table 1: Types of Radish on Basis of Tolerant to Abiotic Stresses Characteristics

S No	Variety Name	Abiotic Stress Tolerance	Key Features
1	Heat Tolerant 1	Heat Tolerance	Performs well in high temperatures, reduced bolting
2	Cold King	Cold Tolerance	Resistant to frost, high root quality in cool climates
3	Drought Warrior	Drought Tolerance	Deep root system, high yield under limited water
4	Salinity Guard	Salinity Tolerance	Grows well in saline soils, good root size and flavour
5	Waterlogged 5	Waterlogging Tolerance	Tolerates excess moisture, minimal root damage

Varietal Achievements

Radish breeding has resulted in the development of several improved varieties that are tailored to specific environmental conditions, market demands, and consumer preferences. These varieties have been bred for traits such as improved yield, disease resistance, abiotic stress tolerance, and root quality.

1. **Pusa Desi Radish:**
   - » **Parentage:** Open-pollinated local selection
   - » **Key Features:** Pusa Desi Radish is an early-maturing variety with long, white roots that have a crisp texture and mild flavor. This variety is particularly well-suited to the cooler months of the year and is known for its high yield and uniform root shape. It is also resistant to common pests such as aphids.
   - » **Adaptation:** Temperate to subtropical climates.
2. **Arka Nishant:**
   - » **Parentage:** Developed through hybridization
   - » **Key Features:** Arka Nishant is a heat-tolerant variety that produces uniform, medium-sized roots with a smooth texture. It has a high resistance to bolting, making it ideal for cultivation in warmer regions. The roots have a mild, peppery flavor and are ideal for both fresh consumption and processing.

- **Adaptation:** Tropical and subtropical climates.

3. **Hisar Daikon:**
   - **Parentage:** Selection from Japanese Daikon lines
   - **Key Features:** Hisar Daikon is a long, white-rooted variety with a sweet, mild flavor and tender texture. It is highly resistant to common radish diseases such as black rot and downy mildew. The variety is particularly popular in both fresh markets and for pickling.
   - **Adaptation:** Temperate to subtropical climates.
4. **Punjab Giant:**
   - **Parentage:** Open-pollinated selection
   - **Key Features:** Punjab Giant is a large-rooted radish variety known for its high yields and long shelf life. It produces large, cylindrical roots that are ideal for both fresh consumption and long-distance transportation. This variety is resistant to soil-borne diseases and performs well in a wide range of soils.
   - **Adaptation:** Subtropical climates, suitable for both spring and autumn planting.
5. **Kalyanpur Purple Radish:**
   - **Parentage:** Hybrid variety
   - **Key Features:** This variety is known for its striking purple skin and white flesh. It is popular for salads due to its unique color and crisp texture. Kalyanpur Purple Radish is highly resistant to root rot and performs well under both drought and mild waterlogging conditions. It is ideal for both smallholder farmers and commercial growers.
   - **Adaptation:** Temperate regions, good tolerance to mild drought conditions.

Table 2: Key Radish Varietal Achievements

S No	Variety Name	Parentage	Key Features	Adaptation
1	Pusa Desi Radish	Open-pollinated	Early-maturing, uniform roots, resistant to pests	Temperate to subtropical
2	Arka Nishant	Hybrid	Heat tolerant, reduced bolting, medium-sized roots	Tropical and subtropical
3	Hisar Daikon	Japanese Daikon lines	Long white roots, disease-resistant, sweet flavor	Temperate to subtropical
4	Punjab Giant	Open-pollinated	Large roots, high yield, resistant to soil diseases	Subtropical, high yield
5	Kalyanpur Purple Radish	Hybrid	Purple skin, crisp texture, root rot resistance	Temperate, drought-tolerant

Technology of Hybrid Seed Production

Producing hybrid seeds in radish involves several steps, including the selection of parent lines, controlled pollination, and careful management to ensure the genetic purity of the hybrid seed. Hybrid radish varieties are developed to improve traits like uniformity, disease resistance, and stress tolerance. Below are the main steps involved in hybrid seed production for radish:

1. **Parent Line Selection:**
   » Hybrid seed production begins with the selection of two genetically distinct parent lines. These parent lines are carefully chosen for their complementary traits. For example, one parent may be selected for its high yield potential, while the other may be chosen for its resistance to diseases or abiotic stresses.
2. **Isolation:**
   » To prevent unwanted cross-pollination, the parent lines are isolated from other radish varieties. Isolation can be achieved either through physical barriers, such as nets, or by planting the parent lines at a considerable distance from other radish crops. This ensures that only the selected parent lines cross-pollinate to produce the desired hybrid.
3. **Hand Emasculation and Pollination:**
   » In hybrid seed production, the male reproductive organs (stamens) of the female parent plant are removed by hand in a process known as emasculation. This is typically done before the flower opens to prevent self-pollination. Once the flowers of the female parent are emasculated, pollen from the male parent is manually applied to the stigma of the female flower. This controlled pollination ensures that the desired genetic traits from both parents are passed on to the hybrid offspring.
4. **Seed Harvesting:**
   » After successful pollination, the fertilized flowers develop siliques (seed pods) containing the hybrid seeds. The siliques are allowed to mature and dry on the plant before they are harvested. The seeds are then extracted, cleaned, and prepared for commercial sale.
5. **Seed Testing and Quality Control:**
   » Hybrid seeds undergo rigorous testing to ensure their germination rate, genetic purity, and uniformity. Field trials are conducted to assess the performance of the hybrid variety under different environmental conditions. The final product is a high-quality hybrid seed that offers superior performance in terms of yield, disease resistance, and abiotic stress tolerance.

Crop Production Techniques

Radish (*Raphanus sativus*) is a fast-growing, cool-season crop that can be grown in various soil types and climates. However, optimal production requires careful management of environmental factors such as temperature, moisture, and soil fertility. Below are the key crop production techniques involved in radish cultivation.

1. **Soil Requirements:**

   » Radish thrives best in well-drained, loose, sandy loam or loamy soil with a pH range of 6.0 to 7.0. Heavy clay soils are not ideal as they can lead to misshapen roots, but they can be improved by adding organic matter. The soil should be rich in organic content to support rapid root development, and adequate moisture retention is essential for maintaining consistent growth.

   » Radish roots can be sensitive to soil compaction, so it's important to prepare the soil properly before planting. Deep tilling and the addition of organic matter such as compost or well-rotted manure can improve soil structure, aeration, and water infiltration. Soil that is too compacted will restrict root growth, leading to stunted or deformed radish roots.

2. **Seed Sowing:**

   » Radish seeds can be directly sown into the field. The seeds should be planted at a depth of 1 to 1.5 cm in rows spaced 25-30 cm apart, with individual plants spaced 5-10 cm apart within the row. This spacing allows enough room for root expansion and ensures good airflow, reducing the risk of disease.

   » The germination period for radish is short, typically 3-7 days, depending on soil temperature and moisture levels. The ideal temperature range for radish seed germination is between 10°C and 20°C. If grown in higher temperatures, radish may bolt, reducing root quality.

3. **Irrigation:**

   » Radish requires consistent moisture throughout its growing cycle. Irrigation should be applied regularly to maintain even soil moisture, especially during the root development phase. Moisture stress can cause the roots to become tough and fibrous, while excess moisture may lead to root cracking or the development of diseases like root rot.

   » Drip irrigation systems are particularly effective in radish production as they deliver water directly to the root zone, minimizing water loss due to evaporation and reducing the incidence of foliar diseases. Overhead irrigation is less ideal because it can lead to leaf wetness, which encourages disease development.

4. **Fertilization:**

   » Radish is a relatively light feeder, but adequate nutrition is essential for good root development. The crop responds well to organic fertilizers, such as compost or well-rotted manure, which supply a slow-release source of nutrients. If using synthetic fertilizers, a balanced application of nitrogen, phosphorus, and potassium (NPK) is recommended.

» Nitrogen promotes vigorous leaf growth, but excessive nitrogen can result in lush foliage at the expense of root formation. Phosphorus is important for root development, while potassium contributes to overall plant health and disease resistance. A general recommendation for fertilization is to apply 50-70 kg/ha of nitrogen, 40-60 kg/ha of phosphorus, and 30-50 kg/ha of potassium, depending on soil fertility and local recommendations.

5. **Weed Management:**

» Weed competition can significantly reduce radish yields, especially during the early stages of growth. Hand weeding is the most common method of weed control in small-scale radish production, while herbicides may be used in larger commercial operations. Mulching with organic materials such as straw or grass clippings can also suppress weed growth and retain soil moisture.

Pest and Disease Management

Radish is susceptible to various pests and diseases that can impact both the quality and yield of the crop. Effective management practices, including the use of resistant varieties, proper crop rotation, and integrated pest management (IPM) strategies, are essential for minimizing crop losses.

1. **Pest Management:**

» **Aphids:** Aphids are common pests of radish, feeding on the sap of the leaves and stems. This can lead to stunted growth, wilting, and the transmission of viral diseases. Aphids can be controlled through biological control agents like ladybugs or lacewings, as well as insecticidal soaps or neem oil.

» **Flea Beetles:** Flea beetles are small, jumping beetles that feed on radish leaves, creating small holes known as "shot-hole" damage. Severe infestations can lead to reduced photosynthesis and poor root development. Flea beetles can be controlled by using row covers to protect young plants or applying insecticidal sprays. Crop rotation can also help reduce flea beetle populations.

» **Root Maggots:** The larvae of the cabbage root fly, known as root maggots, can burrow into radish roots, causing significant damage. Infected roots become spongy and may rot. Preventive measures include using floating row covers, crop rotation, and the application of beneficial nematodes to control larvae in the soil.

» **Cutworms:** Cutworms are nocturnal larvae that feed on the stems of radish plants at ground level, causing wilting and plant death. Control measures include applying biological insecticides like *Bacillus thuringiensis* (Bt), removing plant debris, and using barrier collars around young plants.

2. **Disease Management:**
   - **Downy Mildew (*Peronospora parasitica*):** Downy mildew causes yellowing and necrosis of the leaves, leading to reduced photosynthesis and stunted growth. It thrives in cool, wet conditions and can spread rapidly in dense plantings. Management includes planting resistant varieties, ensuring good air circulation between plants, and applying fungicides when necessary.
   - **Black Rot (*Xanthomonas campestris*):** Black rot is a bacterial disease that causes yellowing and wilting of leaves, starting at the leaf margins. It can lead to significant yield losses if not controlled. Preventive measures include using certified disease-free seeds, practicing crop rotation, and applying copper-based bactericides.
   - **Clubroot (*Plasmodiophorabrassicae*):** Clubroot is a soil-borne disease that causes swelling and deformation of the roots, leading to stunted growth and poor-quality roots. It is most prevalent in acidic soils, so adjusting the soil pH to 7.2 or higher can help reduce the incidence of the disease. Crop rotation and resistant varieties are also effective in managing clubroot.
   - **Powdery Mildew:** Powdery mildew appears as white, powdery spots on the leaves, reducing photosynthesis and weakening the plant. It thrives in warm, dry conditions. Effective control strategies include applying sulfur-based fungicides and ensuring adequate spacing between plants to improve airflow.

Post-Harvest Handling

Proper post-harvest handling is essential to maintain the quality and shelf life of radish. Freshly harvested radish roots are highly perishable and must be handled carefully to prevent mechanical damage, dehydration, and loss of flavor.

1. **Harvesting:**
   - Radish is typically harvested when the roots reach the desired size, which varies depending on the variety. For small, round varieties, harvesting can begin as early as 20-30 days after sowing. Larger varieties like *daikon* may take up to 60 days to mature. Radishes should be harvested before they become overgrown and woody, which can reduce their quality.
   - Harvesting is usually done manually by pulling the roots from the soil. In large commercial operations, mechanical harvesters may be used to speed up the process. Care must be taken to avoid bruising or damaging the roots during harvest, as this can lead to rot during storage.
2. **Cleaning and Grading:**
   - After harvesting, radish roots are cleaned to remove soil and debris. This can be done by rinsing the roots in water or using mechanical brushes. Once cleaned, the roots

are sorted and graded based on size, shape, and color. Uniformity is important for marketability, and roots that are misshapen or damaged are typically discarded or sold at a lower price.

» Grading is particularly important in the fresh market, where consumers prefer radishes that are visually appealing and consistent in size. Radishes for the processing market may be graded less stringently, as appearance is less critical.

3. Storage:

» Radishes are highly perishable and should be stored at low temperatures to maintain their quality. Ideal storage conditions are between 0°C and 4°C with high humidity (90-95%) to prevent dehydration. Under these conditions, radish roots can be stored for up to two weeks without significant loss of quality.

» For longer storage, radishes can be stored in perforated plastic bags or crates with adequate ventilation. Some varieties, such as black radish, have a longer shelf life and can be stored for several months in optimal conditions.

4. Packaging and Transportation:

» Radish roots are typically packed in plastic crates or cartons for transportation to market. Packaging must be sturdy enough to protect the roots from mechanical damage during transit. In the fresh market, radishes are often bundled with their leaves intact, but the leaves should be removed if the radishes are to be stored for an extended period, as they can cause moisture loss from the roots.

» Efficient transportation is crucial for maintaining the freshness of radish, particularly for export markets. Refrigerated transport is commonly used for long-distance shipments to maintain the low temperature and high humidity needed for extended shelf life.

D. Carrot (*Daucus carota*)

Family: Apiaceae

Chromosome number: 2n = 2x = 18

The carrot (*Daucus carota* subsp. *sativus*) is a biennial root vegetable cultivated primarily for its edible taproot. It is one of the most widely grown and consumed vegetables globally, valued for its sweet flavor, crunchy texture, and high nutritional content, particularly its rich supply of beta-carotene, which the body converts into vitamin A. Carrots are also a good source of dietary fiber, antioxidants, potassium, and other vitamins.

Carrots are consumed raw, cooked, or processed into various products such as juices, purees, and snacks. The crop is adaptable to a wide range of climates and is cultivated in temperate, subtropical, and tropical regions. Carrots can vary in color, from the commonly known orange to yellow, red, purple, and white varieties, with different pigments contributing to their nutritional profiles.

Origin and Domestication

Carrots are believed to have originated from wild species found in Central Asia, particularly in present-day Afghanistan. The earliest cultivated carrots were likely purple or yellow in color. These early forms were primarily grown for their aromatic leaves and seeds rather than their roots. Over time, domestication and selective breeding led to the development of larger, sweeter roots.

Orange carrots, the most widely recognized type today, were developed in the 16th and 17th centuries in the Netherlands. The selection for orange roots is believed to have been influenced by Dutch growers, who favored the color in honor of the royal House of Orange.

Domestication of the carrot involved selecting for traits such as root size, sweetness, texture, and color. The development of modern orange carrots is attributed to the selection for high beta-carotene content, which gives the root its distinctive color. Carrot breeding continues today, with efforts focused on improving yield, disease resistance, and nutritional content.

Related Genera

Carrot belongs to the Apiaceae family, which includes several important vegetables and herbs. Some of the related genera and species include:

- **Pastinaca**: The parsnip (*Pastinaca sativa*) is a close relative of carrot and is also grown for its edible root. Parsnips are sweeter than carrots, particularly after exposure to frost. They share similar growing conditions and can be used in breeding programs for cold tolerance or other traits.
- **Petroselinum**: Parsley (*Petroselinum crispum*) is another member of the Apiaceae

family. While grown primarily for its leaves, parsley shares genetic similarities with carrot, and both crops can be affected by similar pests and diseases.

» **Coriandrum**: Coriander (*Coriandrum sativum*) is grown for its leaves (cilantro) and seeds, both of which are widely used in cooking. Although coriander and carrot have different uses, they are closely related and can be grown in similar conditions.

» **Foeniculum**: Fennel (*Foeniculum vulgare*) is a perennial herb with a similar growing habit to carrot. Both crops are biennials grown as annuals for their edible parts. Fennel's anise-flavored bulbs and leaves share some morphological similarities with carrot foliage.

These related genera can sometimes share similar pests and diseases with carrots, making integrated pest management strategies critical for diversified farms.

Related Species

The wild carrot (*Daucus carota* subsp. *carota*), also known as Queen Anne's lace, is the ancestor of the cultivated carrot. Wild carrot is native to Europe and Southwest Asia, and its roots are small, fibrous, and less sweet than those of cultivated varieties. However, wild carrot has been an important genetic resource in carrot breeding programs, particularly for improving disease resistance and stress tolerance.

Other related species of interest in carrot breeding include:

» **Daucus pusillus**: A wild species of carrot native to North America. It has small, inedible roots but may carry genes for drought tolerance.

» **Daucus sahariensis**: A species of wild carrot native to desert regions. It has been used in breeding programs to develop carrot varieties with improved heat and drought tolerance.

» **Daucus glochidiatus**: This species is native to Australia and has been studied for its resistance to pests and diseases.

These wild relatives provide valuable genetic diversity that can be incorporated into carrot breeding programs to improve resistance to biotic and abiotic stresses.

Botanical Varieties of Carrot

Carrot varieties are classified into different botanical types based on their root shape, color, and size. The main botanical varieties include:

1. **Nantes Carrot:**

   » *Description:* Nantes carrots are cylindrical with blunt tips and a smooth, tender texture. They are known for their sweet flavor and crisp flesh.

   » *Root Shape:* Cylindrical

   » *Uses:* Fresh consumption, juicing, and baby carrots.

2. **Chantenay Carrot:**
   - *Description:*Chantenay carrots have short, conical roots and are wider at the top than Nantes carrots. They have a slightly coarser texture but are good for heavy soils.
   - *Root Shape:* Conical
   - *Uses:* Processing, soups, and stews.
3. **Imperator Carrot:**
   - *Description:* Imperator carrots are long and tapered, with a high sugar content. They are the most commonly grown type for the fresh market in the United States.
   - *Root Shape:* Long and tapered
   - *Uses:* Fresh consumption and juicing.
4. **Danvers Carrot:**
   - *Description:*Danvers carrots are medium-length with a conical shape. They are highly adaptable to various soil types, including heavy soils, and have good storage qualities.
   - *Root Shape:* Conical
   - *Uses:* Processing, storage, and fresh market.
5. **Paris Market Carrot:**
   - *Description:* Paris Market carrots are small, round carrots that resemble radishes. They are ideal for shallow or rocky soils where longer carrots cannot develop properly.
   - *Root Shape:* Round
   - *Uses:* Gourmet markets and fresh consumption.
6. **Purple Carrot:**
   - *Description:* Purple carrots are rich in anthocyanins, which give them their deep purple color. They have a slightly peppery flavor compared to orange carrots.
   - *Root Shape:* Various
   - *Uses:* Fresh consumption, specialty markets, and juicing.

Table 1: Major Botanical Varieties of Carrot

S No	Botanical Variety	Root Shape	Color	Uses
1	*Nantes*	Cylindrical	Orange	Fresh, juicing, baby carrots
2	*Chantenay*	Conical	Orange	Processing, soups, stews
3	*Imperator*	Long, tapered	Orange	Fresh, juicing
4	*Danvers*	Conical	Orange	Processing, storage, fresh market
5	*Paris Market*	Round	Orange	Gourmet markets, fresh consumption
6	*Purple Carrot*	Various	Purple	Specialty markets, fresh, juicing

These botanical varieties are suited to different markets and growing conditions, offering diversity in terms of root shape, flavor, and nutritional content. Each variety has specific uses, with Nantes and Imperator types dominating the fresh market, while Danvers and Chantenay are favored for processing.

Growth Habit

Carrot is a biennial plant, meaning it typically completes its life cycle over two growing seasons. However, in commercial cultivation, carrots are usually grown as an annual crop for their roots, which are harvested at the end of the first growing season before flowering occurs. In the first year of growth, the carrot plant develops a rosette of finely divided, feathery leaves, which emerge from a central crown. The plant also produces a thick, fleshy taproot, which stores energy for the plant's second year when it would naturally flower and produce seeds.

The growth of the carrot taproot is influenced by several environmental factors, including soil type, moisture availability, and temperature. Carrots grow best in loose, sandy soils that allow for straight-root development. The ideal temperature range for carrot growth is between 15°C and 20°C. Warmer temperatures can cause the roots to become tough and fibrous, while cooler temperatures enhance the sweetness of the roots due to increased sugar content. Carrots are sensitive to day length, and the length of the growing season will depend on the variety and the climate. Early varieties can be harvested within 60-70 days, while late-maturing varieties may take up to 120 days to reach full maturity.

Floral Morphology

In its second year, the carrot plant produces an upright flowering stalk that can reach heights of 1 to 1.5 meters. The flowers are arranged in compound umbels, a characteristic of the Apiaceae family. Each umbel consists of smaller umbels, or umbellets, which contain numerous tiny, white flowers. The flowers are hermaphroditic, containing both male (stamens) and female (pistils) reproductive organs. Pollination is primarily carried out by insects, particularly bees, which are attracted to the flowers by their nectar. Once pollinated, the flowers develop into small, ribbed fruits, known as schizocarps, which contain the seeds.

Floral Biology

A carrot is a biennial plant, meaning it typically flowers in its second growing season after the root has undergone a period of cold (vernalization). If allowed to grow into its second year, the plant produces a flowering stalk that is taller and branched. The floral structure of carrot is an umbel, a characteristic inflorescence of the Apiaceae family. The umbel is composed of smaller umbels, called umbellets, which are clusters of tiny white flowers. Each umbellet consists of 50–100 flowers, with the terminal flower being slightly larger than the rest.

Carrot flowers are primarily insect-pollinated, and bees are one of the most common pollinators. The flowers are protandrous, meaning the male parts (stamens) mature before the female parts (pistil), which encourages cross-pollination. This floral biology is crucial for maintaining genetic diversity in carrot breeding, especially in hybrid seed production. When the flowers are pollinated, they develop into small, dry fruits called schizocarps, which split into two seeds upon maturity.

Breeding Objectives

The key objectives in carrot breeding include improving root yield, shape, color, and flavor, as well as increasing resistance to pests, diseases, and environmental stresses. With the increasing demand for high-quality produce, breeders are also focusing on enhancing the nutritional content and storage qualities of carrots.

1. **Yield Improvement:**

   » Yield improvement is a primary breeding goal for all vegetable crops, including carrot. High yield is determined by factors such as root size, uniformity, and plant density. Breeding programs aim to develop varieties that produce large, uniform roots while maintaining high-quality flavor and texture.

2. **Root Shape and Color**:

   » Root shape is important for both marketability and ease of harvesting. Carrots with straight, cylindrical roots are preferred, as they are easier to process and package. Breeding efforts focus on reducing root deformities, such as forking, which can occur in heavy or compacted soils. Color is another key trait, with most breeding programs focusing on the improvement of orange carrots, which are rich in beta-carotene. However, there is also interest in developing and improving non-orange varieties, such as purple, red, and yellow carrots, which contain different pigments and offer diverse nutritional benefits.

3. **Flavor and Texture:**

   » The sweetness and crispness of carrots are key factors influencing consumer preference. Breeders select for varieties with high sugar content (primarily sucrose) and a crunchy texture. Breeding for improved flavor also includes reducing the bitterness that can occur in some varieties due to high levels of polyacetylenes.

4. **Disease Resistance**:

   » Carrot is susceptible to a range of diseases, including Alternaria leaf blight (*Alternaria dauci*), Cercospora leaf spot (*Cercosporacarotae*), and bacterial soft rot (*Erwinia carotovora*). Breeding for resistance to these diseases is crucial for reducing crop losses and minimizing the need for chemical control measures. Resistance genes from wild carrot relatives are often used in breeding programs to introduce disease resistance into cultivated varieties.

5. **Abiotic Stress Tolerance:**

   » Carrot production can be severely affected by abiotic stresses such as drought, salinity, and temperature extremes. Breeding for abiotic stress tolerance focuses on selecting varieties that can maintain high yields and root quality under adverse environmental conditions. For example, drought-tolerant varieties may have deeper root systems or increased water-use efficiency, while heat-tolerant varieties are less likely to bolt under high temperatures.

6. **Nutritional Content:**

   » Modern carrot breeding programs emphasize improving the nutritional value of carrots. Orange carrots are a rich source of beta-carotene, which the body converts into vitamin A, while purple and red carrots contain anthocyanins and lycopene, respectively, which are powerful antioxidants. Breeding efforts aim to increase the concentrations of these beneficial compounds to enhance the health benefits of carrot consumption.

7. **Storage and Processing Qualities:**

   » Carrots intended for processing need to have good storage qualities, including resistance to cracking, shriveling, and disease during storage. Breeding for improved storage life focuses on reducing post-harvest losses and maintaining flavor, texture, and color for an extended period. Processing carrots also need to have uniform root size and shape, as well as a high dry matter content, which improves the efficiency of mechanical processing and reduces waste.

Breeding Methods

Carrot breeding employs both conventional and modern techniques to improve desirable traits. The key breeding methods include mass selection, hybridization, and molecular techniques such as marker-assisted selection (MAS).

1. **Mass Selection:**

   » Mass selection is one of the oldest breeding methods used in carrot improvement. In this process, superior plants are selected based on their phenotype, or observable traits, such as root shape, color, and disease resistance. Seeds from these selected plants are then used to grow the next generation. Over several generations, this method can lead to improved populations with the desired characteristics. However, mass selection can be slow and is most effective for improving simple, heritable traits.

2. **Hybridization:**

   » Hybridization involves crossing two genetically distinct parent lines to produce offspring with superior traits, a phenomenon known as heterosis or hybrid vigor. In carrot breeding, hybrid varieties are often preferred because they offer greater uniformity in root shape and size, improved disease resistance, and higher yields. F1 hybrids, which are the first-generation offspring of two inbred parents, are commonly used in commercial carrot production.

3. **Backcross Breeding:**

   » Backcross breeding is used to introduce a specific trait, such as disease resistance, from a donor parent into an elite carrot variety. The donor parent is crossed with the elite variety, and the resulting offspring are then backcrossed with the elite parent for several generations. This process allows the desirable trait to be incorporated into the elite variety while retaining the other superior characteristics of the original parent.

4. **Mutation Breeding:**

   » Mutation breeding involves exposing carrot seeds to chemicals or radiation to induce random genetic mutations. Some of these mutations may result in beneficial traits, such as improved disease resistance or unique root colors. Mutation breeding has been used to develop new carrot varieties with novel characteristics, such as the introduction of high-lycopene carrots.

5. **Marker-Assisted Selection (MAS):**

   » MAS is a modern breeding technique that uses molecular markers to identify plants with desirable traits at an early stage of growth. This method speeds up the breeding process by allowing breeders to select for traits such as disease resistance, yield, and root quality without waiting for the plants to mature. MAS is particularly useful for improving complex traits that are controlled by multiple genes.

6. **Genetic Engineering**:

   » Genetic engineering in carrots has focused on introducing resistance to biotic and abiotic stresses, as well as enhancing the nutritional content of the roots. For example, genetic modification has been used to increase beta-carotene levels in carrots, creating varieties with even higher vitamin A content. While genetically modified carrots are not widely commercialized, research continues in this area.

Breeding for Biotic Stresses

Carrot is susceptible to a range of biotic stresses, including fungal, bacterial, and viral diseases, as well as insect pests. Breeding for resistance to these stresses is essential for reducing crop losses and minimizing the need for chemical control measures.

1. **Alternaria Leaf Blight Resistance:**

   » Alternaria leaf blight, caused by *Alternaria dauci*, is a common fungal disease of carrots, leading to leaf spots, defoliation, and reduced yields. Breeding for resistance to Alternaria involves selecting varieties that can tolerate or resist infection, particularly in humid environments where the disease is most prevalent. Resistant varieties help reduce the need for fungicide applications.

2. **Cercospora Leaf Spot Resistance:**

   » Cercospora leaf spot, caused by *Cercosporacarotae*, is another major fungal disease affecting carrot leaves. It can lead to reduced photosynthesis and stunted growth. Breeding programs focus on developing varieties with resistance to this disease, particularly for regions with warm, humid conditions. Resistance genes from wild carrot relatives have been incorporated into cultivated varieties to improve tolerance to Cercospora leaf spot.

3. **Bacterial Soft Rot Resistance:**

   » Bacterial soft rot, caused by *Erwinia carotovora*, affects the roots of carrots, leading to soft, mushy tissues that rapidly decay. This disease is most problematic during storage or after heavy rainfall. Breeding for resistance to bacterial soft rot involves selecting varieties with improved root structure and enhanced resistance to bacterial infection.

4. **Carrot Fly Resistance:**

   » The carrot fly (*Psilarosae*) is a major pest of carrots, laying its eggs at the base of the plants. The larvae burrow into the roots, causing significant damage and reducing marketability. Breeding for carrot fly resistance focuses on developing varieties with tougher roots or chemical compounds that deter the pest. Some progress has been made in selecting carrot varieties with resistance to carrot fly, particularly through the use of wild relatives.

5. **Virus Resistance:**

   » Carrot is susceptible to several viruses, including the Carrot Motley Dwarf Virus (CMDV) and the Carrot Red Leaf Virus (CRLV), both of which cause stunted growth and poor root development. Breeding for virus resistance is crucial in regions where these viruses are prevalent. Resistance genes from related species or wild carrot varieties are often used to introduce resistance to these viral diseases.

Breeding for Abiotic Stresses

Carrot production is often affected by various abiotic stresses, such as drought, temperature extremes, soil salinity, and waterlogging. Breeding for resistance to these stresses is essential for ensuring stable yields and high-quality roots under suboptimal growing conditions. Below are the key abiotic stresses in carrot cultivation and the breeding strategies used to address them.

1. **Drought Tolerance:**

   » Drought stress can significantly impact carrot yields, particularly in arid and semi-arid regions where water availability is limited. Breeding for drought tolerance involves selecting for varieties that can maintain growth and yield under water-limited conditions. Traits such as deeper root systems, improved water-use efficiency, and the ability to reduce transpiration are important targets in drought-tolerant carrot varieties.

- » Wild relatives of carrot, such as *Daucus sahariensis*, have been used to introduce drought tolerance into commercial carrot varieties. These wild species have evolved in desert environments and possess genetic traits that allow them to thrive in water-scarce conditions. Breeding programs aim to incorporate these traits into cultivated carrots without sacrificing yield or root quality.

2. **Heat Tolerance:**

- » High temperatures can cause bolting (premature flowering) in carrots, which reduces root quality and marketability. Breeding for heat tolerance focuses on selecting varieties that can delay bolting and maintain root development in warm climates. Heat-tolerant carrot varieties are particularly important for regions with extended periods of high temperatures or erratic weather patterns.
- » Heat tolerance is also important for improving the germination rate of carrot seeds in warmer climates. Some modern carrot varieties have been bred to germinate at higher temperatures, allowing for better establishment in tropical and subtropical regions.

3. **Cold Tolerance:**

- » Carrots are a cool-season crop, but exposure to freezing temperatures can cause damage to the roots, resulting in lower yields and poor storage quality. Breeding for cold tolerance involves selecting varieties that can withstand frosts or grow well in cooler climates without bolting or losing root quality.
- » Cold-tolerant varieties are important for winter production in temperate regions and for extending the growing season in areas with short summers. Breeding efforts have focused on improving the cold hardiness of carrots by selecting for traits such as improved root insulation and the ability to withstand low temperatures without suffering from tissue damage.

4. **Salinity Tolerance:**

- » Soil salinity is a major problem in many carrot-growing regions, particularly in coastal areas or regions with poor irrigation practices. High salt concentrations in the soil can reduce carrot growth, cause nutrient imbalances, and lead to poor root development. Breeding for salinity tolerance involves selecting varieties that can exclude excess salts from the root zone or tolerate higher internal salt concentrations without affecting growth.
- » Carrot varieties with salinity tolerance are developed by introducing genes from wild relatives or through selection in saline environments. These varieties are especially valuable in areas with high soil salinity, where conventional carrot varieties would struggle to produce marketable roots.

5. **Waterlogging Tolerance:**

» Waterlogged soils restrict oxygen availability to plant roots, leading to poor root development, reduced yields, and an increased risk of root diseases. Breeding for waterlogging tolerance focuses on selecting carrot varieties that can maintain root function and growth under waterlogged conditions. Waterlogging-tolerant varieties are particularly important for regions with heavy rainfall or poor drainage.

» Breeding programs select for traits such as improved root aeration and the ability to tolerate low-oxygen conditions, which allow carrots to grow in waterlogged soils without suffering from root rot or stunted growth.

Crop Production Techniques

Carrot production requires careful management of soil, water, and environmental factors to ensure high yields and high-quality roots. Below are the essential crop production techniques for growing carrots.

1. **Soil Preparation:**

» Carrots prefer well-drained, loose, sandy loam or loamy soil with a pH range of 6.0 to 6.8. Soil compaction should be avoided as it can lead to misshapen roots and reduced yields. Before planting, the soil should be deeply tilled to a depth of at least 30 cm to encourage straight root growth.

» Organic matter, such as compost or well-rotted manure, should be incorporated into the soil to improve its structure, moisture retention, and fertility. Carrots respond well to soils that are rich in organic matter, as it provides a slow-release source of nutrients that supports root development.

2. **Seed Sowing:**

» Carrot seeds are small and should be sown directly into the field at a shallow depth of 1 to 1.5 cm. Seeds are typically sown in rows spaced 30 to 40 cm apart, with plants spaced 5 to 8 cm apart within the row. This spacing allows enough room for the roots to develop properly without competition for nutrients and water.

» Carrot seeds can be slow to germinate, taking 7 to 21 days depending on soil temperature. The ideal temperature for seed germination is between 10°C and 20°C. During the germination period, the soil should be kept consistently moist to ensure good seedling establishment.

3. **Irrigation:**

» Carrots require consistent moisture throughout their growth cycle, particularly during the root development phase. Inadequate moisture can lead to small, tough, and woody roots, while excess moisture can cause root cracking and increase the risk of root diseases.

- Drip irrigation is the most effective method for carrot production, as it delivers water directly to the root zone while minimizing water loss through evaporation. Overhead irrigation can be used, but care must be taken to avoid wetting the foliage, which can encourage the development of foliar diseases.

4. **Weed Management:**
   - Weed competition can significantly reduce carrot yields, especially during the early stages of growth. Hand weeding is common in small-scale carrot production, while herbicides may be used in commercial operations. Mulching with organic materials, such as straw or grass clippings, can also help suppress weeds and retain soil moisture.
   - Carrots are sensitive to competition from weeds, particularly during the first six weeks of growth when the seedlings are establishing. Ensuring good weed control during this period is critical for achieving high yields.
5. **Fertilization:**
   - Carrots have relatively low nutrient requirements compared to other vegetable crops, but adequate nutrition is essential for good root development. A balanced application of nitrogen, phosphorus, and potassium (NPK) is recommended. Nitrogen promotes leafy growth, but excessive nitrogen can result in lush foliage at the expense of root development.
   - Phosphorus is important for root growth, while potassium contributes to overall plant health and disease resistance. Organic fertilizers, such as compost or well-rotted manure, can be used to provide a slow-release source of nutrients. A typical fertilization regime involves applying 50-60 kg/ha of nitrogen, 30-40 kg/ha of phosphorus, and 40-60 kg/ha of potassium.
6. **Thinning:**
   - Thinning is necessary to ensure that carrot plants have enough space to develop properly. After the seedlings have emerged, they should be thinned to a final spacing of 5-8 cm between plants. Thinning is typically done by hand to avoid disturbing the remaining plants.
   - Crowded carrot plants will produce small, misshapen roots, so it is important to thin the plants early in the growing season.

Post-Harvest Handling

Proper post-harvest handling is critical to maintaining the quality and shelf life of carrots. Freshly harvested carrots are highly perishable and must be handled carefully to prevent mechanical damage, dehydration, and loss of flavor.

1. **Harvesting:**
   - Carrots are typically harvested when the roots reach the desired size, which varies

depending on the variety and market demands. For early-maturing varieties, harvesting can begin as soon as 60-70 days after sowing, while late-maturing varieties may take up to 120 days to reach full maturity.

» Carrots should be harvested before they become overgrown and woody, which can reduce their quality and marketability. In small-scale operations, carrots are harvested manually by pulling the roots from the soil. In larger commercial operations, mechanical harvesters are often used to speed up the process.

2. **Cleaning and Grading:**

» After harvesting, carrots should be cleaned to remove soil and debris. This can be done by rinsing the roots in water or using mechanical brushes. Once cleaned, the roots are sorted and graded based on size, shape, and color. Uniformity is important for marketability, and roots that are misshapen or damaged are typically discarded or sold at a lower price.

» Grading is particularly important for the fresh market, where consumers prefer carrots that are visually appealing and consistent in size. Carrots for the processing market may be graded less stringently, as appearance is less critical.

3. **Storage:**

» Carrots are highly perishable and should be stored at low temperatures to maintain their quality. Ideal storage conditions are between 0°C and 4°C with high humidity (90-95%) to prevent dehydration. Under these conditions, carrots can be stored for several weeks without significant loss of quality.

» For longer storage, carrots should be packed in perforated plastic bags or crates to maintain proper ventilation and prevent the buildup of excess moisture, which can lead to rot. Some carrot varieties, such as those with higher dry matter content, are better suited for long-term storage.

4. **Packaging and Transportation:**

» Carrots are typically packed in plastic crates or cartons for transportation to market. Packaging should be sturdy enough to protect the roots from mechanical damage during transit. In the fresh market, carrots are often sold with their leafy tops intact, but the tops should be removed if the carrots are to be stored for an extended period, as the leaves can cause moisture loss from the roots.

» Efficient transportation is critical for maintaining the freshness of carrots, particularly for export markets. Refrigerated transport is commonly used for long-distance shipments to maintain the low temperature and high humidity needed for extended shelf life.

E. Beetroot (*Beta vulgaris*)

Family: Amaranthaceae (formerly Chenopodiaceae)

Chromosome number: 2n = 2x = 18

Beetroot, commonly referred to as beet, is a biennial root vegetable that is cultivated for its fleshy, edible roots. It is primarily grown for human consumption as well as for animal feed. Beetroot is consumed fresh, processed, or preserved in pickles and juices. It is also used as a natural food coloring agent due to the presence of betalain pigments, responsible for the vibrant red and purple colors. Beetroot is rich in vitamins (particularly folate and vitamin C), minerals (potassium, manganese, and iron), and dietary fiber, offering multiple health benefits. In addition to its nutritional value, beetroot has gained attention for its potential cardiovascular health benefits, largely attributed to the presence of nitrates, which may help in reducing blood pressure.

Origin and Domestication

Beetroot is believed to have originated in the Mediterranean region, where its wild ancestor, *Beta vulgaris* subsp. *maritima*, commonly known as sea beet, is still found along coastlines. The domestication of beetroot likely began with the selection of plants with larger, more tender roots and improved flavor. The cultivation of beets for their roots became widespread during the Roman period, while earlier cultivation focused on the leaves for use in salads.

Historically, two main types of *Beta vulgaris* were cultivated: one for its leaves (chard) and the other for its swollen, fleshy roots (beetroot). Over time, selection and breeding led to the development of various forms of beet, including sugar beet (used for sugar production), fodder beet (for animal feed), and table beet (the common red beetroot).

Related Genera

Beetroot belongs to the Amaranthaceae family, which includes many economically important plants. Some of the related genera and species include:

- **Spinacia:** Spinach (*Spinacia oleracea*) is a leafy vegetable closely related to beetroot. Spinach and beetroot share similar growing conditions and are often cultivated together in crop rotations.
- **Chenopodium:** Quinoa (*Chenopodium quinoa*) is a pseudocereal that belongs to the same family as beetroot. Both plants share similar nutritional profiles, being rich in vitamins and minerals.
- **Amaranthus:** Amaranth is another pseudocereal from the same family. While amaranth is primarily grown for its seeds, it also has edible leaves. Amaranth and beetroot share some similar physiological traits, especially in their ability to tolerate drought and poor soil conditions.

These related genera can be useful in breeding programs for traits such as disease resistance, drought tolerance, and nutritional improvement.

Related Species

The genus *Beta* includes several important species, many of which are valuable for breeding programs aimed at improving disease resistance and stress tolerance in cultivated beets.

1. **Beta vulgaris subsp. maritima (Sea Beet):**
   » The wild ancestor of all cultivated beets, sea beet grows along coastal areas in Europe, the Middle East, and Asia. It is a valuable genetic resource for breeding programs, particularly for improving salinity tolerance and disease resistance in cultivated beets.
2. **Beta vulgaris subsp. cicla (Swiss Chard):**
   » Swiss chard is a leafy form of *Beta vulgaris* grown for its large, tender leaves rather than its roots. It is often used in breeding programs to improve disease resistance and leaf quality in other beet varieties.
3. **Beta patula:**
   » Native to the Madeira archipelago, this species has been used in breeding programs to introduce resistance to beet yellows virus (BYV) and rhizomania, a serious soil-borne disease that affects sugar beet and table beet.
4. **Beta corolliflora:**
   » This wild species is native to Central Asia and has been used to introduce drought tolerance and disease resistance traits into cultivated beet varieties.

Types of Beetroot on Basis of root shape, size and color:

Beetroot comes in several botanical varieties, which differ in root shape, size, and color. The most common varieties are red, but there are also yellow, white, and striped (candy cane) varieties.

1. **Detroit Dark Red:**
   » *Description:* This is the most widely grown variety of red beetroot, known for its deep red color and round, uniform roots. It has a sweet flavor and smooth texture.
   » *Root Shape:* Round
   » *Uses:* Fresh consumption, processing, pickling.
2. **Golden Beet:**
   » *Description:* Golden beet is a yellow variety with a mild, sweet flavor and less earthy taste than red beets. It has a smooth texture and is ideal for salads and roasting.
   » *Root Shape:* Round
   » *Uses:* Fresh consumption, salads, roasting.

3. **Chioggia (Candy Stripe):**
   - *Description:* This Italian heirloom variety is known for its striking pink-and-white concentric rings inside the root. It has a mild, sweet flavor and a slightly softer texture.
   - *Root Shape:* Round
   - *Uses:* Fresh consumption, gourmet markets, salads.
4. **Cylindra:**
   - *Description:*Cylindra beet is a cylindrical variety that produces long, narrow roots. It is favored for its uniform size and ease of slicing for pickling and processing.
   - *Root Shape:* Cylindrical
   - *Uses:* Processing, pickling, fresh consumption.
5. **Albina Vereduna (White Beetroot):**
   - *Description:* Albina Vereduna is a white variety with a milder, less earthy flavor compared to red varieties. It is commonly used in gourmet dishes and specialty markets.
   - *Root Shape:* Round
   - *Uses:* Fresh consumption, gourmet markets.

Table 1: Major Types of Beetroot on Basis of Root Shape, Size and Color

S No	Botanical Variety	Root Shape	Color	Uses
1	*Detroit Dark Red*	Round	Red	Fresh, processing, pickling
2	*Golden Beet*	Round	Yellow	Fresh, salads, roasting
3	*Chioggia (Candy Stripe)*	Round	Red/White	Fresh, gourmet markets, salads
4	*Cylindra*	Cylindrical	Red	Processing, pickling, fresh
5	*Albina Vereduna*	Round	White	Fresh, gourmet dishes

Growth Habit

Beetroot is a biennial plant, meaning it takes two years to complete its life cycle. In the first year, the plant forms a rosette of leaves and a swollen taproot, which stores energy for the second year when it flowers and produces seeds. However, beetroot is typically grown as an annual, with the roots harvested at the end of the first growing season before flowering occurs.

The growth habit of beetroot is influenced by environmental conditions, particularly temperature and day length. Beetroot grows best in cool climates with temperatures between 10°C and 20°C. High temperatures can cause the roots to become woody and develop a bitter flavor, while cold temperatures (below 4°C) can trigger premature bolting.

The root shape and size of beetroot are influenced by soil type and spacing. Loose, well-drained soil is essential for proper root development, while closely spaced plants can result in smaller roots. Carrot-like cylindrical varieties, such as *Cylindra*, are favored in some regions due to their ease of harvest and uniformity in processing.

Floral Morphology

If allowed to grow into its second year, beetroot produces a tall flowering stalk, which can reach heights of 1 to 2 meters. The flowers are small and greenish, arranged in dense clusters called cymes. Beetroot is a wind-pollinated species, and its flowers are typically self-incompatible, meaning cross-pollination between different plants is required for seed production.The flowers are hermaphroditic, containing both male (stamens) and female (pistils) reproductive organs. Once pollinated, the flowers develop into small fruits known as glomerules, which contain multiple seeds. These glomerules are harvested and processed for commercial seed production.

Floral Biology

Beetroot is a wind-pollinated crop that relies on cross-pollination for seed production. It is a self-incompatible species, meaning it cannot fertilize itself, which helps maintain genetic diversity within the population. Cross-pollination occurs when pollen from one plant is carried by the wind to the flowers of another plant. In seed production fields, isolation distances are maintained to prevent unwanted cross-pollination between different varieties or species of beets.

The flowering process in beetroot is influenced by temperature and day length. Vernalization, or exposure to cold temperatures (below 10°C), is required to induce flowering in the second year of growth. Without vernalization, beetroot plants will not flower, making it important to prevent exposure to cold temperatures during the first year of growth when the roots are being harvested.In commercial seed production, beetroot plants are allowed to overwinter and flower in the second year. After pollination, the seed pods (glomerules) mature and are harvested for seed extraction.

Breeding Objectives

Beetroot breeding programs aim to improve traits such as yield, root quality, disease resistance, and abiotic stress tolerance. Below are the primary objectives in beetroot breeding:

1. **Root Yield and Size:**
   » Yield improvement is a key breeding objective for beetroot. High yields are achieved through the selection of varieties that produce large, uniform roots with minimal defects. Root size is particularly important for processing beets, where uniformity is required for efficient slicing and dicing.
2. **Root Shape and Color:**
   » Root shape and color are important for marketability. Consumers prefer smooth, round, or cylindrical roots with vibrant red, yellow, or white flesh, depending on

the variety. Breeding efforts focus on improving the consistency of root shape and enhancing the intensity of root color, particularly in red and yellow varieties.

3. **Disease Resistance:**

   » Beetroot is susceptible to a range of diseases, including Cercospora leaf spot (*Cercosporabeticola*), Rhizoctonia root rot (*Rhizoctonia solani*), and downy mildew (*Peronospora farinosa*). Breeding for disease resistance is crucial for reducing crop losses and minimizing the use of chemical fungicides. Resistance genes from wild relatives are often used in breeding programs to introduce disease resistance into cultivated varieties.

4. **Abiotic Stress Tolerance:**

   » Abiotic stresses such as drought, salinity, and temperature extremes can significantly impact beetroot production. Breeding for abiotic stress tolerance focuses on selecting varieties that can maintain high yields and root quality under challenging environmental conditions. For example, salinity-tolerant varieties are important for coastal farming regions, while drought-tolerant varieties are essential in arid and semi-arid areas.

5. **Flavor and Nutritional Content:**

   » Beetroot is prized for its sweet, earthy flavor and high nutritional content. Breeding programs aim to enhance the flavor profile of beetroot by selecting for varieties with higher sugar content and improved texture. In addition, breeders are working to increase the concentration of betalains, the pigments responsible for the red and yellow colors of beetroot, which have antioxidant properties.

Breeding for Biotic Stresses

Beetroot is vulnerable to several diseases and pests that can reduce crop yields and quality. Breeding for resistance to biotic stresses, such as fungal and bacterial diseases, as well as pest attacks, is crucial for maintaining healthy crops and minimizing losses.

1. **Cercospora Leaf Spot Resistance:**

   » Cercospora leaf spot, caused by the fungus *Cercosporabeticola*, is one of the most common diseases affecting beetroot. The disease manifests as small, circular lesions on the leaves, which eventually coalesce, causing significant defoliation and reducing photosynthesis. This can lead to smaller roots and lower yields.

   » Breeding for resistance to Cercospora leaf spot is a high priority for beetroot growers, particularly in regions with warm, humid conditions. Resistant varieties have been developed through traditional breeding methods, using both natural selection and backcrossing with resistant wild relatives. These varieties exhibit reduced lesion formation and increased tolerance to fungal infection, allowing for healthier plants and higher yields.

2. **Downy Mildew Resistance:**
   - Downy mildew (*Peronospora farinosa*) is a fungal disease that affects beetroot leaves, causing yellowing and necrosis. It thrives in cool, wet conditions and can spread rapidly through the crop if not managed properly. The disease weakens the plants, reducing root development and yield.
   - Resistance to downy mildew is incorporated into beetroot breeding programs by selecting varieties that can tolerate wet conditions and resist fungal invasion. Resistance genes from wild beet species, such as *Beta vulgaris* subsp. *maritima*, are commonly used to introduce downy mildew resistance into commercial varieties.
3. **Rhizoctonia Root Rot Resistance:**
   - Rhizoctonia root rot, caused by *Rhizoctonia solani*, is a soil-borne fungal disease that leads to the decay of beetroot roots, causing them to become spongy and unmarketable. The disease is particularly problematic in poorly drained soils or areas with excessive moisture.
   - Breeding for resistance to Rhizoctonia focuses on selecting varieties with improved root structure and natural tolerance to soil-borne pathogens. This resistance is often combined with good agronomic practices, such as crop rotation and soil drainage improvements, to reduce the impact of the disease.
4. **Beet Cyst Nematode Resistance:**
   - Beet cyst nematode (*Heteroderaschachtii*) is a major pest of beetroot, causing stunted growth and reduced yields. The nematodes invade the roots, forming cysts that damage the root tissues and disrupt water and nutrient uptake.
   - Breeding for resistance to beet cyst nematode involves selecting varieties that can either tolerate or resist nematode invasion. Resistance genes from wild beet species have been successfully incorporated into cultivated varieties, providing growers with nematode-resistant options. Integrated pest management (IPM) strategies, such as crop rotation and the use of resistant varieties, are commonly used to manage nematode populations.
5. **Virus Resistance:**
   - Beetroot is susceptible to several viruses, including beet yellows virus (BYV) and beet curly top virus (BCTV). These viruses are transmitted by insect vectors such as aphids and leafhoppers and can cause stunted growth, yellowing of leaves, and reduced root size.
   - Breeding for virus resistance focuses on selecting varieties that can resist viral infections or reduce the impact of the virus on plant growth. Resistance genes from related species or wild beet varieties are often introduced into commercial varieties through backcrossing or genetic engineering.

Abiotic Stress Tolerance

Abiotic stresses, including drought, salinity, temperature extremes, and poor soil conditions, can negatively impact beetroot production. Breeding for abiotic stress tolerance aims to develop varieties that can thrive under challenging environmental conditions while maintaining high yield and quality.

1. **Drought Tolerance:**
   » Drought stress can lead to smaller, tougher roots in beetroot, reducing both yield and quality. Breeding for drought tolerance involves selecting varieties that can maintain growth and root development under water-limited conditions.

   » Drought-tolerant beetroot varieties are developed by selecting for traits such as deeper root systems, improved water-use efficiency, and the ability to reduce transpiration. These varieties are particularly important in regions with irregular rainfall or limited access to irrigation.

2. **Salinity Tolerance:**
   » Soil salinity is a major problem in many agricultural regions, particularly in coastal areas or regions with poor irrigation management. High salt concentrations in the soil can lead to reduced growth and poor root development in beetroot.

   » Salinity-tolerant varieties of beetroot are bred by selecting plants that can maintain normal growth and yield under saline conditions. Wild relatives of beetroot, such as *Beta vulgaris* subsp. *maritima*, are often used as sources of salinity tolerance. These wild species have evolved to grow in coastal environments and possess genetic traits that allow them to exclude excess salts from the root zone.

3. **Temperature Tolerance**:
   » Extreme temperatures, both hot and cold, can negatively affect beetroot growth. High temperatures can cause bolting (premature flowering), while low temperatures can lead to poor root development and reduced yields.

   » Breeding for temperature tolerance involves selecting varieties that can maintain growth and root quality under temperature extremes. Heat-tolerant varieties are important for regions with hot summers, while cold-tolerant varieties are valuable for extending the growing season in temperate regions.

4. **Waterlogging Tolerance:**
   » Waterlogging, or the saturation of soil with excess water, can lead to oxygen deprivation in beetroot roots, causing root rot and reduced yields. Waterlogging is particularly problematic in regions with heavy rainfall or poor drainage.

   » Breeding for waterlogging tolerance focuses on selecting varieties that can maintain root function and growth under waterlogged conditions. Traits such as improved root aeration and the ability to tolerate low-oxygen conditions are important for waterlogging-tolerant varieties.

Crop Production Techniques

Successful beetroot cultivation requires proper management of soil, water, and environmental factors. Below are the key crop production techniques involved in beetroot cultivation.

1. **Soil Preparation:**
   - Beetroot grows best in well-drained, loamy soils with a pH range of 6.0 to 7.5. The soil should be rich in organic matter to promote healthy root development. Heavy clay soils should be avoided, as they can lead to misshapen roots.
   - Before planting, the soil should be deeply tilled to a depth of at least 25-30 cm to loosen the soil and allow for proper root development. Incorporating compost or well-rotted manure into the soil improves fertility and water retention, supporting healthy root growth.
2. **Seed Sowing**:
   - Beetroot seeds are small and can be sown directly into the field. Seeds are typically sown at a depth of 1.5 to 2 cm in rows spaced 25-30 cm apart, with individual plants spaced 8-10 cm apart within the row.
   - Beetroot seeds germinate best at temperatures between 10°C and 25°C, with optimal germination occurring within 5 to 10 days. Adequate soil moisture is critical during germination and seedling establishment, so irrigation should be applied if rainfall is insufficient.
3. **Irrigation:**
   - Consistent soil moisture is essential for beetroot, particularly during the root development phase. Water stress can lead to poor root formation and tough, woody roots, while excessive moisture can cause root cracking and increase the risk of root rot.
   - Drip irrigation is the preferred method for beetroot cultivation, as it delivers water directly to the root zone while minimizing water loss through evaporation. Overhead irrigation is less ideal, as it can lead to leaf wetness and the development of foliar diseases.
4. **Weed Management:**
   - Weed competition can significantly reduce beetroot yields, particularly during the early stages of growth. Hand weeding is common in small-scale beetroot production, while herbicides may be used in commercial operations.
   - Mulching with organic materials, such as straw or grass clippings, can help suppress weed growth and retain soil moisture. Mulching also improves soil structure and provides nutrients as it decomposes.

5. **Fertilization:**

   » Beetroot requires balanced fertilization to support healthy root development. Nitrogen, phosphorus, and potassium are the primary nutrients required. However, excessive nitrogen can lead to lush foliage at the expense of root development.

   » A typical fertilization regime involves applying 40-60 kg/ha of nitrogen, 30-50 kg/ha of phosphorus, and 60-80 kg/ha of potassium, depending on soil fertility. Organic fertilizers, such as compost or well-rotted manure, can also be used to improve soil fertility and provide a slow-release source of nutrients.

6. **Thinning:**

   » Thinning is necessary to ensure that beetroot plants have enough space to develop properly. After the seedlings have emerged, they should be thinned to a final spacing of 8-10 cm between plants. Crowded plants will produce smaller roots, so thinning should be done early in the growing season.

Post-Harvest Handling

Proper post-harvest handling is critical to maintaining the quality and shelf life of beetroot. Freshly harvested beetroot is highly perishable and must be handled carefully to prevent mechanical damage, dehydration, and loss of flavor.

1. **Harvesting:**

   » Beetroot is typically harvested when the roots reach the desired size, which varies depending on the variety and market demands. For small, early-maturing varieties, harvesting can begin as soon as 50-60 days after sowing, while larger varieties may take 90-120 days to mature.

   » Beetroot should be harvested before the roots become woody, as this reduces their quality and marketability. Harvesting is usually done manually by pulling the roots from the soil, although mechanical harvesters are used in larger commercial operations.

2. **Cleaning and Grading:**

   » After harvesting, beetroot roots should be cleaned to remove soil and debris. This can be done by rinsing the roots in water or using mechanical brushes. Once cleaned, the roots are sorted and graded based on size, shape, and color.

   » Uniformity is important for marketability, and roots that are misshapen or damaged are typically discarded or sold at a lower price. Grading is especially important for the fresh market, where consumers prefer smooth, uniformly sized roots.

3. **Storage:**

    » Beetroot is highly perishable and should be stored at low temperatures to maintain quality. Ideal storage conditions are between 0°C and 4°C with high humidity (90-95%) to prevent dehydration.

    » For longer storage, beetroot should be packed in perforated plastic bags or crates to maintain proper ventilation and prevent moisture buildup, which can lead to rot. Some beetroot varieties, such as those with higher dry matter content, are better suited for long-term storage.

4. **Packaging and Transportation:**

    » Beetroot is typically packed in plastic crates or cartons for transportation to market. The packaging should be sturdy enough to protect the roots from mechanical damage during transit. Efficient transportation is essential for maintaining the freshness of beetroot, particularly for export markets.

Solanaceous Vegetable Breeding

A. Potato (*Solanum tuberosum*)

Family: Solanaceae

Chromosome number: 2n = 4x = 48

Potato is one of the world's most important food crops, cultivated for its underground tubers. These tubers are an excellent source of energy due to their high starch content and are a major part of diets across the globe. In addition to carbohydrates, potatoes provide essential nutrients such as vitamin C, vitamin B6, potassium, and dietary fiber. The crop's ability to grow in various environments and its adaptability to different farming systems has made it a staple in many countries. Potatoes are used in multiple forms: fresh potatoes are cooked and consumed directly; processed potatoes are turned into products such as French fries, potato chips, and potato flour; and they are used in many processed foods like soups, stews, and snacks. Given its versatility, the potato industry plays a critical role in both food security and economic development worldwide.

Origin and Domestication

The potato originated in the Andes Mountains of South America, particularly in modern-day Peru and northwest Bolivia. The crop was domesticated around 7,000–10,000 years ago by indigenous people who relied on it as a major food source. Wild relatives of the potato (*Solanum spp.*) are still found in the region, growing in diverse environments from coastal deserts to high-altitude plateaus. Domestication involved selecting traits such as larger tubers, better taste, and reduced levels of glycoalkaloids (compounds that can be toxic in high concentrations). Over centuries, local farmers developed hundreds of varieties adapted to specific microclimates and agricultural practices.

The Spanish brought the potato to Europe in the 16th century, where it initially met resistance but gradually became accepted as a staple food due to its high productivity and nutritional value. From Europe, the potato spread across the world, becoming a key crop in temperate regions, including North America, Asia, and Africa.

Related Genera and Species

The potato belongs to the Solanaceae family, which includes several important food crops such as tomatoes, eggplants, and peppers. It is closely related to various species within the *Solanum* genus, some of which have been used in breeding programs to improve resistance to diseases and environmental stresses.

1. **Tomato (*Solanum lycopersicum*):**
   » Like potatoes, tomatoes are part of the *Solanum* genus and share similar growth

requirements. Both crops are affected by many of the same pests and diseases, such as late blight (*Phytophthora infestans*), making tomato a valuable model crop for research that also benefits potato breeding.

2. **Eggplant (*Solanum melongena*):**

   » Eggplant, another member of the Solanaceae family, is not as closely related to the potato as the tomato, but it shares some of the same physiological traits, such as flowering and fruit development mechanisms. It is grown in similar climates and soil conditions.

3. **Wild Potato Relatives:**

   » The wild relatives of cultivated potato (*Solanum tuberosum*) are found in South America. These species, such as *Solanum demissum*, *Solanum acaule*, and *Solanum bulbocastanum*, are often used in breeding programs for their disease resistance traits, particularly resistance to late blight and potato virus Y (PVY).

Different Types of Potato on Basis of Morphological Characteristics

Potato varieties are classified into different types based on their morphological characteristics, such as tuber shape, skin color, flesh color, and culinary uses. The major types of potatoes are:

1. **Round White Potatoes:**

   » *Description:* These potatoes have smooth, thin, white to tan skins and white flesh. They are commonly used for boiling, frying, and as all-purpose cooking potatoes.

   » *Uses:* Boiling, frying, salads, and general-purpose cooking.

2. **Russet Potatoes:**

   » *Description:* Russet potatoes have rough, brown skin and white flesh. They are high in starch and have a dry, mealy texture, making them ideal for baking and frying.

   » *Uses:* Baking, French fries, mashed potatoes.

3. **Red Potatoes:**

   » *Description:* Red potatoes have smooth, thin red skins and white or yellow flesh. They have a waxy texture and hold their shape well when cooked, making them perfect for salads and soups.

   » *Uses:* Salads, soups, boiling, roasting.

4. **Fingerling Potatoes:**

   » *Description:* Fingerling potatoes are small, elongated tubers with thin skin. They come in a variety of colors, including yellow, red, and purple. Their small size and nutty flavor make them popular in gourmet cooking.

   » *Uses:* Roasting, grilling, salads.

5. **Yellow Potatoes**:

   » *Description:* Yellow potatoes, including varieties like Yukon Gold, have golden skin and flesh. They are slightly waxy, making them suitable for boiling, mashing, and roasting.

   » *Uses:* Mashing, roasting, grilling.

6. **Purple and Blue Potatoes**:

   » *Description:* These potatoes have dark purple or blue skin and flesh, rich in antioxidants called anthocyanins. They are often used in specialty dishes due to their striking color.

   » *Uses:* Roasting, mashing, gourmet dishes.

Table 1: Major Type of Potato on Basis of Colour and Uses

S. No	Botanical Variety	Tuber Shape	Skin Color	Flesh Color	Uses
1	Round White Potato	Round	White	White	Boiling, frying, salads, general-purpose
2	Russet Potato	Oval	Brown	White	Baking, French fries, mashed potatoes
3	Red Potato	Round	Red	White	Salads, soups, boiling, roasting
4	Fingerling Potato	Elongated	Various	Various	Roasting, grilling, salads
5	Yellow Potato	Round	Yellow	Yellow	Mashing, roasting, grilling
6	Purple/Blue Potato	Round/Oval	Purple/Blue	Purple/Blue	Roasting, gourmet dishes, mashing

Growth Habit

Potatoes are herbaceous perennial plants that are usually grown as annuals for their tubers. The plant produces a fibrous root system and underground stolons from which the tubers develop. The tubers serve as storage organs for nutrients, enabling the plant to survive unfavorable conditions.

The aerial part of the potato plant consists of stems, leaves, and flowers. The leaves are compound, made up of leaflets that are pinnately arranged along the stem. Potato plants produce small, white, pink, or purple flowers, followed by small, green, tomato-like fruit, which contain seeds. However, potatoes are primarily propagated vegetatively from tubers (seed potatoes), rather than from botanical seeds.

Potato growth is divided into five stages:

1. **Sprout Development:**

   » This stage begins when the seed tuber starts sprouting. The growth of sprouts depends on temperature, moisture, and soil conditions. Warmer temperatures and good soil moisture promote rapid sprout emergence.

2. **Vegetative Growth:**

   » During this stage, the plant produces stems, leaves, and stolons. Healthy vegetative growth is critical for photosynthesis and the development of the tubers. The size of the plant canopy determines how much energy the plant can store in the tubers.

3. **Tuber Initiation:**

   » Tuber initiation occurs when the stolons begin to form small swellings that will eventually become mature tubers. This phase is influenced by day length and temperature. Cool temperatures and short days are favorable for tuber formation.

4. **Tuber Bulking:**

   » In this stage, the tubers grow rapidly, accumulating carbohydrates and other nutrients. Adequate moisture and nutrients are essential during tuber bulking to maximize yield and tuber size. Nutrient management, particularly nitrogen, plays a key role in this phase.

5. **Maturation:**

   » The maturation stage begins when the plant canopy starts to die back, and the tubers reach their full size. The skin of the tubers toughens during this stage, which is important for preventing damage during harvest and storage.

Floral Morphology

The potato plant produces flowers that are typically five-petaled and come in various colors, including white, pink, and purple. The flowers are hermaphroditic, containing both male (stamens) and female (pistil) reproductive organs. Pollination is primarily insect-mediated, although potatoes are often vegetatively propagated and rarely grown for seed production. The fruit that develops after pollination is a small, green, tomato-like berry that contains numerous seeds. These botanical seeds can be used for breeding purposes, but the commercial potato crop is propagated from tubers. Since potatoes reproduce vegetatively, their flowers and fruit have limited commercial significance, except in breeding programs.

Breeding Objectives

Breeding programs for potatoes focus on improving various traits, such as yield, disease resistance, stress tolerance, and tuber quality. The main objectives of potato breeding include:

1. **Yield Improvement:**

   » One of the primary objectives of potato breeding is to increase yield. This involves selecting varieties that produce a higher number of tubers per plant, larger tuber sizes, and greater overall biomass. Breeding for higher yield also involves selecting for resistance to diseases and environmental stresses, as these factors can significantly reduce yield.

2. **Disease Resistance:**
    » Potatoes are susceptible to numerous diseases, including late blight (*Phytophthora infestans*), early blight (*Alternaria solani*), and potato virus Y (PVY). Breeding for resistance to these diseases is a top priority, as disease outbreaks can devastate potato crops. Wild relatives of the potato are often used as sources of resistance genes.
3. **Abiotic Stress Tolerance:**
    » Potatoes are sensitive to abiotic stresses such as drought, salinity, and extreme temperatures. Breeding programs aim to develop varieties that can maintain productivity under these challenging conditions. Drought-tolerant varieties, for example, may have deeper root systems or better water-use efficiency.
4. **Tuber Quality**:
    » Tuber quality is critical for both fresh consumption and processing. Breeding efforts focus on improving traits such as tuber size, shape, skin color, flesh color, and texture. Varieties with high dry matter content are preferred for processing into chips and fries, while varieties with smooth skin and uniform shape are favored for the fresh market.
5. **Nutritional Content:**
    » Enhancing the nutritional value of potatoes is another breeding objective. Potatoes are naturally rich in carbohydrates, but breeding programs also focus on increasing the levels of vitamins, minerals, and antioxidants, such as vitamin C, potassium, and anthocyanins.

Breeding Methods in Potato

Breeding methods for potatoes aim to improve yield, disease resistance, tolerance to abiotic stresses, and tuber quality. Potato breeding faces unique challenges due to the crop's tetraploid nature ($4x = 48$), which complicates the inheritance of traits.

Conventional Breeding Methods

1. **Introduction:**
    » This method involves the introduction of improved potato varieties from one ecological region to another for testing and evaluation. These introductions are evaluated for adaptability, yield performance, and resistance to local diseases and environmental stresses.
    » Examples of successful introductions include varieties like *Kufri Jyoti*, *Kufri Sindhuri*, and *Kufri Pukhraj*, which have gained widespread acceptance among farmers for their high yield and disease resistance.
2. **Selection:**
    » Potato breeding often employs both mass selection and individual plant selection.

Superior plants are selected based on desirable characteristics such as higher yield, disease resistance, and tuber quality. The selection process aims to make the most of genetic diversity available in the germplasm collection.

» Selections are made from both local and exotic varieties.

Source Selections:

S No	Variety Name	Source	Features
1	Kufri Jyoti	Indian Agricultural Research Institute (IARI)	Early maturing, disease resistant
2	Kufri Sindhuri	Central Potato Research Institute (CPRI)	Red skin, tolerant to drought
3	Kufri Pukhraj	IARI	High yielding, large tubers, heat tolerant
4	Atlantic	Introduced from the USA	Suitable for processing, high dry matter

3. **Hybridization (Heterosis Breeding):**

   » Hybridization is a major technique in potato breeding. It involves crossing two or more parents with complementary traits to produce a hybrid with superior qualities (heterosis). Potato hybrids are primarily bred for increased yield, resistance to pests and diseases, and improved tuber quality.

   » Successful hybrids, such as *Kufri Chipsona-1* and *Kufri Chipsona-3*, have been developed to meet specific processing needs (such as chipping).

Examples of Potato Hybrids:

S No	Hybrid Name	Parentage	Features
1	Kufri Chipsona-1	Kufri Jyoti × Atlantic	High dry matter, suitable for chips
2	Kufri Chipsona-3	Kufri Jyoti × Atlantic	Low reducing sugars, ideal for processing
3	Kufri Sutlej	Kufri Jyoti × Phureja	Early bulking, high yield
4	Kufri Bahar	Kufri Jyoti × Kufri Sindhuri	Drought-tolerant, good for tropical areas

4. **Pedigree Method:**

   » In this method, selected plants from a hybrid population are selfed and evaluated in subsequent generations to select for the best-performing lines. Pedigree selection is a slow but effective method for fixing desired traits in potato breeding.

   » Selection begins from F1 hybrids, and evaluations continue until pure lines are developed (usually till F5 or F6).

5. **Bulk Method:**

   » This method is commonly used when breeding for complex traits, such as resistance to multiple diseases. In bulk breeding, the entire hybrid population is grown together, and selections are made after a few generations. It allows for the recombination of genes that control polygenic traits.

6. **Backcross Method:**
   » The backcross method is used to introduce specific traits, such as disease resistance, from a wild species or landrace into a cultivated variety. The cultivated variety is repeatedly crossed with the wild species, followed by selection for the desired trait.
   » This method has been successfully used to develop varieties resistant to late blight (*Phytophthora infestans*) and potato cyst nematodes (*Globodera spp.*).
7. **Mutation Breeding:**
   » Induced mutations have been used to create novel potato varieties with improved traits. For instance, *Kufri Kanchan* is a mutant developed from *Kufri Jyoti* using gamma irradiation. Mutation breeding is effective for developing varieties with enhanced disease resistance and improved tuber quality.

Non-Conventional Breeding Methods

1. **Tissue Culture:**
   » Tissue culture techniques are widely used in potato breeding for the rapid multiplication of disease-free planting material. Micropropagation allows for the production of virus-free seed potatoes, which is critical for maintaining healthy crops.
   » In addition, somatic hybridization via protoplast fusion has been employed to combine traits from different potato species that cannot be crossed by conventional means. This has led to the development of hybrids with resistance to various pathogens and abiotic stresses.
2. **Genetic Engineering (Recombinant DNA Technology):**
   » Genetic engineering has emerged as a powerful tool in potato breeding. Transgenic potatoes with resistance to insects, viruses, and herbicides have been developed using genes from other organisms.
   » The most notable example is the development of genetically modified potatoes resistant to the Colorado potato beetle (*Leptinotarsa decemlineata*). These potatoes contain a gene from *Bacillus thuringiensis* (Bt), which produces a protein toxic to the beetle larvae but safe for humans and animals.

Breeding for Biotic Stresses

Biotic stresses, such as diseases and pests, are a significant challenge in potato cultivation. Breeding for resistance to these stresses is essential for ensuring sustainable production. Below are some of the key biotic stresses targeted in potato breeding programs:

Table: Biotic Stress Resistance in Potato Varieties

S No	Disease/Pest	Source Species	Resistant Varieties
1	Late Blight	*Solanum demissum*	Kufri Jyoti, Kufri Bahar
2	Early Blight	*Solanum tuberosum* (wild relatives)	Kufri Sindhuri, Kufri Badshah
3	Potato Virus Y (PVY)	*Solanum stoloniferum*	Kufri Ashoka, Kufri Arun
4	Potato Cyst Nematode	*Solanum vernei, Solanum andigenum*	Kufri Swarna, Kufri Kanchan

1. **Late Blight Resistance:**

   » Late blight caused by *Phytophthora infestans* is one of the most destructive diseases affecting potato crops. Breeding for resistance has focused on introducing resistance (R) genes from wild species like *Solanum demissum* into cultivated varieties. However, due to the pathogen's ability to evolve, continuous efforts are needed to develop new resistant varieties.

2. **Early Blight Resistance:**

   » Early blight, caused by *Alternaria solani*, is another major fungal disease. Resistance to early blight has been introduced into potato varieties through the use of polygenic traits that enhance the plant's natural defense mechanisms.

3. **Virus Resistance**:

   » Potato is susceptible to several viral diseases, including potato virus Y (PVY) and potato leaf roll virus (PLRV). Breeding for viral resistance often involves the incorporation of resistance genes from wild relatives like *Solanum stoloniferum* and *Solanum andigenum.*

4. **Nematode Resistance:**

   » Potato cyst nematodes (*Globodera spp.*) cause significant yield losses in potato cultivation. Resistance genes from wild potato relatives like *Solanum vernei* have been incorporated into cultivated varieties to combat this pest.

Breeding for Abiotic Stresses

Abiotic stresses, including drought, heat, and salinity, significantly affect potato yields, particularly in regions with harsh climates. Breeding programs focus on developing varieties that can thrive in these adverse conditions.

Table: Abiotic Stress Resistant Potato Varieties

S No	Abiotic Stress	Tolerant Species	Resistant Varieties
1	Drought	*Solanum tuberosum, Solanum acaule*	Kufri Pukhraj, Kufri Bahar
2	Heat	*Solanum tuberosum*	Kufri Surya, Kufri Kanchan
3	Salinity	*Solanum bulbocastanum*	Kufri Naveen, Kufri Swarna

1. **Drought Tolerance:**
   - Potato is highly sensitive to drought conditions, which can cause reduced tuber formation and lower yields. Breeding for drought tolerance involves selecting varieties with deeper root systems and improved water-use efficiency. Drought-tolerant varieties like *Kufri Pukhraj* and *Kufri Bahar* are widely grown in rainfed regions.
2. **Heat Tolerance:**
   - High temperatures during the growing season can reduce potato yields and affect tuber quality. Heat-tolerant varieties like *Kufri Surya* and *Kufri Kanchan* have been developed for tropical and subtropical regions where temperatures often exceed the optimal range for potato growth.
3. **Salinity Tolerance:**
   - Soil salinity is a growing problem in many potato-producing regions, particularly in coastal areas and regions with poor irrigation practices. Salinity-tolerant varieties like *Kufri Naveen* have been developed to withstand high salt concentrations in the soil, ensuring stable yields under saline conditions.

Crop Production Techniques

Efficient crop production techniques are essential for achieving high potato yields and maintaining tuber quality. Proper management of soil, water, and nutrient inputs, as well as pest and disease control, are critical to successful potato cultivation.

1. **Soil Preparation:**
   - Potatoes grow best in well-drained, sandy loam soils with a pH range of 5.0 to 6.5. The soil should be loose and free from stones and debris to allow for proper tuber development. Compacted soils can lead to misshapen tubers and reduce yield.
   - Before planting, the soil should be deeply tilled to a depth of 30-40 cm to ensure good aeration and root penetration. Organic matter, such as compost or well-rotted manure, can be incorporated into the soil to improve its structure, moisture retention, and fertility.
2. Se**ed Tuber Selection and Plantin**g:
   - Potatoes are propagated vegetatively using seed tubers, which are small tubers or pieces of larger tubers that contain at least one "eye" or bud. Seed tubers should be certified disease-free and stored in cool, dark conditions before planting.
   - Planting is typically done in rows spaced 60-90 cm apart, with seed tubers placed 15-20 cm deep and spaced 25-30 cm apart within the row. Proper spacing is essential to ensure that the plants have enough room to develop healthy tubers.
3. **Irrigation:**
   - Potatoes require consistent moisture throughout their growing cycle, particularly

during the tuber initiation and bulking stages. Water stress during these stages can lead to smaller, lower-quality tubers.

» Drip irrigation or furrow irrigation is commonly used in potato cultivation. Drip irrigation is preferred for its water efficiency and ability to deliver moisture directly to the root zone. Overhead irrigation should be avoided, as it can increase the risk of foliar diseases such as late blight.

4. **Weed Management**:

» Weeds compete with potato plants for nutrients, water, and sunlight, reducing yields. Weed control is particularly important during the early stages of growth, when potato plants are still small and less competitive.

» Mechanical cultivation, hand weeding, and the use of herbicides are common methods of weed control in potato fields. Mulching with organic materials, such as straw or grass clippings, can also help suppress weeds and retain soil moisture.

5. **Fertilization:**

» Potatoes are nutrient-intensive and require balanced fertilization to support healthy growth and high yields. Nitrogen, phosphorus, and potassium are the primary nutrients required by potatoes.

» A typical fertilization regime involves applying 100-150 kg/ha of nitrogen, 80-100 kg/ha of phosphorus, and 150-200 kg/ha of potassium, depending on soil fertility. Organic fertilizers, such as compost or manure, can also be used to improve soil fertility and provide a slow-release source of nutrients.

6. **Hilling:**

» Hilling is the process of mounding soil around the base of potato plants as they grow. This encourages tuber development by protecting them from sunlight, which can cause greening and make the tubers inedible.

» Hilling also helps improve soil drainage and reduces the risk of tubers being exposed to the air, which can lead to spoilage. It is typically done two to three times during the growing season, as the plants reach a height of 15-20 cm.

Post-Harvest Handling

Proper post-harvest handling is essential to maintaining the quality and shelf life of potatoes. Freshly harvested potatoes are highly perishable and must be handled carefully to prevent mechanical damage, dehydration, and sprouting.

1. **Harvesting:**

» Potatoes are typically harvested when the foliage begins to die back, indicating that the tubers have reached full maturity. Harvesting is usually done 90-120 days after planting, depending on the variety and growing conditions.

» In small-scale farming operations, potatoes are harvested manually by lifting the tubers out of the soil using a fork or spade. In larger commercial operations, mechanical harvesters are used to dig up the tubers and separate them from the soil.

2. **Curing and Storage**:

» After harvesting, potatoes should be cured for 10-14 days at temperatures of 12-18°C with high humidity (85-95%) to allow the skins to toughen and any minor wounds to heal. Curing helps prevent spoilage during storage and improves the shelf life of the tubers.

» Potatoes should be stored in cool, dark conditions at temperatures between 4-10°C to prevent sprouting and spoilage. Exposure to light should be minimized to prevent greening, which occurs when chlorophyll develops in the tuber skin, producing toxic solanine compounds.

3. **Grading and Packaging:**

» After curing, potatoes are sorted and graded based on size, shape, and skin quality. Uniformity is important for marketability, with premium grades being sold for fresh consumption and lower grades used for processing or animal feed.

» Potatoes are typically packed in ventilated crates or sacks for transportation to market. The packaging should be sturdy enough to protect the tubers from mechanical damage during transit.

4. **Transportation:**

» Potatoes are transported to market or storage facilities using trucks or rail. Proper ventilation is essential during transportation to prevent the buildup of moisture and heat, which can lead to rot or sprouting.

B. Brinjal (*Solanum melongena*)

Family: Solanaceae

Chromosome Number: 2n = 2x = 24

Brinjal, commonly known as eggplant or aubergine, is a significant vegetable crop belonging to the Solanaceae family. It is cultivated for its fleshy, spongy fruit, which is used extensively in a variety of culinary dishes across Asia, Europe, and the Americas. The fruit comes in diverse shapes, sizes, and colors, ranging from small, round, and purple to long, slender, and green. In addition to its versatility in cooking, brinjal has gained recognition for its high nutritional content, being a good source of dietary fiber, vitamins (particularly B vitamins), and minerals such as potassium and magnesium.

Brinjal is also rich in antioxidants, including phenolic compounds like anthocyanins, which contribute to the pigmentation of the purple varieties. These antioxidants play a role in reducing oxidative stress in the body and promoting cardiovascular health. Due to its low caloric value and high fiber content, brinjal is commonly recommended in weight management diets.The cultivation of brinjal is widespread in tropical and subtropical regions, with India, China, and the Philippines being the leading producers. The crop thrives in warm climates and is cultivated both in open fields and under protected conditions such as greenhouses. It is a perennial plant in its native habitat but is typically grown as an annual in most agricultural systems.

Origin and Domestication

Brinjal is believed to have originated in South Asia, specifically in the regions encompassing India and Burma (now Myanmar). It is closely related to several wild species of *Solanum* that still exist in the region, which have been used in breeding programs to improve traits such as disease resistance and abiotic stress tolerance. Ancient texts, including Indian Sanskrit manuscripts, and historical Chinese records from as far back as the 5th century BC, reference brinjal, indicating its long history of cultivation and domestication in the region.

The domestication of brinjal involved selecting for larger fruit size, reduced bitterness, and improved palatability. The wild relatives of brinjal, such as *Solanum incanum*, still produce small, bitter fruits that are unsuitable for consumption. Early farmers selected plants with desirable traits, eventually developing the cultivated forms of *Solanum melongena* that we see today. In addition to India and Southeast Asia, brinjal was introduced to Europe and the Mediterranean region by traders during the Middle Ages, where it became an important crop in Mediterranean cuisine.

Related Genera and Species

Brinjal belongs to the Solanaceae family, which includes other economically significant crops like tomatoes, potatoes, and peppers. Within the genus *Solanum*, brinjal is closely related to several other species, some of which have been used in breeding programs for improving various agronomic traits.

1. **Tomato (*Solanum lycopersicum*):**
   - » Brinjal shares many physiological traits with the tomato, including similar flowering patterns and responses to environmental stresses. Both crops are susceptible to similar pests and diseases, which has prompted joint research and breeding efforts to develop resistant varieties.
2. **Wild Relatives of Brinjal:**
   - » Wild relatives of brinjal, such as *Solanum incanum*, *Solanum insanum*, and *Solanum torvum*, play an essential role in breeding programs. These species possess genes for traits like resistance to bacterial wilt, drought tolerance, and salt tolerance, which are being introgressed into cultivated brinjal varieties through backcrossing and hybridization techniques.
   - » ***Solanum torvum*** is especially important as it is widely used as a rootstock for grafting in brinjal cultivation to manage soil-borne diseases and nematode problems.
3. **Other Related Species:**
   - » The close relationship between brinjal and its wild relatives has allowed breeders to introduce useful traits from species like *Solanum aethiopicum* (scarlet eggplant), which is grown in Africa and is resistant to several biotic stresses, and *Solanum macrocarpon* (African eggplant), which contributes traits like drought tolerance and improved fruit size.

Botanical Varieties of Brinjal

Brinjal exhibits considerable variation in fruit shape, size, and color, which has led to the classification of several botanical varieties. These varieties differ in their market preferences, culinary uses, and adaptability to growing conditions.

Table 1: Major Botanical Varieties of Brinjal

S No	**Variety Name**	**Fruit Shape**	**Fruit Color**	**Region Grown**	**Uses**
1	Long Purple	Cylindrical	Dark Purple	India, Southeast Asia	Frying, grilling
2	Round Green	Round	Light Green	Thailand, Vietnam	Stir-frying, soups
3	Black Beauty	Oval	Black	Europe, North America	Roasting, baking

4	White Eggplant	Egg-shaped	White	Mediterranean, Middle East	Grilling, stuffing
5	Thai Green Pea Eggplant	Small, round	Green	Thailand, Malaysia	Curries, stir-frying
6	Japanese Long Eggplant	Slender	Purple	Japan, Korea	Steaming, stir-frying

1. **Long Purple Varieties:**
   » These varieties are characterized by long, slender fruits with smooth, glossy skin. They are widely grown in India, Southeast Asia, and parts of China. Long purple varieties are often used in frying and grilling due to their tender flesh and mild flavor.
2. **Round Green Varieties:**
   » Round green brinjals are popular in Southeast Asia, particularly in Thailand and Vietnam. These small, round fruits are used extensively in local cuisine, especially in stir-fried dishes and curries. They are known for their slightly bitter taste and firm texture.
3. **Black Beauty and Oval Varieties:**
   » *Black Beauty* is a well-known variety grown in Europe and North America. These large, oval-shaped fruits are prized for their smooth, shiny skin and mild, creamy flesh. They are often used in Mediterranean cuisine for dishes like eggplant parmesan and baba ghanoush.
4. **White Eggplant Varieties:**
   » White eggplant, with its egg-shaped, creamy-white fruits, is popular in the Mediterranean and Middle Eastern regions. It has a slightly firmer texture than purple varieties and is often used in grilling, stuffing, and as a meat substitute in vegetarian dishes.
5. **Thai Green Pea Eggplant:**
   » These small, round fruits are a staple in Southeast Asian cooking, particularly in Thai and Vietnamese cuisines. Thai green eggplant has a unique crunchy texture and slightly bitter flavor, making it perfect for use in curries and stir-fries.
6. **Japanese Long Eggplant:**
   » Japanese eggplants are slender, purple-skinned varieties with a mild flavor and tender flesh. They are commonly used in Japanese and Korean cuisines, often steamed or stir-fried and served with soy-based sauces.

Growth Habit of Brinjal

Brinjal is a perennial plant in its native tropical environment, but it is grown as an annual in most agricultural systems. It grows as a bushy, upright herb, reaching heights of 0.5 to

1.5 meters, depending on the variety and growing conditions. The plant has a fibrous root system that spreads widely in well-drained soils, supporting the development of large fruits.

The leaves of the brinjal plant are broad, with a rough, hairy texture, and are often lobed. The plant produces violet or white flowers that are hermaphroditic, containing both male and female reproductive organs. These flowers are typically borne singly or in clusters, and pollination is predominantly self-pollinating, although insect pollination can also occur.

Floral Morphology

The flowers of brinjal are solanaceous, meaning they are similar in structure to other members of the Solanaceae family, like tomatoes and peppers. Each flower is borne on a short pedicel and consists of five to six fused petals, which form a star-like appearance. The petals are usually violet, though some varieties exhibit white or pink petals.

The stamens are yellow, forming a cone around the style, which is located in the center of the flower. The ovary is superior, meaning it is located above the point where the petals and sepals attach. Once pollinated, the ovary develops into the fleshy fruit that is harvested as brinjal.

Anthesis (the opening of the flower) typically occurs in the morning, and the flowers remain receptive to pollination for about 24-36 hours. Under normal conditions, brinjal flowers self-pollinate, but cross-pollination can occur with the help of pollinators like bees. Following successful pollination, fruit set begins, and the fruit matures over 25-30 days, depending on the variety and environmental conditions.

Breeding Methods in Brinjal

Brinjal breeding aims to improve several key traits, including resistance to biotic and abiotic stresses, yield, fruit quality, and adaptability to different growing environments. Due to its long cultivation history and extensive variation in fruit characteristics, brinjal offers ample opportunities for breeders to develop improved varieties through both conventional and modern techniques.

Conventional Breeding Methods

1. **Introduction:**

   » In the introduction method, seeds from brinjal varieties in one ecological region are introduced into another for testing and evaluation. This helps determine the adaptability of introduced varieties to new environmental conditions, including resistance to pests, diseases, and abiotic stresses like drought and salinity.

   » Successful introductions have led to the spread of many brinjal varieties across continents. For example, the *Black Beauty* variety was initially bred in Europe and North America but has since been adopted in Asia and Africa due to its favorable fruit quality and yield.

2. **Selection:**

» Selection is a traditional breeding method in brinjal that involves identifying superior plants from genetically diverse populations. In mass selection, breeders select individual plants exhibiting desirable traits such as high yield, disease resistance, or superior fruit quality. These selected plants are then propagated to form the next generation.

» **Individual Plant Selection (IPS):** This involves selecting chance variants or superior plants from a mixed population, which are then propagated and evaluated over several generations to stabilize the desired traits. Selections can be based on fruit size, shape, color, and resistance to pests like the brinjal shoot and fruit borer.

Source Selections in Brinjal:

S No	Variety Name	Source	Features
1	Arka Navneet	Indian Institute of Horticultural Research (IIHR)	High yield, round purple fruits
2	Pusa Kranti	Indian Agricultural Research Institute (IARI)	High yield, medium-long fruits, purple
3	Pusa Purple Cluster	IARI	Early-maturing, cluster-bearing fruits
4	Pant Samrat	G.B. Pant University of Agriculture	Long, cylindrical fruits, high yield
5	Arka Nidhi	IIHR	Resistant to shoot and fruit borer

3. **Hybridization (Heterosis Breeding):**

» Hybridization is widely used to combine favorable traits from different parental lines. Heterosis or hybrid vigor is exploited in brinjal to improve yield, fruit quality, and resistance to pests and diseases. Hybrids typically show enhanced performance compared to their parental lines.

» Hybrid varieties such as *Pusa Hybrid 6* and *Mahyco Hybrid* have been developed for commercial cultivation. These hybrids produce uniform fruits with improved disease resistance and better marketability.

Examples of Brinjal Hybrids:

S No	Hybrid Name	Parentage	Features
1	Pusa Hybrid 6	Pusa Kranti × Local Selection	High yield, early maturing, purple fruits
2	Mahyco Hybrid	Arka Nidhi × Local Line	Resistant to shoot and fruit borer
3	Arka Navneet Hybrid	IIHR Line × Local Line	High yield, round fruits
4	Pant Hybrid-1	Pant Samrat × Local Long Variety	Long cylindrical fruits, high yield

4. **Pedigree Method:**
   - The pedigree method is commonly employed in brinjal breeding, especially when breeding for complex traits like disease resistance and yield. In this method, plants from hybrid populations are selfed and their offspring evaluated over successive generations. Selection is continued for desirable traits, including fruit quality and pest resistance.
5. **Backcross Method:**
   - The backcross method is used when breeders want to introduce a specific trait, such as resistance to bacterial wilt, into a well-established brinjal variety. In this process, a resistant variety is repeatedly crossed with the recurrent parent variety to retain the original characteristics of the variety while introducing the desired resistance trait.
6. **Mutation Breeding:**
   - Mutation breeding involves inducing genetic changes using physical or chemical mutagens to create new variations in brinjal. This method has led to the development of unique varieties with improved resistance to pests and diseases, as well as enhanced fruit quality. For example, varieties like *Pant Rituraj* have been developed through induced mutation to show early maturity and pest tolerance.

Non-Conventional Breeding Methods

1. **Tissue Culture:**
   - Tissue culture is employed in brinjal for the rapid multiplication of disease-free plants. Micropropagation allows the production of large numbers of uniform plants in a short period, which is particularly useful for maintaining pure lines in hybrid seed production. Somatic hybridization has also been explored in brinjal, allowing the combination of traits from different *Solanum* species.
2. **Genetic Engineering (Recombinant DNA Technology):**
   - Genetic engineering has emerged as a key tool in modern brinjal breeding. Genetically modified (GM) brinjal varieties, such as Bt brinjal, have been developed to confer resistance to major pests like the brinjal shoot and fruit borer (*Leucinodesorbonalis*). Bt brinjal contains a gene from the bacterium *Bacillus thuringiensis* (Bt), which produces a protein toxic to the borer larvae but safe for human consumption. This technology has reduced the need for chemical pesticides and increased yields in regions affected by heavy pest infestations.

Breeding for Biotic Stresses

Brinjal is susceptible to a wide range of biotic stresses, including fungal, bacterial, and viral diseases, as well as insect pests. Breeding for resistance to these stresses is a priority in ensuring sustainable brinjal production.

Table: Biotic Stress Resistance in Brinjal Varieties

S No	Disease/Pest	Source Species	Resistant Varieties
1	Bacterial Wilt	*Solanum torvum*	Pusa Purple Round, Arka Nidhi
2	Brinjal Shoot & Fruit Borer	*Solanum melongena* (wild relatives)	Pusa Hybrid 6, Arka Navneet
3	Phomopsis Blight	*Solanum integrifolium*	Pant Rituraj, Arka Nidhi
4	Little Leaf Disease	*Solanum aethiopicum*	Pusa Purple Cluster, Pusa Kranti

1. **Bacterial Wilt Resistance:**

   » Bacterial wilt caused by *Ralstonia solanacearum* is one of the most destructive diseases in brinjal cultivation. Breeding for resistance to bacterial wilt has involved the use of wild relatives like *Solanum torvum* as rootstocks for grafting. Grafted plants have shown considerable resistance to bacterial wilt and are widely adopted in regions with high disease pressure.

2. **Brinjal Shoot and Fruit Borer Resistance:**

   » The brinjal shoot and fruit borer (*Leucinodesorbonalis*) is the most significant pest of brinjal, causing severe damage to both shoots and fruits. Breeding programs focus on developing varieties that are resistant to this pest through the use of wild relatives or genetic engineering techniques such as Bt brinjal.

3. **Phomopsis Blight Resistance:**

   » Phomopsis blight caused by *Phomopsis vexans* is a fungal disease that affects brinjal plants, leading to leaf spots, fruit rot, and stem cankers. Resistance to Phomopsis blight has been introduced into commercial varieties by selecting for tolerance in wild relatives like *Solanum integrifolium.*

4. **Little Leaf Disease Resistance:**

   » Little leaf disease is caused by phytoplasma infections, leading to stunted growth and reduced fruit set. Resistance to little leaf disease has been identified in African brinjal species like *Solanum aethiopicum*, and these traits have been introgressed into cultivated brinjal varieties.

Breeding for Abiotic Stresses

Abiotic stresses, such as drought, salinity, and extreme temperatures, pose significant challenges in brinjal cultivation. Breeding programs focus on developing varieties that can tolerate these conditions without compromising yield or fruit quality.

Table: Abiotic Stress Resistant Brinjal Varieties

S No	Abiotic Stress	Tolerant Species	Resistant Varieties
1	Drought	*Solanum torvum, Solanum aethiopicum*	Arka Nidhi, Pusa Kranti
2	Salinity	*Solanum melongena* (wild types)	Pant Samrat, Pusa Purple Cluster
3	Heat Tolerance	*Solanum melongena*	Pant Hybrid-1, Arka Navneet

1. **Drought Tolerance:**
   » Drought tolerance is an important trait for brinjal grown in arid and semi-arid regions. Varieties such as *Arka Nidhi* and *Pusa Kranti* have been bred for improved water-use efficiency and deeper root systems, allowing them to maintain productivity under water-limited conditions.
2. **Salinity Tolerance:**
   » Soil salinity is a growing concern in coastal regions where brinjal is cultivated. Breeding programs have focused on selecting varieties that can withstand high soil salt concentrations without exhibiting reductions in yield or fruit quality. Varieties like *Pant Samrat* and *Pusa Purple Cluster* are examples of salt-tolerant brinjal cultivars.
3. **Heat Tolerance:**
   » High temperatures can negatively impact brinjal fruit set and quality. Heat-tolerant varieties like *Pant Hybrid-1* and *Arka Navneet* have been developed to perform well under hot conditions, making them suitable for cultivation in tropical regions.

Floral Biology of Brinjal

Brinjal exhibits a typical floral morphology of the Solanaceae family, with perfect, hermaphroditic flowers that contain both male and female reproductive organs. The flower structure plays a significant role in pollination and fruit set, impacting the yield of the crop. The knowledge of floral biology is essential for breeding programs and seed production techniques.

1. **Flower Structure:**
   » The flowers of brinjal are generally violet to light purple, although some varieties exhibit white or pinkish hues. They are arranged in solitary or cymose inflorescences, with each flower comprising five petals fused at the base to form a star-shaped corolla.
   » The calyx, consisting of five sepals, is persistent and enlarges as the fruit develops. This persistent calyx helps protect the developing fruit from pests and diseases.
2. **Stamens and Anthers:**
   » Brinjal flowers have five stamens with yellow anthers, which are closely packed together, forming a cone around the central style. The anthers release pollen through longitudinal slits, and pollination typically occurs within 24-36 hours after anthesis.
   » In most varieties, the stigma is positioned slightly below or at the same level as

the anthers, promoting self-pollination. However, cross-pollination can also occur, especially with the assistance of pollinators such as bees.

3. **Ovary and Style:**
   - The ovary of brinjal is superior, with a single locule containing numerous ovules. Following successful pollination and fertilization, the ovary develops into the fleshy fruit. The style is slender and elongates after anthesis, with the stigma becoming receptive to pollen during this period.
   - The floral structure is generally adapted for self-pollination, although cross-pollination rates can vary depending on environmental factors and the presence of pollinators.
4. **Pollination and Fruit Set:**
   - Pollination in brinjal typically occurs in the morning, between 6 AM and 10 AM, when the flowers are fully open, and the pollen is viable. Under optimal conditions, fruit set occurs readily after self-pollination, but hand pollination is sometimes employed in seed production programs to enhance fruit set in controlled environments.
   - The presence of adequate pollinators, such as honeybees or bumblebees, can increase the fruit set rate in open-field conditions. In protected cultivation, techniques like gentle vibration of the flower trusses or manual pollination are used to ensure effective pollination.

Breeding Objectives for Brinjal

Brinjal breeders focus on several key objectives, aiming to improve both yield and quality while addressing the challenges posed by pests, diseases, and environmental stresses. The primary breeding objectives include improving fruit characteristics, enhancing biotic and abiotic stress resistance, and increasing overall productivity.

1. **Improved Yield:**
   - Yield improvement is a central objective in brinjal breeding. Breeders aim to develop varieties that produce high marketable yields under diverse agro-climatic conditions. Factors influencing yield include the number of fruits per plant, fruit size, and resistance to environmental stresses.
   - Yield is also influenced by the plant's ability to resist common pests and diseases, which can drastically reduce production. High-yielding hybrids, such as *Pusa Hybrid 6* and *Arka Navneet*, have been developed to address these concerns.
2. **Fruit Quality:**
   - Fruit quality is a critical aspect of brinjal breeding, especially as consumer preferences vary across different regions. Breeders work to improve fruit size, shape, color, and texture to meet market demands.

» Varieties with longer shelf life and resistance to post-harvest diseases are also preferred by farmers and retailers. Other important fruit quality traits include high nutritional content, minimal bitterness, and firm flesh that withstands transport.

3. **Resistance to Biotic Stresses:**

» Breeding for resistance to pests such as the brinjal shoot and fruit borer, bacterial wilt, and Phomopsis blight is a priority. These biotic stresses can significantly reduce yields and fruit quality. Modern breeding programs have successfully introduced resistance to these diseases using both conventional and genetic engineering methods, such as Bt brinjal.

Table: Biotic Stress-Resistant Varieties

S No	Disease/Pest	Source Species	Resistant Varieties
1	Bacterial Wilt	*Solanum torvum*	Pusa Purple Round, Arka Nidhi
2	Brinjal Shoot and Fruit Borer	Bt gene introgression	Bt Brinjal, Mahyco Hybrid
3	Phomopsis Blight	*Solanum integrifolium*	Pant Rituraj, Arka Nidhi
4	Little Leaf Disease	*Solanum aethiopicum*	Pusa Purple Cluster, Pusa Kranti

4. **Resistance to Abiotic Stresses:**

» Abiotic stresses such as drought, salinity, and temperature extremes can negatively impact brinjal production. Breeding programs focus on developing varieties that can thrive under these challenging conditions.

» Varieties with improved drought tolerance, salinity tolerance, and resistance to high temperatures are particularly important for cultivation in arid and semi-arid regions.

Table: Abiotic Stress-Resistant Brinjal Varieties

S No	Abiotic Stress	Tolerant Species	Resistant Varieties
1	Drought	*Solanum torvum, Solanum aethiopicum*	Arka Nidhi, Pusa Kranti
2	Salinity	*Solanum melongena* (wild types)	Pant Samrat, Pusa Purple Cluster
3	Heat Tolerance	*Solanum melongena*	Pant Hybrid-1, Arka Navneet

5. **Nutritional Quality and Biofortification:**

» Brinjal is an important source of dietary fiber, vitamins, and minerals. Breeding programs have started focusing on biofortification to increase the content of essential nutrients such as vitamins A, C, and antioxidants like anthocyanins in purple-skinned varieties.

» Biofortified brinjal varieties have higher nutritional value, making them more appealing to health-conscious consumers. These varieties also offer potential solutions to address micronutrient deficiencies in developing countries.

Post-Harvest Handling and Storage of Brinjal

Proper post-harvest handling is crucial to maintaining the quality and shelf life of brinjal, as the fruit is highly perishable. Poor handling can lead to significant losses due to mechanical damage, disease, and spoilage. Post-harvest management practices include proper harvesting techniques, storage conditions, and transportation methods to minimize losses.

1. **Harvesting Techniques:**
   - » Brinjal fruits are harvested when they reach full maturity but before they become overripe. Depending on the variety, fruits can be harvested at different stages of maturity to suit market demands. Early-maturing varieties are typically harvested within 50-70 days after transplanting, while late-maturing varieties may take up to 90 days.
   - » Harvesting is typically done manually using a sharp knife or scissors to avoid damaging the plant. Care must be taken not to bruise or scratch the fruits during harvest, as this can lead to post-harvest diseases.
2. **Curing and Grading:**
   - » After harvesting, brinjal fruits are cured in a cool, shaded area to allow any minor mechanical injuries to heal. This reduces the likelihood of rot during storage. Curing usually takes 24-48 hours.
   - » Grading is performed based on fruit size, color, and quality. Uniformly sized fruits with glossy skin are preferred for the fresh market. Grading also helps to remove any damaged or diseased fruits before they enter the supply chain.
3. **Storage Conditions:**
   - » Brinjal is highly sensitive to chilling injury and should not be stored at temperatures below 10°C. Ideal storage conditions for brinjal are between 10°C and 12°C with high relative humidity (85-90%). Under these conditions, brinjal can be stored for up to two weeks.
   - » In large-scale production systems, refrigerated storage units are used to extend the shelf life of brinjal fruits. Modified atmosphere packaging (MAP) has also been explored as a way to reduce spoilage and extend freshness.
4. **Transportation:**
   - » Brinjal fruits are highly perishable and must be transported quickly and under optimal conditions to reduce post-harvest losses. Proper packaging in crates or ventilated cartons is essential to prevent mechanical damage during transport.
   - » Transportation in cool trucks is recommended for long distances to maintain the freshness of the fruit. For local markets, brinjal can be transported in non-refrigerated vehicles, but care must be taken to avoid exposure to direct sunlight or high temperatures.

C. Tomato (*Solanum lycopersicum*)

Family: Solanaceae

Chromosome Number: 2n = 2x = 24

Tomato is one of the most widely cultivated vegetable crops globally, used both as fresh produce and in processed forms like sauces, ketchup, and canned goods. Known for its high content of vitamins, especially Vitamin C, A, and K, as well as minerals like potassium and magnesium, tomato is a key contributor to human nutrition. The fruit contains powerful antioxidants, including lycopene, which has been studied for its potential health benefits, particularly in reducing the risk of cancer and cardiovascular diseases. India ranks second in tomato production globally, following China, with increasing demand both in domestic markets and for export.

Origin and Domestication

Tomato is believed to have originated in the Andean region of South America, particularly in areas now part of Peru, Ecuador, and northern Chile. Wild relatives of the cultivated tomato (*Solanum pimpinellifolium* and *Solanum peruvianum*) can still be found in these regions, offering genetic diversity valuable for modern breeding programs. Domestication likely began in Mexico, where early cultivators were selected for larger fruit size and improved flavor. Over time, tomato spread to Europe through Spanish explorers in the 16th century and later to Asia and Africa, where it became an integral part of regional cuisines. The process of domestication brought significant changes to the size, shape, and color of the tomato fruit. In wild species, the fruit is typically small and round, while modern cultivars exhibit a variety of shapes, including round, oblong, and pear-shaped fruits. Domesticated varieties also show reduced bitterness and increased sugar content, enhancing their appeal for human consumption.

Related Genera and Species

Tomato belongs to the *Solanum* genus, which also includes other important crops like potato (*Solanum tuberosum*) and eggplant (*Solanum melongena*). The tomato plant shares many genetic and physiological characteristics with its relatives, making it possible to transfer useful traits between species through traditional and modern breeding techniques. Several wild species of *Solanum* have been utilized in tomato breeding programs to introduce traits like disease resistance, drought tolerance, and improved fruit quality.

Table: Wild Relatives of Tomato and Their Uses in Breeding

S No	Species Name	Key Traits	Contribution to Breeding
1	*Solanum pimpinellifolium*	Resistance to late blight and powdery mildew	Disease resistance, improved fruit set
2	*Solanum peruvianum*	Drought and salt tolerance	Abiotic stress tolerance
3	*Solanum cheesmaniae*	Heat tolerance, pest resistance	Heat and pest tolerance
4	*Solanum chilense*	Cold tolerance	Cold and drought resistance

1. **Wild Relatives of Tomato:**

 » *Solanum pimpinellifolium* is a small-fruited species found in coastal areas of Peru and Ecuador. It has been widely used in breeding programs to introduce resistance to fungal diseases like late blight and powdery mildew.

 » *Solanum peruvianum*, a drought-tolerant species, has contributed genes for improving tomato's resistance to saline and arid conditions.

 » *Solanum cheesmaniae* is native to the Galapagos Islands and has been used to breed heat-tolerant tomatoes that can withstand higher temperatures, making them suitable for cultivation in tropical and subtropical regions.

Botanical Varieties of Tomato

Tomato exhibits significant morphological diversity, with fruits varying in shape, size, color, and flavor. Several botanical varieties have been classified based on these characteristics, offering distinct advantages for different uses, whether for fresh consumption or processing.

Table: Major Types of Tomato Varieties on Basis of Fruit Shape, Colour and Their Uses

S No	Variety Name	Fruit Shape	Fruit Color	Region Grown	Uses
1	*Solanum lycopersicum var. commune*	Round	Red	Global	Fresh consumption, processing
2	*Solanum lycopersicum var. pyriforme*	Pear-shaped	Red/ Yellow	Mediterranean	Sauces, canning
3	*Solanum lycopersicum var. cerasiforme*	Cherry	Red/ Yellow	Europe, North America	Salads, garnishing
4	*Solanum lycopersicum var. grandifolium*	Large-leaved	Red	India, Southeast Asia	Processing, fresh market
5	*Solanum lycopersicum var. validum*	Upright	Red	North America, Europe	Fresh consumption, cooking

1. **Round Varieties (*Solanum lycopersicum var. commune*):**
   - This is the most commonly grown type of tomato, characterized by its round shape and red skin. It is used extensively for both fresh consumption and processing into products like ketchup and sauces.
2. **Pear-Shaped Varieties (*Solanum lycopersicum var. pyriforme*):**
   - Pear-shaped tomatoes are popular in Mediterranean cuisine, especially for making sauces and canned products. They tend to have a high pulp-to-seed ratio, making them ideal for processing.
3. **Cherry Tomato Varieties (*Solanum lycopersicum var. cerasiforme*):**
   - Cherry tomatoes are small, round, and come in a variety of colors, including red, yellow, and orange. They are often used in salads and as garnishes due to their sweet flavor and attractive appearance.
4. **Large-Leaved Varieties (*Solanum lycopersicum var. grandifolium*):**
   - These varieties are characterized by their large leaves and medium to large fruits. They are popular in India and Southeast Asia, where they are used for both fresh consumption and processing.
5. **Upright Varieties (*Solanum lycopersicum var. validum*):**
   - These varieties grow upright and produce firm, medium-sized fruits. They are commonly cultivated in North America and Europe for fresh consumption and cooking.

Growth Habit of Tomato

Tomato plants exhibit different growth habits, which influence their cultivation methods and suitability for different growing environments. Understanding the growth habit of tomato varieties is essential for maximizing yield and managing the crop effectively.

1. **Determinate vs. Indeterminate Growth:**
   - **Determinate Varieties:** In determinate tomato varieties, the plants grow to a fixed height and stop growing once they produce a terminal flower cluster. These plants tend to be bushier, with fruits ripening simultaneously over a short period, making them ideal for mechanical harvesting and processing.
   - **Indeterminate Varieties:** Indeterminate varieties continue to grow and produce fruit throughout the growing season. These plants have a more vine-like growth habit, and the fruits ripen over an extended period. Indeterminate tomatoes are preferred for fresh market production because they offer a continuous harvest.
2. **Semi-Determinate Varieties:**
   - Semi-determinate varieties exhibit characteristics of both determinate and indeterminate plants. They produce multiple fruit clusters along the stem, like

indeterminate plants, but eventually terminate in a flower cluster, limiting further growth. These varieties offer a balance between continuous fruit production and ease of management.

Floral Biology of Tomato

Tomato flowers are solanaceous, exhibiting a structure typical of the Solanaceae family. The knowledge of floral biology is crucial for breeding programs and for understanding pollination mechanisms that affect fruit set.

1. **Flower Structure:**
   » The flowers of the tomato plant are small, yellow, and borne in clusters called inflorescences. Each flower consists of five to six petals that form a star-shaped corolla. The flowers are perfect (hermaphroditic), containing both male and female reproductive organs.
2. **Stamens and Anthers:**
   » The stamens of tomato flowers are fused together to form a cone around the pistil. The anthers release pollen through longitudinal slits, which typically occurs in the early morning hours. The flower structure is primarily adapted for self-pollination, though cross-pollination can occur with the help of pollinators like bees.
3. **Ovary and Style:**
   » The ovary of the tomato flower is superior, and the style elongates as the flower matures. Pollination usually occurs soon after the flower opens, and successful pollination leads to the development of the fruit.
4. **Pollination and Fruit Set:**
   » Pollination in tomatoes is usually self-pollination, with the pollen falling from the anther cone onto the stigma. However, cross-pollination can enhance genetic diversity, especially in breeding programs. Hand pollination is sometimes employed in greenhouse settings to improve fruit set, especially in the absence of natural pollinators.

Breeding Methods in Tomato

Tomato breeding focuses on enhancing yield, improving fruit quality, and increasing resistance to both biotic and abiotic stresses. Over the years, several breeding techniques have been employed to develop superior tomato varieties that meet the demands of both farmers and consumers.

Conventional Breeding Methods

1. **Introduction:**
   » Introduction is one of the oldest methods of crop improvement, where varieties from one region are introduced into another for evaluation. Several tomato introductions,

such as *Roma* and *Marglobe*, became successful due to their adaptability to new environments. These varieties are often tested for their performance under local conditions, including their resistance to diseases, pests, and climate stresses.

2. **Selection:**

   » In tomato breeding, both mass selection and individual plant selection are widely used. This involves selecting superior plants from a genetically diverse population based on desirable traits such as fruit size, yield, resistance to pests, and adaptability to environmental conditions. The selected plants are propagated and further evaluated in subsequent generations.

Source Selections in Tomato:

S No	Variety Name	Source	Features
1	Pusa Ruby	IARI	Indeterminate, early, firm, uniform ripening
2	Pusa Early Dwarf	IARI	Determinate, early maturing, medium size
3	Arka Meghali	IIHR	Resistant to bacterial wilt, good fruit size
4	Pant T3	GBPUA&T, Pantnagar	Semi-determinate, round, smooth fruit
5	Arka Vikas	IIHR	Tolerant to moisture stress, high yield

3. **Hybridization (Heterosis Breeding):**

   » Hybridization is a widely used method to combine favorable traits from two or more parents. Hybrid varieties are often developed to exhibit heterosis or hybrid vigor, resulting in increased yield, uniformity in fruit size, and improved resistance to pests and diseases. In tomatoes, hybrids like *Pusa Hybrid 4* and *Arka Rakshak* have been developed to meet the market's demand for high-yielding, disease-resistant varieties.

Examples of Hybrid Varieties:

S No	Hybrid Name	Parentage	Features
1	Pusa Hybrid 4	Pusa Early Dwarf × Local Line	High yield, determinate, good fruit set
2	Arka Rakshak	IIHR Hybrid	Resistant to bacterial wilt, early blight
3	Pant Hybrid 2	Pant Early × Local Line	High yield, heat tolerant
4	Arka Vishal	IIHR Hybrid	Large fruit, good for processing

4. **Pedigree Method:**

   » Pedigree selection involves crossing two parents with desirable traits, followed by selection in the segregating generations (F2, F3, etc.). In this method, breeders track the inheritance of specific traits through successive generations. For example, breeding for disease resistance, like resistance to Fusarium wilt, is often achieved using the pedigree method.

5. **Bulk Method:**

   » In the bulk method, a large number of plants from a segregating population are grown together without selection, allowing for natural selection to act on the population. This method is useful when breeding for complex traits such as yield, where multiple genes are involved.

6. **Backcross Method:**

   » The backcross method is used to transfer a specific desirable trait from a donor parent to an established variety. In tomato, this method has been effectively used to introduce disease resistance traits, such as resistance to root-knot nematodes or bacterial wilt, into otherwise high-yielding commercial varieties.

7. **Mutation Breeding:**

   » Mutation breeding involves using physical or chemical mutagens to create new genetic variations. In tomato, induced mutations have led to the development of varieties like *Pusa Lal Meeruti*, which is resistant to multiple diseases and offers improved yield and quality.

Non-Conventional Breeding Methods

1. **Tissue Culture:**

   » Tissue culture techniques, including micropropagation, are widely used for producing disease-free planting material and for maintaining pure lines in hybrid seed production. Tomato tissue culture is particularly useful for producing uniform plants that are free from viral infections. Somatic hybridization through protoplast fusion has also been explored in tomato breeding to combine desirable traits from different species.

2. **Genetic Engineering (Recombinant DNA Technology):**

   » Genetic engineering has allowed breeders to introduce specific traits into tomato varieties that are difficult to achieve through conventional breeding. For example, Bt tomatoes have been developed by introducing the *Bacillus thuringiensis* (Bt) gene, which provides resistance to insects like the tomato fruit borer (*Helicoverpaarmigera*). This reduces the need for chemical pesticides and improves yield.

Breeding for Biotic Stresses in Tomato

Tomato is highly susceptible to a wide range of biotic stresses, including fungal, bacterial, and viral diseases, as well as insect pests. Breeding for resistance to these stresses is critical to ensuring stable yields and high-quality produce.

Table: Biotic Stress-Resistant Varieties

S No	Disease/Pest	Source Species	Resistant Varieties
1	Late Blight	*Solanum pimpinellifolium*	Arka Rakshak, Pant T3
2	Fusarium Wilt	*Solanum lycopersicum*	Pusa Ruby, Arka Meghali
3	Tomato Leaf Curl Virus	*Solanum peruvianum*	Pusa Early Dwarf, Arka Abha
4	Root-Knot Nematode	*Solanum lycopersicum* (wild types)	Arka Vikas, Pusa Hybrid 2
5	Bacterial Wilt	*Solanum torvum*	Arka Meghali, Pant Hybrid 1

1. **Late Blight Resistance:**

   » Late blight, caused by *Phytophthora infestans*, is one of the most devastating diseases of tomato. Resistance to late blight has been introduced into commercial varieties using genes from wild relatives like *Solanum pimpinellifolium* and *Solanum peruvianum*. Breeding for durable resistance remains a priority as the pathogen evolves rapidly.

2. **Fusarium Wilt Resistance:**

   » Fusarium wilt, caused by *Fusarium oxysporum*, is a soil-borne fungal disease that affects tomato roots. Resistant varieties like *Pusa Ruby* and *Arka Meghali* have been developed through conventional breeding methods. Resistance to Fusarium wilt is often linked to the presence of specific R genes, which confer immunity to the fungus.

3. **Tomato Leaf Curl Virus (ToLCV) Resistance:**

   » ToLCV is a major viral disease affecting tomatoes in tropical and subtropical regions. The virus is transmitted by whiteflies and can cause significant yield losses. Breeding for resistance has involved using wild species like *Solanum peruvianum* as donors of resistance genes. Varieties like *Pusa Early Dwarf* and *Arka Abha* have shown strong resistance to ToLCV.

4. **Root-Knot Nematode Resistance:**

   » Root-knot nematodes (*Meloidogyne spp.*) are microscopic parasites that infect the roots of tomato plants, causing galls and reducing water and nutrient uptake. Resistant varieties like *Arka Vikas* and *Pusa Hybrid 2* have been developed using both conventional breeding and biotechnological approaches.

5. **Bacterial Wilt Resistance:**

   » Bacterial wilt, caused by *Ralstonia solanacearum*, is another significant disease in tomato cultivation, particularly in warm and humid regions. Wild relatives like *Solanum torvum* have been used to introduce resistance genes into cultivated varieties like *Arka Meghali* and *Pant Hybrid 1*.

Breeding for Abiotic Stresses in Tomato

Abiotic stresses such as drought, salinity, and extreme temperatures have a profound impact on tomato yields, especially in regions prone to adverse weather conditions. Breeding for abiotic stress tolerance is essential for ensuring stable yields in such environments.

Table: Abiotic Stress-Resistant Varieties

S No	Abiotic Stress	Tolerant Species	Resistant Varieties
1	Drought	*Solanum peruvianum, Solanum cheesmaniae*	Arka Vikas, Pusa Sadabahar
2	Heat Tolerance	*Solanum pimpinellifolium*	Pant Hybrid 2, Arka Rakshak
3	Salinity	*Solanum cheesmaniae, Solanum chilense*	Arka Meghali, Pusa Ruby

1. **Drought Tolerance:**
   » Drought tolerance is a key breeding objective in tomato, especially for cultivation in arid and semi-arid regions. Varieties like *Arka Vikas* and *Pusa Sadabahar* have been bred to exhibit improved water-use efficiency, allowing them to maintain yields under water-limited conditions.

2. **Heat Tolerance:**
   » High temperatures can significantly affect tomato fruit set, leading to reduced yields. Heat-tolerant varieties like *Pant Hybrid 2* and *Arka Rakshak* have been developed using genes from wild species such as *Solanum pimpinellifolium*. These varieties perform well under tropical and subtropical conditions.

3. **Salinity Tolerance:**
   » Saline soils can hinder tomato growth and reduce fruit quality. Breeding for salinity tolerance has focused on selecting varieties that can withstand high soil salt concentrations. Varieties like *Arka Meghali* and *Pusa Ruby* have shown good tolerance to saline conditions without compromising yield.

Breeding for Quality Traits in Tomato

In tomato breeding, fruit quality traits are a key focus, especially for varieties grown for fresh consumption, processing, and export markets. Consumers often prioritize characteristics such as flavor, color, texture, and nutritional content, while processors look for tomatoes with high solids content and resistance to mechanical damage during harvesting and transportation. Breeding for improved quality traits ensures market competitiveness and consumer satisfaction.

1. **Flavor and Taste:**
   » The flavor of tomatoes is influenced by a balance between sugar and acid content. Breeding programs aim to develop varieties with an optimal sugar-to-acid ratio, enhancing both sweetness and tanginess. Soluble solids, which include sugars like

glucose and fructose, contribute to the sweetness, while organic acids such as citric acid determine the tartness of the fruit.

- » Genetic variation in sugar and acid content among tomato varieties has allowed breeders to select for improved flavor profiles. Heirloom tomatoes like *Brandywine* are known for their rich flavor, while modern hybrids such as *Pusa Hybrid 4* offer a balance of sweetness and acidity that appeals to both fresh market consumers and processing industries.

2. **Color:**

- » The red color of tomatoes is primarily due to the pigment lycopene, a powerful antioxidant with potential health benefits. Breeding for high lycopene content has become a priority in the development of nutritionally enhanced tomatoes. Varieties like *Pusa Gaurav* are rich in lycopene and are often recommended for use in health-conscious markets.
- » Yellow and orange tomatoes, which contain lower levels of lycopene but are high in beta-carotene (a precursor to Vitamin A), are also popular in niche markets. Breeding programs have successfully developed varieties with diverse fruit colors, such as *Pant Bahar* (yellow) and *Tangerine* (orange), to cater to consumer preferences for novelty and visual appeal.

3. **Texture and Firmness:**

- » Tomato texture is important for both fresh consumption and processing. Consumers prefer firm, juicy tomatoes for salads and sandwiches, while processors require varieties with thicker flesh and higher dry matter content to withstand mechanical handling and long-distance transport.
- » Firmness is closely related to the presence of specific genes like the *Nor* gene, which controls fruit ripening. Varieties with non-ripening or delayed-ripening traits, such as *Arka Vikas*, are favored in the processing industry as they have a longer shelf life and are less prone to bruising during handling.

4. **Nutritional Content:**

- » Tomatoes are a rich source of vitamins (A, C, and K) and minerals (potassium, calcium). In recent years, biofortification efforts have been directed at increasing the levels of these essential nutrients. Varieties like *Pusa Hybrid 1* have been developed with enhanced Vitamin C content, while breeding for higher beta-carotene and lycopene content is ongoing.
- » The inclusion of biofortified tomatoes in diets can help address micronutrient deficiencies, particularly in developing countries where tomatoes are a staple food.

5. **Processing Quality:**

    » Tomatoes destined for processing must have specific traits that make them suitable for canning, sauce-making, or drying. These traits include high soluble solids content (Brix), thick flesh, and reduced seed content. Processing tomatoes such as *Roma* and *Pusa Ruby* are bred to meet these requirements, ensuring efficient production of tomato paste, sauces, and canned products.

Table: Key Quality Traits in Tomato Varieties

S No	Variety Name	Key Quality Traits	Main Use
1	Pusa Gaurav	High lycopene content, firm texture	Fresh market, health food
2	Arka Vikas	Thick flesh, high solids	Processing, long-distance transport
3	Pant Bahar	Yellow color, sweet flavor	Niche market, fresh consumption
4	Pusa Hybrid 1	High Vitamin C, good fruit set	Fresh market, health food
5	Roma	High Brix content, low seed count	Canning, sauce-making

Post-Harvest Management of Tomato

Tomato is a highly perishable crop, and post-harvest losses can be significant if proper management practices are not followed. Post-harvest handling, storage, and transportation are crucial for maintaining fruit quality, reducing spoilage, and ensuring profitability for farmers and distributors.

1. **Harvesting Techniques:**

    » Tomatoes should be harvested at the correct stage of maturity to ensure optimal flavor and shelf life. For fresh market tomatoes, the fruits are typically harvested at the breaker stage, when they begin to change color from green to red. For processing tomatoes, harvesting is done when the fruit reaches full maturity to maximize solids content.

    » Harvesting is often done manually, especially in smallholder farms, to minimize mechanical damage. However, in large-scale operations, mechanical harvesters are used, particularly for determinate varieties bred for simultaneous fruit ripening.

2. **Grading and Sorting:**

    » After harvesting, tomatoes are sorted based on size, color, and quality. Grading ensures that uniform fruits are packed for the market, which increases their appeal to consumers and reduces the chances of post-harvest diseases spreading from damaged or diseased fruits.

    » Sorting is typically done by hand in smaller farms, but larger operations employ automated grading machines that can sort tomatoes according to size and color using advanced optical technologies.

3. **Storage and Curing:**

 » Tomatoes are highly sensitive to temperature fluctuations, and improper storage can lead to chilling injuries or accelerated ripening. The optimal storage temperature for tomatoes is between 10°C and 13°C with high relative humidity (85-90%). This helps extend shelf life without compromising fruit flavor or texture.

 » Tomatoes should not be stored at temperatures below 10°C, as this can cause chilling injury, leading to reduced flavor and texture. In larger operations, tomatoes are often stored in cold storage facilities or under modified atmosphere packaging (MAP) to reduce ethylene production and delay ripening.

4. **Packaging and Transportation:**

 » Proper packaging is essential to prevent mechanical damage and extend the shelf life of tomatoes during transportation. Tomatoes for the fresh market are typically packed in ventilated cartons or plastic crates that allow air circulation and prevent bruising.

 » For long-distance transportation, temperature-controlled trucks are used to maintain the optimal storage environment. Tomatoes destined for processing are usually transported in bulk containers, where they are less susceptible to damage due to their firmer texture and thicker skin.

5. **Minimizing Post-Harvest Losses:**

 » Post-harvest losses in tomato production can reach up to 30% in some regions, primarily due to improper handling, storage, and transportation. Training farmers and distributors in proper post-harvest techniques can significantly reduce these losses and improve overall profitability.

 » Innovations in packaging, such as biodegradable films that can reduce moisture loss and delay ripening, are being developed to address these challenges. Additionally, the use of ethylene inhibitors during storage can help control ripening and reduce spoilage.

Table: Post-Harvest Management Techniques for Tomato

S No	Post-Harvest Stage	Best Practices	Impact on Quality
1	Harvesting	Harvest at breaker or mature stage	Enhances shelf life, better flavor
2	Grading and Sorting	Remove damaged or diseased fruits	Reduces spoilage, improves market appeal
3	Storage	Store at 10-13°C, 85-90% humidity	Extends shelf life, prevents chilling injury
4	Packaging	Use ventilated cartons or crates	Minimizes mechanical damage
5	Transportation	Temperature-controlled trucks	Maintains quality over long distances

Tomato Varietal Development and Seed Production

Tomato varietal development focuses on combining traits like high yield, disease resistance, and superior fruit quality. Seed production is a critical component of the tomato industry, as hybrid seeds are widely used for commercial cultivation.

1. **Varietal Development:**

   » Developing a new tomato variety involves several years of breeding and testing. Breeders select parental lines with desirable traits, cross them, and evaluate the resulting hybrids over multiple growing seasons. The goal is to develop stable varieties that perform consistently across different environments.

   » Modern breeding programs incorporate molecular markers to accelerate the selection process. Marker-assisted selection (MAS) allows breeders to identify plants with desired genetic traits early in the breeding process, reducing the time required to develop new varieties.

2. **Hybrid Seed Production:**

   » Hybrid seeds are produced by crossing two genetically distinct parental lines. The seeds from this cross (F1 generation) exhibit hybrid vigor, resulting in plants with improved performance over the parents. Hybrid seed production requires strict isolation of the parental lines to prevent cross-pollination and ensure seed purity.

   » Seed production is typically done using hand emasculation and pollination techniques in controlled environments. In recent years, cytoplasmic male sterility (CMS) has been employed to facilitate hybrid seed production without the need for manual emasculation.

3. **Open-Pollinated Varieties (OPVs):**

   » While hybrids dominate the commercial tomato market, open-pollinated varieties (OPVs) are still popular among small-scale farmers who prefer to save their seeds. OPVs are more genetically diverse than hybrids, offering greater adaptability to local growing conditions.

   » Varieties like *Pusa Ruby* and *Arka Vikas* are popular OPVs that have been cultivated for decades, known for their consistency in yield and adaptability to various climates.

Table: Key Tomato Varieties for Different Uses

S No	Variety Name	Main Use	Key Traits
1	Pusa Ruby	Fresh market, processing	Early maturing, uniform ripening
2	Arka Vikas	Processing, long-distance transport	Thick flesh, high solids
3	Pant Hybrid 2	Fresh market, heat-tolerant regions	Heat tolerance, good fruit size
4	Pusa Hybrid 4	Fresh market, high yield	High yield, good flavor
5	Roma	Canning, sauce-making	High Brix, low seed content

B. Chilli (*Capsicum spp.*)

Family: Solanaceae

Chromosome Number: 2n = 2x = 24 (Capsicum annuum)

Chilli, also known as hot pepper, is one of the most important spice crops globally. It is widely cultivated for its pungent fruits, which are used as a condiment, in food flavoring, and in various culinary preparations. Chilli fruits are rich in vitamins A, C, and E, and contain capsaicin, the compound responsible for the pungency. India is one of the leading producers of chilli, contributing significantly to global production. Varieties of chilli differ in terms of size, shape, color, pungency, and flavor, catering to both fresh consumption and the processing industry.

Origin and Domestication

Chillies originated in the Americas, with evidence of their use dating back over 6,000 years in regions such as Mexico and Peru. The domestication of chilli is believed to have taken place in the southern part of Mexico and parts of Central America. After the arrival of Christopher Columbus in the New World, chillies were introduced to Europe and eventually spread to Asia, Africa, and other parts of the world.

The domestication process has led to significant diversification in chilli fruit characteristics, including variations in fruit size, color, shape, and capsaicin content. The five major domesticated species of chilli include:

- *Capsicum annuum*: The most widely cultivated species, including bell peppers and many hot peppers.
- *Capsicum frutescens*: Includes hot peppers such as tabasco.
- *Capsicum chinense*: Known for varieties like habanero, which are among the hottest chillies.
- *Capsicum baccatum*: Mainly cultivated in South America.
- *Capsicum pubescens*: Known for its hairy leaves and distinct fruit characteristics.

Types of Chilli on Basis of Fruit Characteristics and Use

Chilli exhibits significant diversity in fruit morphology, ranging from small, highly pungent varieties to large, mild bell peppers. Botanical classification of chilli is based on characteristics such as fruit size, shape, and pungency levels.

Table: Major Botanical Varieties of Chilli

S No	Variety Name	Fruit Characteristics	Region Grown	Main Use
1	*Capsicum annuum var. annuum*	Bell-shaped, mild, large fruits	Global	Fresh market, processing
2	*Capsicum annuum var. glabriusculum*	Small, highly pungent fruits	Mexico, South America	Hot sauces, condiments
3	*Capsicum chinense*	Small, highly pungent, wrinkled	Caribbean, Mexico	Hot sauces, pickling
4	*Capsicum frutescens*	Medium-sized, very pungent	Southeast Asia, Africa	Fresh consumption, sauces
5	*Capsicum baccatum*	Medium-sized, moderately pungent	South America	Fresh market, sauces

Growth Habit of Chilli

Chilli plants exhibit a bushy growth habit, with varying heights depending on the species and variety. The growth habit significantly influences yield and the management practices adopted for cultivation.

1. **Annual vs. Perennial Growth:**

   » In tropical and subtropical regions, chilli is often grown as a perennial crop, with plants surviving for multiple seasons. In temperate regions, however, chilli is usually grown as an annual crop due to its sensitivity to frost and cold temperatures.

   » Perennial chilli plants can produce fruit over several years, making them valuable in home gardens and small-scale farming systems. Annual chilli varieties, on the other hand, are typically planted for commercial production, where high yields are expected in a single growing season.

2. **Determinate vs. Indeterminate Varieties:**

   » **Determinate Varieties:** These varieties have a compact, bushy growth habit, with fruit set occurring over a short period. Determinate varieties are often preferred for mechanical harvesting due to their uniform fruit ripening. They are widely used in commercial chilli production.

   » **Indeterminate Varieties:** Indeterminate varieties continue to grow and flower throughout the growing season, producing fruits over an extended period. These varieties are often cultivated for fresh markets and home gardens, where continuous harvesting is desirable.

3. **Canopy Management:**

   » Canopy management practices, such as pruning and thinning, are important for maintaining plant health and maximizing yield. Proper canopy management improves air circulation, reduces the risk of fungal diseases, and promotes better fruit set.

Floral Biology of Chilli

Understanding the floral biology of chilli is crucial for breeding programs and seed production. Chilli flowers are typically solitary, borne at the axils of leaves, and exhibit a range of colors, including white, green, and purple, depending on the variety.

1. **Flower Structure:**
   » Chilli flowers are hermaphroditic, containing both male (stamens) and female (pistil) reproductive organs. Each flower consists of five petals that form a radial, star-shaped corolla. The calyx is green and persistent, enlarging as the fruit develops.
   » The flowers are typically self-pollinated, although cross-pollination can occur, especially in the presence of pollinators like bees.
2. **Stamens and Anthers:**
   » The stamens consist of filaments and anthers, which produce pollen. Pollen release (dehiscence) typically occurs in the morning, and self-pollination occurs as the stigma is positioned close to the anthers.
3. **Pollination and Fruit Set:**
   » Pollination is mostly autogamous (self-pollination), but cross-pollination can also occur, especially in open-field conditions where pollinators are present. Hand pollination is sometimes employed in seed production programs to ensure controlled crosses.
   » Successful pollination leads to fruit set, and the ovary develops into the characteristic fruit of chilli. Poor pollination, caused by unfavorable weather conditions or lack of pollinators, can result in low fruit set and yield.

Breeding Methods in Chilli

Chilli breeding focuses on improving yield, fruit quality, resistance to biotic and abiotic stresses, and the development of varieties suited for specific uses such as fresh consumption, processing, or spice production.

Conventional Breeding Methods

1. **Introduction:**
   » In this method, chilli varieties from one region are introduced to another for evaluation and potential adaptation. Introductions are tested for their performance in terms of yield, resistance to pests and diseases, and adaptability to local climatic conditions.
2. **Selection:**
   » Selection involves identifying superior plants from a genetically diverse population and propagating them to develop stable varieties. Selection is based on traits like fruit size, shape, pungency, disease resistance, and adaptability to environmental conditions.

Source Selections in Chilli:

S No	Variety Name	Source	Key Features
1	Pusa Jwala	IARI	Highly pungent, early maturing
2	Arka Meghana	IIHR	Resistant to bacterial wilt
3	Pant C1	GBPUA&T	High yield, medium pungency
4	Co 4	TNAU	High yield, good for processing
5	Pusa Sadabahar	IARI	Continuous fruiting, highly pungent

3. **Hybridization (Heterosis Breeding):**

   » Hybridization is used to combine desirable traits from different parental lines. Heterosis or hybrid vigor is exploited to improve yield, resistance to pests and diseases, and fruit quality. Hybrid varieties such as *Pusa Hybrid 1* and *Arka Suphal* have been developed for commercial production.

Examples of Chilli Hybrids:

S No	Hybrid Name	Parentage	Features
1	Pusa Hybrid 1	Pusa Jwala × Local Line	High yield, early maturity, pungent
2	Arka Suphal	IIHR Hybrid	Resistant to bacterial wilt, good fruit set
3	Pant Hybrid C2	Pant C1 × Local Line	High yield, heat-tolerant

4. **Pedigree Method:**

   » In pedigree selection, breeders select plants from segregating populations (F2, F3, etc.) based on desirable traits. This method is particularly effective when breeding for specific traits like disease resistance or improved fruit quality.

5. **Mutation Breeding:**

   » Mutation breeding is used to create genetic variability by inducing mutations through chemical or physical mutagens. Several chilli varieties with improved disease resistance and fruit characteristics have been developed through induced mutation.

Non-Conventional Breeding Methods

1. **Tissue Culture:**

   » Tissue culture techniques are widely used for producing disease-free chilli plants and for propagating elite lines in hybrid seed production. Micropropagation ensures uniformity in planting material and allows rapid multiplication of selected genotypes.

2. **Genetic Engineering (Recombinant DNA Technology):**

   » Genetic engineering offers new possibilities for improving chilli varieties, particularly in introducing resistance to pests and diseases. Bt chilli, for example, incorporates the *Bacillus thuringiensis* gene to provide resistance against insect pests like the fruit borer.

Breeding for Biotic Stresses in Chilli

Chilli plants are prone to several pests and diseases that can significantly reduce yield and fruit quality. Breeding for resistance to biotic stresses is a priority in chilli improvement programs.

Table: Biotic Stress-Resistant Varieties

S No	Disease/Pest	Source Species	Resistant Varieties
1	Bacterial Wilt	*Capsicum chinense*	Arka Meghana, Pant C1
2	Chilli Mosaic Virus	*Capsicum annuum* (wild types)	Pusa Sadabahar, Arka Suphal
3	Anthracnose	*Capsicum annuum* (wild types)	Co 4, Pusa Jwala
4	Fruit Borer	*Capsicum chinense*	Pusa Hybrid 1, Arka Suphal
5	Powdery Mildew	*Capsicum baccatum*	Pusa Sadabahar, Pant C1

1. **Bacterial Wilt Resistance:**
   - Bacterial wilt, caused by *Ralstonia solanacearum*, is a serious disease affecting chilli production. Resistance has been introduced through breeding programs, using wild relatives like *Capsicum chinense* as donors of resistance genes. Varieties like *Arka Meghana* and *Pant C1* have shown strong resistance to bacterial wilt.
2. **Chilli Mosaic Virus Resistance:**
   - Chilli mosaic virus (ChiMV) is a major viral disease that causes significant yield losses. Breeding for resistance involves selecting resistant genotypes from wild relatives of *Capsicum annuum*. Varieties like *Pusa Sadabahar* and *Arka Suphal* have demonstrated high levels of resistance to this virus.

Breeding for Abiotic Stresses in Chilli

Chilli crops are vulnerable to various abiotic stresses such as drought, high temperature, salinity, and waterlogging, which can significantly impact yield and quality. Breeding for abiotic stress tolerance is essential for ensuring stable chilli production in diverse climatic conditions, particularly in regions prone to environmental extremes.

Table: Abiotic Stress-Resistant Varieties

S No	Abiotic Stress	Tolerant Species	Resistant Varieties
1	Drought	*Capsicum annuum* (wild types)	Arka Meghana, Pusa Sadabahar
2	High Temperature	*Capsicum chinense*	Pant Hybrid C2, Arka Suphal
3	Salinity	*Capsicum frutescens*	Pusa Hybrid 1, Arka Vikas
4	Waterlogging	*Capsicum annuum*	Pant C1, Pusa Jwala

1. **Drought Tolerance:**
   - Drought stress reduces plant growth, fruit set, and overall yield in chilli, especially

in arid and semi-arid regions. Breeding for drought tolerance focuses on selecting varieties that have deeper root systems, better water-use efficiency, and higher osmotic adjustment capacity. Varieties like *Arka Meghana* and *Pusa Sadabahar* have been bred for improved drought tolerance, allowing them to perform well under water-limited conditions.

2. **High Temperature Tolerance:**

   » High temperatures, particularly during flowering and fruit set, can lead to poor pollination, fruit drop, and reduced yields in chilli. Heat-tolerant varieties such as *Pant Hybrid C2* and *Arka Suphal* have been developed to withstand elevated temperatures. These varieties maintain good fruit set and overall productivity even when grown in hot climates.

3. **Salinity Tolerance:**

   » Chilli crops are sensitive to soil salinity, which can result in reduced plant growth, poor fruit quality, and yield losses. Breeding for salinity tolerance involves selecting plants that can maintain ion homeostasis and exclude excess sodium from sensitive tissues. Varieties like *Pusa Hybrid 1* and *Arka Vikas* have shown moderate tolerance to saline conditions, making them suitable for cultivation in regions with saline soils.

4. **Waterlogging Tolerance:**

   » Excessive soil moisture or waterlogging can severely affect chilli plants, leading to root rot, poor nutrient uptake, and reduced yield. Breeding for waterlogging tolerance involves selecting varieties that can maintain root health and oxygen supply under flooded conditions. Varieties like *Pant C1* and *Pusa Jwala* have been bred to tolerate temporary waterlogging, allowing for better performance in areas with heavy rainfall.

Post-Harvest Management of Chilli

Post-harvest management of chilli is crucial for preserving fruit quality, reducing spoilage, and ensuring profitability for farmers and traders. Chilli fruits are perishable, and improper handling can lead to significant losses due to physical damage, microbial infections, and moisture loss.

1. **Harvesting Techniques:**

   » Chilli fruits should be harvested at the correct stage of maturity to ensure maximum flavor, color, and pungency. For fresh market chilli, fruits are typically harvested when they are fully mature but still firm. For processing purposes, fruits may be harvested at the red-ripe stage to ensure high color and capsaicin content.

   » Harvesting is usually done manually to minimize damage to the delicate fruits, especially in smallholder farms. In commercial-scale operations, mechanical harvesters may be used for varieties bred for uniform ripening.

2. **Grading and Sorting:**
   » After harvesting, chilli fruits are sorted based on size, color, shape, and overall quality. Grading ensures that uniform fruits are packed for the market, increasing their appeal to consumers and reducing the chances of spoilage due to damaged or overripe fruits.
   » In large-scale operations, automated grading machines are used to sort chillies by size and color, ensuring consistency in packaging and reducing labor costs.
3. **Drying and Storage:**
   » Drying is an important post-harvest process for chilli, especially for varieties intended for the spice industry. Sun drying is the most common method, but mechanical drying can be employed to ensure consistent moisture levels and avoid contamination.
   » Proper drying reduces the moisture content of the fruit to a safe level (typically below 10%), preventing mold growth and spoilage during storage. Chillies can be stored in ventilated containers or sacks, in cool, dry conditions to maintain quality.
4. **Packaging and Transportation:**
   » Proper packaging is essential to prevent mechanical damage and extend the shelf life of chilli during transportation. Chillies for the fresh market are often packed in ventilated crates or cartons to prevent bruising and moisture buildup.
   » For long-distance transportation, temperature-controlled storage is used to maintain fruit quality. For dried chillies, proper packaging with moisture-resistant materials helps maintain flavor, color, and pungency during transport.
5. **Minimizing Post-Harvest Losses:**
   » Post-harvest losses in chilli can be reduced through proper handling, drying, and storage practices. The use of ethylene inhibitors during storage can help control ripening and extend shelf life. In addition, improved packaging materials, such as breathable plastic films, can reduce moisture loss and prevent microbial growth during storage and transportation.

Table: Post-Harvest Management Techniques for Chilli

S No	Post-Harvest Stage	Best Practices	Impact on Quality
1	Harvesting	Harvest at mature stage, handle gently	Preserves flavor, color, and pungency
2	Grading and Sorting	Remove damaged or overripe fruits	Reduces spoilage, improves market value
3	Drying	Sun drying or mechanical drying	Reduces moisture, prevents mold growth
4	Packaging	Use ventilated crates or moisture-proof packaging	Minimizes mechanical damage, retains quality

5	Transportation	Temperature-controlled storage for long distances	Maintains quality during transport

Chilli Varietal Development and Seed Production

Varietal development in chilli focuses on improving traits like yield, pungency, resistance to pests and diseases, and tolerance to abiotic stresses. The seed production process for chilli involves both open-pollinated varieties (OPVs) and hybrid seeds, with hybrid varieties being more commonly used in commercial production due to their superior performance.

1. **Varietal Development:**
   » Developing a new chilli variety involves crossing different parental lines to combine desirable traits. Breeders evaluate the resulting hybrids over several growing seasons to ensure that the desired traits, such as yield, pungency, and disease resistance, are stable.

   » Molecular markers are often used in modern breeding programs to accelerate the selection process. Marker-assisted selection (MAS) helps identify plants with specific genetic traits, allowing breeders to select superior lines earlier in the breeding cycle.
2. **Hybrid Seed Production:**
   » Hybrid seeds are produced by crossing two genetically distinct parental lines. Hybrid varieties exhibit heterosis (hybrid vigor), resulting in improved performance over the parents in terms of yield, uniformity, and resistance to diseases.

   » The production of hybrid seeds requires the isolation of the parental lines to prevent cross-pollination. In hybrid seed production, hand emasculation and pollination are often used, although cytoplasmic male sterility (CMS) is also employed to facilitate hybrid seed production without manual emasculation.
3. **Open-Pollinated Varieties (OPVs):**
   » While hybrid seeds dominate the commercial chilli market, OPVs are still popular among small-scale farmers due to their adaptability and the ability to save seeds for future planting. OPVs offer greater genetic diversity, which makes them more resilient to environmental changes and biotic stresses.

Table: Key Chilli Varieties for Different Uses

S No	Variety Name	Main Use	Key Traits
1	Pusa Jwala	Fresh market, spice production	Highly pungent, early maturing
2	Arka Suphal	Processing, resistant to diseases	Bacterial wilt resistance, high yield
3	Pant C1	Fresh market, home gardens	Medium pungency, good yield
4	Pusa Hybrid 1	Fresh market, high yield	High yield, good fruit set
5	Co 4	Processing, spice industry	High yield, good for drying

Quality Traits in Chilli

Breeding for quality traits in chilli involves improving flavor, pungency, color, and nutritional content. These traits determine the market value of chilli, whether for fresh consumption, processing, or use as a spice.

1. **Pungency:**
   - Pungency in chilli is determined by the capsaicin content, which varies among species and varieties. Breeding for higher or lower pungency depends on the target market. For example, mild varieties like *Bell Pepper* are bred for fresh consumption, while highly pungent varieties like *Pusa Jwala* are favored for spice production.
2. **Color:**
   - The red color of chilli is due to carotenoids, primarily capsanthin and capsorubin. Breeding for high color content is important for both the fresh market and processing industries, as vibrant color is a key quality attribute. Varieties like *Pusa Sadabahar* and *Arka Suphal* have been bred for intense red color, which makes them suitable for spice production.
3. **Nutritional Content:**
   - Chilli fruits are rich in vitamins A and C, as well as other antioxidants like capsanthin. Breeding programs aim to improve the nutritional content of chilli by selecting varieties with higher concentrations of these beneficial compounds. Biofortified varieties with enhanced vitamin and mineral content are also being developed to address micronutrient deficiencies.

E. Bell Pepper (*Capsicum annuum*)

Family: Solanaceae

Chromosome Number: 2n = 2x = 24

Bell pepper, also known as sweet pepper or capsicum, is an important vegetable crop cultivated for its large, non-pungent fruits, which are commonly used in salads, stir-fries, and as stuffing. The fruit is available in a variety of colors, including green, red, yellow, and orange, depending on the ripeness and variety. Bell pepper is rich in vitamins A and C, as well as antioxidants like capsanthin, making it a nutritionally valuable crop. In global production, bell peppers are widely grown in both open fields and under protected conditions such as greenhouses.

Origin and Domestication

Bell peppers, like other members of the *Capsicum* genus, originated in the Americas. They are believed to have been domesticated in Central and South America thousands of years ago. Initially, wild ancestors of bell pepper produced small, highly pungent fruits. Through domestication and selection, larger, non-pungent fruits were developed, leading to the modern bell pepper varieties we see today.

The spread of bell pepper around the world occurred during the Columbian Exchange, where they were introduced to Europe, Asia, and Africa. The absence of capsaicin in bell peppers makes them distinct from hot peppers, and they have become a staple in global cuisine due to their versatility and mild flavor.

Botanical Varieties of Bell Pepper

Bell peppers are classified into several varieties based on fruit size, color, shape, and other characteristics. Most commonly, bell peppers are categorized by their color at maturity, which can vary from green to red, yellow, orange, and even purple.

Table: Major Botanical Varieties of Bell Pepper

S No	Variety Name	Fruit Characteristics	Region Grown	Main Use
1	*Capsicum annuum var. grossum*	Large, blocky, non-pungent fruits	Global	Fresh consumption, processing
2	*Capsicum annuum var. annuum*	Medium-sized, sweet fruits	Europe, Asia	Fresh market
3	*Capsicum annuum var. microcarpum*	Small, sweet fruits	South America	Pickling, fresh consumption

Growth Habit of Bell Pepper

Bell pepper plants exhibit a compact, bushy growth habit, with a height ranging from 50 to 100 cm, depending on the variety and growing conditions. Understanding the growth habit is important for crop management, particularly in greenhouse environments where space optimization is crucial.

1. **Annual vs. Perennial Growth:**
   - Bell pepper is typically grown as an annual crop, especially in temperate regions where frost sensitivity limits its lifespan. In tropical and subtropical regions, bell pepper can be cultivated as a short-lived perennial, producing fruits for several seasons.
   - When grown as a perennial, the plants tend to become woody at the base over time, and their productivity declines after a few seasons. As an annual crop, bell pepper is planted afresh every season to ensure high yields.
2. **Canopy Management:**
   - Proper canopy management, including pruning and staking, is necessary to ensure good light penetration and air circulation within the plant canopy. This helps reduce the incidence of diseases like powdery mildew and improves fruit set and quality.
   - In greenhouse cultivation, bell pepper plants are often trained using vertical support systems to optimize space and light exposure.

Floral Biology of Bell Pepper

Bell pepper flowers are perfect, meaning they contain both male (stamens) and female (pistil) reproductive organs. Flowers are generally white, borne singly in the leaf axils, and capable of self-pollination. However, cross-pollination can occur in open-field conditions, especially with the help of insects like bees.

1. **Flower Structure:**
   - Each flower consists of five white petals, five green sepals, and five stamens surrounding a single pistil. The flowers are relatively small and usually measure about 1.5 to 2 cm in diameter.
   - The ovary is superior and develops into the fleshy fruit of the bell pepper. The calyx is persistent and remains attached to the fruit throughout its development.
2. **Pollination and Fruit Set:**
   - Pollination in bell peppers is primarily self-pollination, although cross-pollination can occur in the presence of insect pollinators. Greenhouse-grown bell peppers may require manual pollination or the use of bumblebees to ensure good fruit set.
   - Fruit set can be influenced by environmental factors such as temperature and humidity. Optimal pollination and fruit set occur when temperatures range between 18°C and 24°C. Excessively high temperatures can lead to flower drop and poor fruit set.

Breeding Methods in Bell Pepper

Breeding programs for bell pepper aim to improve yield, fruit quality, disease resistance, and tolerance to abiotic stresses. Both conventional and non-conventional breeding methods are used to develop superior bell pepper varieties.

Conventional Breeding Methods

1. **Introduction:**

   » The introduction of bell pepper varieties from one ecological zone to another has led to the adaptation of many introduced varieties to local conditions. For example, varieties introduced from Europe have been tested and adapted for cultivation in Asian and African countries.

2. **Selection**:

   » Selection plays a critical role in bell pepper breeding, where superior plants are selected from a genetically diverse population. The selection process is based on traits like fruit size, shape, color, resistance to pests and diseases, and adaptability to different growing conditions.

Source Selections in Bell Pepper:

S No	Variety Name	Source	Key Features
1	California Wonder	USA	Large, blocky fruits, widely adaptable
2	Yolo Wonder	USA	High yield, disease resistance
3	Arka Gaurav	India (IIHR)	Resistant to bacterial wilt, good yield
4	Bharat	India (PAU)	Tolerant to heat, high yield
5	Wonder King	Europe	High fruit quality, suitable for export

3. **Hybridization (Heterosis Breeding):**

   » Hybridization is widely used to develop bell pepper hybrids with improved traits. Hybrids exhibit hybrid vigor (heterosis), which results in increased yields, better fruit quality, and enhanced resistance to diseases.

Examples of Bell Pepper Hybrids:

S No	Hybrid Name	Parentage	Features
1	Pusa Hybrid 4	California Wonder × Local Line	High yield, large blocky fruits
2	Arka Mohini	IIHR Hybrid	Resistant to bacterial wilt, early maturity
3	Yolo King Hybrid	Yolo Wonder × Local Line	High fruit set, uniform size

4. **Pedigree Method:**

   » Pedigree breeding is commonly used in bell pepper to track the inheritance of traits such as disease resistance and fruit quality. Single plant selection is initiated in the F2 generation and continues through successive generations until stable, uniform lines are developed.

5. **Mutation Breeding:**

   » Mutation breeding has been employed in bell pepper to induce new genetic variations that can improve traits like disease resistance, fruit color, and yield. For example, the use of chemical mutagens has led to the development of improved bell pepper varieties with unique colors and shapes.

Non-Conventional Breeding Methods

1. **Tissue Culture:**

   » Tissue culture techniques are used to propagate disease-free planting material and for the rapid multiplication of elite genotypes. In bell pepper, somatic embryogenesis and micropropagation are commonly employed to produce uniform, true-to-type plants.

2. **Genetic Engineering (Recombinant DNA Technology):**

   » Genetic engineering offers potential in bell pepper breeding by introducing genes that confer resistance to diseases, pests, and abiotic stresses. For example, transgenic bell pepper plants with resistance to viral diseases have been developed using recombinant DNA technology.

Breeding for Biotic Stresses in Bell Pepper

Bell pepper plants are susceptible to various diseases and pests, including bacterial wilt, powdery mildew, and fruit borers. Breeding for resistance to these biotic stresses is essential for ensuring high yields and quality fruits.

Table: Biotic Stress-Resistant Varieties

S No	Disease/Pest	Source Species	Resistant Varieties
1	Bacterial Wilt	*Capsicum chinense*	Arka Mohini, Yolo Wonder
2	Powdery Mildew	*Capsicum annuum* (wild types)	Arka Gaurav, Bharat
3	Anthracnose	*Capsicum annuum*	California Wonder, Wonder King
4	Fruit Borer	*Capsicum chinense*	Pusa Hybrid 4, Arka Gaurav
5	Tobacco Mosaic Virus	*Capsicum frutescens*	Arka Mohini, Yolo King Hybrid

1. **Bacterial Wilt Resistance:**

   » Bacterial wilt is a major disease affecting bell pepper production, particularly in warm, humid regions. Breeding programs have focused on developing resistant varieties by utilizing wild relatives such as *Capsicum chinense*. Varieties like *Arka Mohini* and *Yolo Wonder* exhibit good resistance to bacterial wilt.

2. Powdery Mildew Resistance:

- Powdery mildew is a fungal disease that can severely affect bell pepper plants, leading to reduced yields. Resistant varieties such as *Arka Gaurav* and *Bharat* have been developed through selection and hybridization to reduce the impact of this disease.

Breeding for Abiotic Stresses in Bell Pepper

Abiotic stresses like drought, high temperature, and salinity can affect bell pepper growth and yield. Breeding for tolerance to these stresses is essential to ensure stable production in regions with harsh environmental conditions.

Table: Abiotic Stress-Resistant Varieties

S No	Abiotic Stress	Tolerant Species	Resistant Varieties
1	Drought	*Capsicum chinense*	Arka Mohini, Bharat
2	High Temperature	*Capsicum frutescens*	Yolo King Hybrid, Pusa Hybrid 4
3	Salinity	*Capsicum annuum*	Wonder King, California Wonder
4	Waterlogging	*Capsicum chinense*	Arka Gaurav, Bharat

Post-Harvest Management of Bell Pepper

Bell pepper is a perishable crop, and post-harvest management plays a critical role in maintaining fruit quality and reducing losses. Proper handling, storage, and transportation are essential for preserving the freshness, color, and texture of bell pepper fruits.

1. Harvesting Techniques:

- Bell peppers should be harvested at the appropriate stage of maturity to ensure optimal flavor, texture, and color. Green bell peppers are typically harvested before full ripening, while red, yellow, and orange varieties are harvested at full maturity.
- Manual harvesting is preferred to avoid bruising and damage to the fruits. Fruits should be handled carefully to prevent mechanical injury.

2. Grading and Sorting:

- After harvesting, bell peppers are graded based on size, color, and overall quality. Grading ensures that only uniform, high-quality fruits are packed for the market, improving their market value.

3. Storage and Transportation:

- Bell peppers are highly sensitive to temperature fluctuations, and improper storage can lead to chilling injury or accelerated ripening. The optimal storage temperature for bell peppers is between 7°C and 10°C with high relative humidity (90-95%).
- For long-distance transportation, bell peppers are packed in ventilated crates or cartons to maintain their quality during transport.

Bulb and Tuberous Vegetables

A. Onion (*Allium cepa*)

Family: Amaryllidaceae

Chromosome Number: 2n = 2x = 16

Onion (*Allium cepa*) is a biennial plant, often cultivated as an annual, and is one of the most important vegetable crops grown worldwide. It is a member of the *Allium* genus, which includes garlic, leeks, and shallots. Onions are known for their distinctive flavor, which varies from mild to pungent depending on the variety. They are a staple ingredient in cuisines across the globe and are used in various forms, including fresh, dried, and processed. Onions are rich in sulfur compounds, which contribute to their pungency and have been shown to offer health benefits, such as reducing the risk of certain cancers, lowering blood pressure, and supporting heart health. They are also a good source of vitamins C and B6, manganese, and dietary fiber. The bulb is the edible part of the onion and is formed by the concentric layers of modified leaves that store nutrients.

Origin and Domestication

Onion is believed to have originated in Central Asia, specifically in the regions of modern-day Iran and Pakistan. Evidence suggests that it was cultivated as early as 5,000 years ago. The plant spread to ancient civilizations in Egypt, Greece, and Rome, where it became a staple food due to its storage capabilities and ease of transport. Onions were particularly valued in ancient Egypt, where they were used as offerings and placed in the tombs of pharaohs. The bulb's circular structure was symbolic of eternity to the ancient Egyptians.

Domestication has led to the development of various onion types, including red, white, and yellow onions, as well as specialized varieties such as scallions and shallots. Early domesticated onions were likely smaller and more pungent than the varieties cultivated today, but selective breeding has led to improvements in size, flavor, and shelf life. Modern onion breeding focuses on enhancing yield, disease resistance, and tolerance to environmental stresses.

Types of Onion on Basis of Bulb Characteristics and Use

Onions can be classified into several botanical varieties based on bulb size, shape, color, and pungency. These characteristics are influenced by both genetics and environmental factors, such as day length, which determines whether a variety is short-day or long-day.

Table: Major Botanical Varieties of Onion

S No	Variety Name	Bulb Characteristics	Region Grown	Main Use
1	*Allium cepa var. cepa*	Large, globe-shaped bulbs	Global	Fresh market, processing
2	*Allium cepa var. aggregatum*	Smaller bulbs, less pungent	Southeast Asia, India	Fresh consumption, pickling
3	*Allium cepa var. proliferum*	Produces bulbils at the top	Temperate regions	Specialty onions, fresh market

1. ***Allium cepa var. cepa* (Globe Onion):**
   » This is the most common variety of onion, grown worldwide for its large, globe-shaped bulbs. The bulbs can be white, yellow, or red, depending on the cultivar, and range in size from small to large. Globe onions are typically grown for fresh market consumption and processing, with many varieties being well-suited for long-term storage.
2. ***Allium cepa var. aggregatum* (Shallot or Multiplier Onion):**
   » This variety is characterized by smaller bulbs that grow in clusters. Shallots have a milder flavor compared to globe onions and are highly valued in culinary applications for their delicate taste. They are commonly grown in Southeast Asia and India and are often used in pickling or as a gourmet ingredient in dishes.
3. ***Allium cepa var. proliferum* (Tree Onion):**
   » Known for its unique growth habit, this variety produces bulbils at the top of the plant instead of flowers. These bulbils can be harvested and replanted or used as food. Tree onions are grown mainly in temperate regions and are considered a specialty crop.

Growth Habit of Onion

Onion is a biennial plant that typically completes its life cycle over two years. However, it is commonly cultivated as an annual for bulb production. The plant produces a rosette of hollow, cylindrical leaves in the first year, followed by the development of a bulb at the base. The bulb serves as a storage organ, allowing the plant to survive adverse environmental conditions. In the second year, the plant bolts, sending up a flower stalk, and produces seeds.

1. **Bulb Formation:**
   » Bulb formation in onions is triggered by day length. Short-day onions initiate bulb development when exposed to 10 to 12 hours of daylight, while long-day onions require 14 to 16 hours of daylight. Day-neutral varieties, which are less sensitive to photoperiod, can be grown in a wider range of latitudes.

   » The size and shape of the bulb are influenced by both genetic and environmental factors. Onion bulbs are typically globular or slightly flattened, with layers of fleshy scales that store water and nutrients. These scales are covered by a thin, papery outer skin that helps protect the bulb during storage.

2. **Root System:**

   » Onions have a shallow, fibrous root system that extends laterally rather than deep into the soil. This shallow rooting makes onions particularly sensitive to water stress and nutrient deficiencies. As a result, proper soil moisture management and fertilization are critical for ensuring optimal growth and bulb development.

3. **Flowering and Seed Production:**

   » In its second year, the onion plant sends up a tall, hollow flower stalk called a scape, which can reach heights of up to 1.5 meters. The scape terminates in an umbel, a rounded cluster of small white or pink flowers. Pollination is primarily carried out by insects, particularly bees.

   » Onion seeds are small, black, and angular in shape. For seed production, onions are typically grown as biennials, with the bulbs harvested in the first year and replanted in the second year to produce seeds.

Floral Biology of Onion

Understanding the floral biology of onion is essential for breeding and seed production. Onion flowers are perfect, meaning they contain both male (stamens) and female (pistil) reproductive organs. The flowers are borne in large, rounded clusters called umbels, which can contain hundreds of individual flowers. Each flower is small and white, with six tepals (undifferentiated petals and sepals), six stamens, and a single pistil.

1. **Pollination:**

   » Onions are primarily cross-pollinated, although self-pollination can also occur. Cross-pollination is facilitated by insects, particularly bees, which transfer pollen from one flower to another. Because onions are prone to inbreeding depression, maintaining genetic diversity through cross-pollination is important for seed production.

2. **Flowering and Seed Set:**

   » Onion plants flower in their second year of growth when grown as biennials. The timing of flowering is influenced by environmental factors, particularly temperature and day length. Cool temperatures followed by warm, long days are optimal for flower initiation and seed set.

   » After pollination, the flowers develop into small seed capsules, each containing multiple seeds. The seeds mature within 45 to 60 days after pollination, at which point the umbels can be harvested for seed extraction.

Breeding Methods in Onion

Breeding in onion focuses on improving yield, bulb quality, disease resistance, and tolerance to environmental stresses. Both conventional and non-conventional breeding methods are employed to develop superior onion varieties that cater to the diverse needs of farmers and consumers.

Conventional Breeding Methods

1. **Introduction:**

   » Introduction involves the evaluation of onion varieties introduced from different ecological zones. By assessing their adaptability to local growing conditions, promising varieties are identified and incorporated into breeding programs. For example, varieties introduced from Europe may be adapted for cultivation in tropical regions by selecting for traits like heat tolerance.

2. **Selection:**

   » Selection is a widely used method in onion breeding, where superior plants are selected from genetically diverse populations based on desirable traits such as bulb size, shape, color, and resistance to diseases. In mass selection, a group of plants with similar traits is selected, while in individual plant selection, outstanding plants are chosen for further propagation.

Source Selections in Onion:

S No	Variety Name	Source	Key Features
1	Pusa Red	India (IARI)	High yield, red bulbs, long storage
2	Arka Kalyan	India (IIHR)	Resistant to bolting, early maturity
3	Texas Grano	USA	Large bulbs, mild flavor
4	Red Creole	USA	Heat tolerant, red bulbs
5	Bombay Yellow	India (Local Landrace)	High pungency, good storage

3. **Hybridization (Heterosis Breeding):**

   » Hybridization is used to combine desirable traits from different parental lines, such as disease resistance, high yield, and improved bulb quality. Hybrid varieties often exhibit hybrid vigor (heterosis), resulting in improved performance over their parents. Hybrid onion varieties are widely used for commercial cultivation due to their uniformity and high yield.

Examples of Onion Hybrids:

S No	Hybrid Name	Parentage	Features
1	Pusa Hybrid 3	Pusa Red × Local Line	High yield, good storage quality
2	Arka Niketan	IIHR Hybrid	Early maturity, uniform bulbs
3	Texas Hybrid	Texas Grano × Local Line	Large bulbs, mild flavor

Breeding for Biotic and Abiotic Stresses in Onion

Onion crops are susceptible to a range of biotic stresses, including fungal, bacterial, and viral diseases, as well as abiotic stresses like drought, salinity, and heat. Breeding for resistance to these stresses is crucial for improving onion productivity and reducing losses.

Table: Biotic Stress-Resistant Varieties

S No	Disease/Pest	Source Species	Resistant Varieties
1	Downy Mildew	*Allium cepa* (wild types)	Arka Niketan, Pusa Red
2	Purple Blotch	*Allium cepa*	Texas Grano, Bombay Yellow
3	Fusarium Basal Rot	*Allium cepa*	Red Creole, Arka Kalyan

1. **Downy Mildew Resistance:**

   » Downy mildew is a common fungal disease affecting onion crops, leading to leaf chlorosis, reduced bulb size, and yield losses. Resistant varieties like *Arka Niketan* and *Pusa Red* have been developed through breeding programs to mitigate the impact of this disease.

2. **Purple Blotch Resistance:**

   » Purple blotch, caused by *Alternaria porri*, is another serious fungal disease in onions. Breeding for resistance involves selecting varieties that can withstand infection and exhibit reduced disease severity. Varieties like *Texas Grano* and *Bombay Yellow* have shown resistance to purple blotch.

Post-Harvest Management of Onion

Onions are highly perishable, and proper post-harvest management is essential to maintain quality and extend shelf life. Post-harvest losses can be significant, especially in regions with inadequate storage facilities.

1. **Harvesting Techniques:**

   » Onions should be harvested when the tops begin to dry and fall over. Harvesting at the right stage of maturity is crucial to ensure good storage quality. Manual harvesting is commonly practiced, although mechanical harvesters are used in large-scale operations.

2. **Curing and Drying:**

   » After harvesting, onions are cured by drying them in the field or in well-ventilated storage sheds. Curing allows the outer skin to become dry and papery, which helps protect the bulb during storage. Proper curing reduces moisture content and prevents mold growth.

3. **Storage:**

   » Onions are stored in cool, dry conditions with good ventilation. The ideal storage temperature is between 0°C and 4°C, with relative humidity levels of 65-70%. Proper storage conditions can extend the shelf life of onions for several months.

Table: Post-Harvest Management Techniques for Onion

S No	Post-Harvest Stage	Best Practices	Impact on Quality
1	Harvesting	Harvest at full maturity, handle carefully	Preserves bulb quality and flavor
2	Curing and Drying	Cure in a well-ventilated area for 1-2 weeks	Reduces moisture content, prevents spoilage
3	Storage	Store in cool, dry conditions with ventilation	Extends shelf life, prevents sprouting

B. Garlic (*Allium sativum*)

Family: Amaryllidaceae

Chromosome Number: 2n = 2x = 16

Garlic is an ancient bulbous plant, known for its strong flavor and medicinal properties. It is widely used as a spice, condiment, and medicinal herb in various cultures. Garlic is rich in sulfur-containing compounds such as allicin, which gives it its pungent odor and numerous health benefits, including antibacterial, antiviral, and antifungal properties. It is also a source of vitamins C and B6, manganese, and fiber. Garlic is consumed in various forms: raw, cooked, or processed into products like garlic powder, garlic oil, and garlic paste. Garlic is grown worldwide, with significant production in China, India, and other temperate and subtropical regions. Unlike onion, garlic does not typically form seeds and is propagated vegetatively through its cloves, making breeding programs for garlic unique and challenging.

Origin and Domestication

Garlic is believed to have originated in Central Asia, with wild relatives still found in regions such as Kazakhstan, Uzbekistan, and Turkmenistan. It has been cultivated for over 5,000 years and was widely used in ancient civilizations, including Egypt, Greece, and Rome, for both culinary and medicinal purposes. Historical records show that garlic was used as a remedy for various ailments, from respiratory infections to digestive problems, and even as an offering in ancient Egyptian tombs.

The domestication of garlic involved the selection of clones that exhibited desirable traits such as larger bulbs, higher pungency, and better storage capabilities. Unlike many other crops, garlic does not produce viable seeds; thus, all domesticated garlic varieties are propagated vegetatively through their cloves. This method of propagation has resulted in a narrow genetic base for garlic, which presents challenges in breeding programs aimed at improving traits such as disease resistance and yield.

Types of Garlic on Basis of Bulb Characteristics and Use

Garlic is classified into two main botanical types: **Hardneck Garlic (*Allium sativum var. ophioscorodon*)** and **Softneck Garlic (*Allium sativum var. sativum*)**. These two types differ in terms of their growth habits, bulb formation, and storage characteristics.

Table: Major Botanical Varieties of Garlic

S No	Variety Name	Bulb Characteristics	Region Grown	Main Use
1	*Allium sativum var. sativum*	Large, white or purple bulbs	Global	Fresh market, processing
2	*Allium sativum var. ophioscorodon*	Hardneck, forms flower stalks	Temperate regions	Specialty garlic, fresh market

1. ***Allium sativum var. sativum* (Softneck Garlic):**
   - This variety is the most commonly cultivated type of garlic and is characterized by its soft, pliable neck, which allows it to be braided for storage. Softneck garlic typically produces larger bulbs with more cloves, making it the preferred choice for commercial production. It is grown in a wide range of climates, from temperate to subtropical regions, and is known for its longer shelf life compared to hardneck garlic. Common varieties include *California Early* and *California Late.*
2. ***Allium sativum var. ophioscorodon* (Hardneck Garlic):**
   - Hardneck garlic varieties are characterized by their hard, stiff flower stalks (scapes) that form during the growing season. These varieties tend to have fewer but larger cloves per bulb and are known for their complex flavors, which can range from mild to pungent. Hardneck garlic is less suited for long-term storage compared to softneck varieties, but it is highly prized by gourmet chefs and specialty markets. Varieties such as *Rocambole* and *Purple Stripe* are examples of hardneck garlic.

Growth Habit of Garlic

Garlic is a perennial plant that is typically cultivated as an annual for bulb production. The plant grows from individual cloves, each of which produces a new garlic plant with a bulb composed of multiple cloves. Garlic plants form a rosette of narrow, flat leaves, with the bulb developing underground. In hardneck varieties, the plant produces a central flowering stalk, or scape, that curls before straightening and forming bulbils at the top.

1. **Bulb Formation:**
   - Garlic bulbs develop underground and are composed of individual cloves surrounded by papery scales. The size and number of cloves per bulb are influenced by environmental factors such as day length, temperature, and soil conditions. Bulb formation in garlic is triggered by cool temperatures followed by a period of warmth, which allows the plant to switch from vegetative growth to bulb development.
2. **Root System:**
   - Garlic has a shallow, fibrous root system that extends laterally rather than deep into the soil. This shallow rooting system makes garlic sensitive to water stress and nutrient deficiencies. Irrigation and fertilization practices must be carefully managed to ensure optimal bulb development and high yields.
3. **Flowering and Bulbil Production:**
   - In hardneck garlic varieties, the plant produces a flowering stalk, or scape, that emerges from the center of the bulb. The scape eventually forms bulbils, which are small, aerial garlic cloves that can be used for propagation. However, garlic typically does not produce viable seeds, and commercial garlic production relies on the vegetative propagation of cloves.

Floral Biology of Garlic

Unlike many other crops, garlic rarely flowers, and even when it does, the flowers typically do not produce viable seeds. Instead, garlic reproduces vegetatively through its cloves or bulbils (in the case of hardneck varieties). Understanding the floral biology of garlic is important for breeding and propagation purposes, especially when developing new varieties through vegetative selection.

1. **Floral Structure:**

   » In hardneck garlic varieties, the plant produces a scape that terminates in an umbel, a rounded cluster of small, white or pinkish flowers. These flowers are sterile in most cases, meaning that they do not produce seeds. Instead, the umbel often contains bulbils, which are small, clove-like structures that can be used for propagation.

2. **Bulbil Production:**

   » Bulbils are small, aerial garlic cloves that form at the top of the flowering stalk in hardneck garlic varieties. These bulbils can be harvested and planted to grow new garlic plants, although it typically takes several years for bulbil-propagated garlic to reach full bulb size. Bulbil propagation is an alternative to clove propagation, offering a means of maintaining genetic diversity in garlic populations.

3. **Vegetative Propagation:**

   » Since garlic does not produce viable seeds, it is propagated vegetatively through its cloves. Each clove is capable of developing into a new garlic plant, producing a bulb identical to the parent plant. This method of propagation allows for the maintenance of specific garlic varieties over long periods, but it also limits genetic diversity within garlic populations.

Breeding Methods in Garlic

Breeding garlic presents unique challenges due to its sterile nature and reliance on vegetative propagation. Most garlic breeding programs focus on selecting superior clones with desirable traits such as larger bulb size, higher yield, disease resistance, and improved storage quality.

Conventional Breeding Methods

1. **Introduction:**

   » Introduction involves evaluating garlic varieties introduced from different regions to identify those that perform well in local growing conditions. Varieties introduced from Europe and Asia, for example, have been tested and adapted for cultivation in North America and other regions. An introduction helps broaden the genetic base of garlic, even though it relies on vegetative propagation.

2. **Selection:**

» Selection is the primary breeding method for garlic, as it does not produce viable seeds. Breeders select the best-performing plants based on traits such as bulb size, shape, clove number, disease resistance, and adaptability to different environmental conditions. Once a superior plant is identified, it is propagated vegetatively through its cloves or bulbils.

Source Selections in Garlic:

S No	Variety Name	Source	Key Features
1	California Early	USA	High yield, large bulbs, long storage
2	Rocambole	Europe	Hardneck, excellent flavor, gourmet
3	Creole Red	Mexico	Red-skinned, heat tolerant
4	Purple Stripe	Europe	Hardneck, good for baking
5	Silverwhite	Italy	Softneck, good for braiding

3. **Clonal Selection:**

» Clonal selection is a process by which individual plants with desirable traits are selected and propagated vegetatively to create new garlic varieties. Clonal selection helps maintain the uniformity of traits such as bulb size, flavor, and disease resistance. This method is especially important in garlic breeding, as it ensures that the selected traits are passed on to future generations.

4. **Mutation Breeding:**

» Mutation breeding involves the use of chemical mutagens or radiation to induce genetic variations in garlic. While this method has been less commonly used in garlic breeding, it has the potential to create new garlic varieties with improved traits such as disease resistance and bulb size.

Non-Conventional Breeding Methods

1. **Tissue Culture:**

» Tissue culture techniques are used to propagate disease-free planting material and rapidly multiply elite garlic clones. In garlic, tissue culture can be used to produce virus-free plants, which is important for maintaining the health and productivity of garlic crops. Tissue culture also allows for the preservation of genetic material and the rapid multiplication of new varieties.

2. **Genetic Engineering (Recombinant DNA Technology):**

» Genetic engineering offers potential in garlic breeding by introducing genes that confer resistance to diseases, pests, and environmental stresses. While genetic engineering in garlic is still in its early stages, it holds promise for the development of improved garlic varieties with enhanced traits.

Breeding for Biotic Stresses in Garlic

Garlic crops are susceptible to various biotic stresses, including fungal diseases such as white rot, downy mildew, and fusarium basal rot. Breeding for resistance to these biotic stresses is essential for ensuring high yields and bulb quality.

Table: Biotic Stress-Resistant Varieties

S No	Disease/Pest	Source Species	Resistant Varieties
1	White Rot	*Allium sativum* (wild types)	California Early, Silverwhite
2	Fusarium Basal Rot	*Allium sativum*	Rocambole, Purple Stripe
3	Nematodes	*Allium sativum*	Creole Red, California Late

1. **White Rot Resistance:**
   - White rot is a serious fungal disease affecting garlic, leading to reduced yields and poor bulb quality. Resistant varieties like *California Early* and *Silverwhite* have been developed through clonal selection to combat this disease.
2. **Fusarium Basal Rot Resistance:**
   - Fusarium basal rot is caused by the soilborne fungus *Fusarium oxysporum* and can result in significant crop losses. Resistant varieties like *Rocambole* and *Purple Stripe* have been selected for their ability to tolerate this disease.

Post-Harvest Management of Garlic

Proper post-harvest management is critical for preserving garlic quality and extending shelf life. Garlic is a highly perishable crop, and improper handling can result in post-harvest losses due to moisture loss, sprouting, and disease.

1. **Harvesting Techniques:**
   - Garlic should be harvested when the leaves begin to turn yellow and dry out. This indicates that the bulbs have reached full maturity. Manual harvesting is commonly practiced, although mechanical harvesters may be used in large-scale operations.
2. **Curing and Drying:**
   - After harvesting, garlic bulbs must be cured by drying them in well-ventilated areas for several weeks. Curing allows the outer layers of the bulb to dry and harden, which helps protect the cloves during storage. Proper curing reduces moisture content and prevents the growth of mold and bacteria.

3. **Storage:**

» Garlic is best stored in cool, dry conditions with good air circulation. The ideal storage temperature is between 0°C and 4°C, with low humidity to prevent sprouting and moisture loss. Proper storage conditions can extend the shelf life of garlic for up to 8-12 months.

Table: Post-Harvest Management Techniques for Garlic

S No	Post-Harvest Stage	Best Practices	Impact on Quality
1	Harvesting	Harvest at full maturity, handle carefully	Preserves bulb size and flavor
2	Curing and Drying	Cure in a dry, well-ventilated area for 2-3 weeks	Reduces moisture, prevents spoilage
3	Storage	Store in cool, dry conditions with ventilation	Extends shelf life, prevents sprouting

C. Sweet Potato (*Ipomoea batatas*)

Family: Convolvulaceae

Chromosome Number: 2n = 6x = 90

Sweet potato is a highly nutritious, tuberous root crop that is cultivated worldwide, especially in tropical and subtropical regions. It is an important food source, providing dietary fiber, vitamins (particularly vitamin A), and minerals. Sweet potatoes come in a variety of colors, including orange, purple, white, and yellow, depending on the variety and growing conditions. The crop is known for its versatility, being consumed in both fresh and processed forms. It is also valued for its adaptability to diverse growing conditions and its ability to produce high yields in marginal soils.

Sweet potatoes are often confused with yams, though they belong to entirely different plant families. Sweet potato roots are primarily used as food, but the leaves and young shoots are also edible and are consumed as a leafy green vegetable in some cultures. The plant is typically propagated vegetatively through vine cuttings or slips, though seed production is possible under certain conditions.

Origin and Domestication

Sweet potato is believed to have originated in Central and South America, with evidence suggesting that it was domesticated by indigenous peoples over 5,000 years ago. Archaeological findings in Peru have shown the presence of cultivated sweet potatoes dating back to 2500 BCE. The crop spread to Polynesia, Asia, and Africa via human migration and trade, and it is now grown in over 100 countries worldwide.

The domestication of sweet potatoes involved selecting varieties with desirable traits such as larger tuber size, increased sweetness, and better storage capabilities. Today, sweet potato varieties differ greatly in terms of flesh and skin color, texture, sweetness, and starch content. Selective breeding has also led to the development of varieties with enhanced nutritional content, particularly those high in beta-carotene (a precursor to vitamin A).

Major Types of Sweet Potato on Basis of Tuber Characteristics and Use

Sweet potato varieties are classified based on tuber skin and flesh color, shape, sweetness, and starch content. Each variety is suited to different culinary uses, such as fresh consumption, baking, boiling, or industrial processing into starch, flour, or ethanol.

Table: Major Types of Sweet Potato on Basis of Tuber Characteristics and Use

S No	Variety Name	**Tuber Characteristics**	**Region Grown**	**Main Use**
1	*Ipomoea batatas var. batatas*	Large, orange-fleshed tubers	Global	Fresh consumption, processing
2	*Ipomoea batatas var. purpurea*	Purple-skinned, purple-fleshed	Southeast Asia, Japan	Specialty market, processing
3	*Ipomoea batatas var. alba*	White-fleshed, starchy tubers	Africa, South America	Boiling, starch production

1. ***Ipomoea batatas var. batatas*** **(Orange-Fleshed Sweet Potato):**

   » This is the most commonly cultivated variety, characterized by its large, elongated tubers with orange flesh. Orange-fleshed sweet potatoes are rich in beta-carotene, making them an important source of vitamin A, particularly in regions where vitamin A deficiency is prevalent. They are widely grown for fresh consumption, baking, and processing into products like chips and fries.

2. ***Ipomoea batatas var. purpurea*** **(Purple-Fleshed Sweet Potato):**

   » Purple-fleshed sweet potatoes are known for their vibrant color, which is due to the presence of anthocyanins, powerful antioxidants. These varieties are popular in Southeast Asia and Japan, where they are used in both fresh consumption and processed into specialty products like purple sweet potato flour and noodles. They are also gaining popularity in the health food market due to their antioxidant properties.

3. ***Ipomoea batatas var. alba*** **(White-Fleshed Sweet Potato):**

   » White-fleshed sweet potatoes are more starchy and less sweet than orange or purple varieties. They are commonly grown in Africa and South America and are primarily used for boiling or as a source of starch for industrial applications. These varieties are often used in the production of sweet potato flour, which is used in baking and as a thickening agent.

Growth Habit of Sweet Potato

Sweet potato is a herbaceous, perennial plant that is typically grown as an annual for its tuberous roots. The plant produces long, trailing vines that spread across the soil surface, with adventitious roots developing at the nodes. These roots thicken into storage roots, or tubers, which are harvested for consumption. The growth habit of sweet potato makes it well-suited for cultivation in both smallholder and commercial farming systems.

1. **Vine Growth:**

   » Sweet potato plants produce trailing vines that can extend up to several meters in length. The vines are typically prostrate, but in some cases, they may climb or twine around nearby supports. The length and vigor of vine growth are influenced by environmental factors such as temperature, soil fertility, and water availability. In regions with ample rainfall and warm temperatures, sweet potato vines grow vigorously, covering the ground and suppressing weeds.

2. **Root System:**

   » Sweet potatoes have a shallow root system, with adventitious roots forming at the nodes of the vines. These roots penetrate the soil and develop into storage roots, which are the edible tubers. The plant's shallow root system makes it susceptible to drought stress, but it also allows the plant to tolerate poor soils and grow in a wide range of soil types, from sandy to clayey soils.

3. **Flowering and Seed Production**:

   » Sweet potato plants can produce small, funnel-shaped flowers that are typically purple or white. The flowers are borne singly or in clusters and are pollinated by insects, particularly bees. However, sweet potato rarely produces viable seeds under normal growing conditions, and commercial propagation is almost exclusively done vegetatively through vine cuttings or slips. In some breeding programs, controlled pollination is used to produce seeds for hybrid development.

Floral Biology of Sweet Potato

Sweet potato flowers are perfect, meaning they contain both male (stamens) and female (pistil) reproductive organs. The flowers resemble those of morning glories (which are also part of the *Convolvulaceae* family) and are usually purple, pink, or white in color. Although sweet potatoes can flower under optimal conditions, viable seed production is rare, and the crop is typically propagated vegetatively.

1. **Flower Structure:**

   » Each sweet potato flower consists of five fused petals forming a tubular corolla, five stamens attached to the corolla, and a single pistil with an ovary. The flowers are generally small and delicate, measuring about 2 to 4 cm in diameter. Flowering is more common in certain varieties and under specific environmental conditions, such as long days and cool nights.

2. **Pollination and Seed Set:**

   » Pollination in sweet potato is primarily insect-mediated, with bees playing a significant role in transferring pollen from one flower to another. However, even with successful pollination, seed production is infrequent. In breeding programs, controlled hand pollination is used to cross varieties and produce seeds for hybrid development.

3. **Vegetative Propagation**:

   » Since sweet potato rarely produces viable seeds, commercial propagation is done vegetatively. Vine cuttings or slips are taken from the mother plant and transplanted into the field to grow new plants. This method of propagation allows for the rapid multiplication of superior varieties and ensures the uniformity of traits such as tuber size, shape, and color.

Breeding Methods in Sweet Potato

Breeding sweet potato is challenging due to its complex polyploidy (hexaploid, $2n = 6x = 90$) and limited seed production. However, breeding programs focus on improving traits such as yield, disease resistance, tuber quality, and adaptation to diverse environmental conditions.

Conventional Breeding Methods

1. **Introduction:**

   » The introduction of sweet potato varieties from different ecological zones plays an important role in expanding the genetic base of the crop. Breeding programs evaluate introduced varieties for their adaptability to local conditions and their potential to improve yield and resistance to pests and diseases.

2. **Selection:**

   » Selection is one of the most important breeding methods for sweet potato. Farmers and breeders select plants with desirable traits such as higher tuber yields, better storage quality, improved sweetness, and resistance to diseases. In regions where sweet potato is a staple crop, local landraces are often selected for their adaptability to local growing conditions.

Source Selections in Sweet Potato:

S No	Variety Name	Source	Key Features
1	Beauregard	USA	High yield, orange flesh, disease-resistant
2	Jewel	USA	Early maturity, sweet flavour
3	Tainung 65	Taiwan	Purple skin, high starch content
4	Resisto	USA	High beta-carotene content, nematode-resistant
5	Ejumula	Uganda	High beta-carotene, drought-tolerant

3. **Hybridization (Heterosis Breeding):**

   » Hybridization involves crossing different sweet potato varieties to combine desirable traits from both parents. For example, breeders may cross a variety with high yield potential with one that has resistance to pests or diseases. The resulting hybrids often exhibit hybrid vigor, with improved performance over the parent varieties.

4. **Polyploid Breeding:**

   » Due to sweet potato's polyploid nature, polyploid breeding techniques are used to manipulate the number of chromosomes in breeding lines. This allows breeders to introduce new genetic variations and develop superior hybrids with improved traits such as enhanced yield, better tuber quality, and resistance to abiotic stresses.

Non-Conventional Breeding Methods

1. **Tissue Culture:**

   » Tissue culture techniques are used to propagate disease-free sweet potato planting material and to rapidly multiply elite clones. This method is particularly important for producing virus-free plants and maintaining the genetic integrity of selected varieties. Tissue culture also enables the conservation of genetic diversity in breeding programs.

2. **Genetic Engineering (Recombinant DNA Technology):**

   » Genetic engineering holds promise in sweet potato breeding by allowing the introduction of genes that confer resistance to diseases, pests, and abiotic stresses. Although genetic engineering in sweet potato is still in the experimental stage, it has the potential to develop varieties with enhanced traits such as improved nutritional content and stress tolerance.

Breeding for Biotic and Abiotic Stresses in Sweet Potato

Sweet potato is susceptible to a wide range of biotic stresses, including fungal and bacterial diseases, as well as pests such as sweet potato weevil. Abiotic stresses like drought, salinity, and heat also affect sweet potato production, particularly in regions with poor soils or erratic rainfall.

Table: Biotic Stress-Resistant Varieties

S No	Disease/Pest	Source Species	Resistant Varieties
1	Sweet Potato Weevil	*Ipomoea batatas* (wild types)	Tainung 65, Resisto
2	Root-Knot Nematodes	*Ipomoea batatas*	Beauregard, Jewel
3	Fusarium Wilt	*Ipomoea batatas*	Ejumula, Jewel

1. **Sweet Potato Weevil Resistance**:

   » Sweet potato weevil is a major pest affecting tuber quality and yield. Resistant varieties like *Tainung 65* and *Resisto* have been developed to combat this pest, reducing the need for chemical control measures.

2. **Root-Knot Nematode Resistance:**

   » Root-knot nematodes can cause significant damage to sweet potato crops by stunting plant growth and reducing tuber size. Varieties like *Beauregard* and *Jewel* have shown resistance to nematode infestations, improving yields in nematode-infested fields.

Post-Harvest Management of Sweet Potato

Proper post-harvest management is crucial for maintaining the quality and extending the shelf life of sweet potatoes. Post-harvest losses are often caused by improper handling, storage conditions, and pest infestations.

1. **Harvesting Techniques:**
   » Sweet potatoes should be harvested when the leaves begin to yellow and die back, indicating that the tubers have reached full maturity. Careful manual harvesting is necessary to avoid damaging the delicate tuber skin, which can lead to post-harvest rotting.
2. **Curing and Drying:**
   » After harvesting, sweet potatoes should be cured in warm, humid conditions (around 25-30°C with 85-90% humidity) for 5-10 days. Curing allows small wounds on the tuber skin to heal, reducing the risk of rot during storage. It also enhances the flavor and sweetness of the tubers.
3. **Storage**:
   » Sweet potatoes should be stored in cool, dry conditions with adequate ventilation. The optimal storage temperature is 13-16°C with 85-90% humidity. Under proper storage conditions, sweet potatoes can be stored for several months without significant loss of quality.

Table: Post-Harvest Management Techniques for Sweet Potato

S No	Post-Harvest Stage	Best Practices	Impact on Quality
1	Harvesting	Harvest at full maturity, handle carefully	Preserves tuber quality, reduces damage
2	Curing and Drying	Cure in warm, humid conditions for 5-10 days	Enhances sweetness, prevents rotting
3	Storage	Store in cool, dry conditions with ventilation	Extends shelf life, reduces sprouting

Breeding of Cucurbitaceous Vegetables

A. Pumpkin (*Cucurbita spp.*)

Family: Cucurbitaceae

Chromosome Number: 2n = 2x = 40 (varies depending on species)

Pumpkins are cultivated for their edible fruit, seeds, and leaves. They are integral to food systems in many regions, providing essential nutrients like beta-carotene, vitamin C, and potassium. In addition to food uses, pumpkins are grown for their ornamental and cultural significance, particularly in festivals such as Halloween. The diversity in pumpkin species and varieties makes them versatile for various culinary, medicinal, and industrial purposes. Some varieties have thick flesh, making them ideal for soups and pies, while others are known for producing high-quality seeds or oil. Pumpkins are native to North and South America, and their domestication and cultivation have spread globally. The adaptability of pumpkin plants to a wide range of soils and climates, including arid and temperate regions, makes them a valuable crop worldwide. They are primarily cultivated in the USA, India, China, and Mexico, though they are grown in many other countries for both domestic use and export markets.

Origin and Domestication

Pumpkins have a long history of domestication, believed to have started in the Americas over 7,500 years ago. Archaeological evidence from Mexico indicates that ancient peoples used early forms of pumpkins, which were likely bitter and small. Through domestication, humans selected varieties with larger fruits, less bitterness, and increased sweetness. The domesticated species that evolved include *Cucurbita pepo*, *C. maxima*, and *C. moschata*, which are still cultivated today. Domestication of pumpkins was intertwined with the development of maize and beans, forming the "Three Sisters" agricultural system used by Native American tribes. This system utilized the complementary growing habits of the three crops—maize provided structure for beans to climb, while pumpkins spread across the ground, suppressing weeds and retaining moisture.

Botanical Varieties of Pumpkin

Pumpkins exhibit great diversity in terms of fruit size, color, shape, and flavor, depending on the species and variety. The major botanical varieties belong to the three key species *Cucurbita pepo*, *Cucurbita maxima*, and *Cucurbita moschata*. Each has unique characteristics and uses, ranging from fresh consumption to processing for oils or seeds.

Table: Major Types of Pumpkin on Basis of Fruit Characteristics and Use

S No	Variety Name	Species	Fruit Characteristics	Region Grown	Main Use
1	*Cucurbita pepo*	*C. pepo*	Small to medium, ribbed fruits	Global	Fresh consumption, ornamental
2	*Cucurbita maxima*	*C. maxima*	Large, round fruits with thick flesh	Temperate regions	Cooking, processing
3	*Cucurbita moschata*	*C. moschata*	Elongated or pear-shaped fruits	Tropical and subtropical regions	Fresh consumption, pies

1. ***Cucurbita pepo*** **(Field Pumpkin, Halloween Pumpkin):**

    » *Cucurbita pepo* varieties are well-known for their bright orange color and ribbed skin, particularly used for ornamental purposes such as carving for Halloween. However, many cultivars within this species are also consumed as food. The skin is generally tough, and the flesh is often less sweet than other pumpkin species, making them better suited for decoration or animal feed. The varieties within this species include summer squashes like zucchini and winter squashes like acorn squash, making *C. pepo* extremely diverse.

2. ***Cucurbita maxima*** **(Giant Pumpkin, Hubbard Squash):**

    » *Cucurbita maxima*produces some of the largest pumpkins, often grown for competitions. This species is characterized by its dense, sweet flesh, making it ideal for pies, soups, and purées. *C. maxima* is grown primarily in temperate regions and requires a longer growing season compared to *C. pepo*. Varieties such as *Big Max*, *Hubbard*, and *Buttercup* are highly prized for both culinary and processing purposes, as they produce large quantities of storable, nutrient-dense flesh.

3. ***Cucurbita moschata*** **(Butternut Squash, Cheese Pumpkin):**

    » This species is highly adaptable to warm and humid climates, thriving in tropical and subtropical regions. Its varieties, such as butternut squash, are popular for their smooth, sweet, and nutty flavor, which makes them perfect for pies, roasted dishes, and purées. *C. moschata* is also known for its excellent storage qualities, allowing the fruits to last for several months after harvest. Varieties like *Musquee de Provence* are popular in French cuisine, especially for making traditional pumpkin pies and soups.

Growth Habit of Pumpkin

Pumpkins are herbaceous annual plants that grow as sprawling vines. Their growth habit allows them to cover large areas, making them ideal for ground cover crops, which help suppress weeds and conserve soil moisture. Understanding the growth characteristics of pumpkin plants is essential for successful cultivation, particularly in terms of vine management, irrigation, and pest control.

1. **Vine Growth:**
   » Pumpkin plants produce long, prostrate vines that can spread up to 10 meters in length. The vines are fast-growing, and depending on the variety, they may exhibit different growth habits—some may be bushy, while others spread across the ground. The large, lobed leaves help protect the developing fruit from sun damage, while the tendrils allow the plant to anchor itself as it grows. The vigorous growth of pumpkin vines can lead to rapid ground cover, which is beneficial for reducing weed competition but requires ample space in the field.
2. **Root System:**
   » Pumpkins have a deep, fibrous root system that allows them to extract nutrients and water from deeper soil layers. The primary root, or taproot, can reach depths of up to 90 cm, with lateral roots extending horizontally to support the sprawling vines. The deep root system enables pumpkins to be somewhat drought-tolerant, but they require consistent moisture during the fruit-setting stage to ensure proper development.
3. **Flowering and Pollination:**
   » Pumpkin plants are monoecious, meaning they produce separate male and female flowers on the same plant. Male flowers typically appear first and are more abundant, while female flowers appear later and contain a small, immature fruit at the base. Pollination is primarily carried out by bees, which transfer pollen from the male flowers to the female flowers. Poor pollination can result in misshapen or small fruits, making bee activity essential for a successful pumpkin crop. Introducing beehives to pumpkin fields or ensuring the presence of pollinators can enhance fruit set and yield.
4. **Fruit Development**:
   » After successful pollination, the fruit begins to develop from the ovary of the female flower. The size and shape of the pumpkin depend on the variety and growing conditions. The fruits can range from small, decorative pumpkins to massive fruits weighing over 1,000 pounds. Most pumpkins take between 90 and 120 days to mature, with color change and hardening of the skin indicating that the fruit is ready for harvest.

Floral Biology of Pumpkin

Pumpkin flowers are large, bright yellow, and funnel-shaped, making them attractive to pollinators. Understanding the floral biology of pumpkins is crucial for improving fruit set and ensuring high yields, particularly in commercial cultivation systems that rely on natural pollination by insects.

1. **Male and Female Flowers:**
   » Pumpkin plants produce male and female flowers separately. Male flowers are borne

on long peduncles and contain pollen-producing stamens, while female flowers have a small, immature fruit at their base, which grows into a pumpkin after pollination. The ratio of male to female flowers can vary, but male flowers usually appear first. Ensuring a sufficient number of both flower types is essential for achieving good fruit set.

2. **Pollination:**

   » Pollination in pumpkins is primarily mediated by bees. Without sufficient pollination, the developing fruit will either not form or will be deformed. In large-scale production systems, pollinator-friendly practices such as introducing managed bee colonies or planting nectar-rich flowers around pumpkin fields can help ensure adequate pollination. It is also important to limit the use of pesticides during flowering to protect pollinators.

3. **Fruit Set and Development:**

   » The fruit develops from the fertilized ovary of the female flower. Depending on the variety, fruit development can take several weeks to several months. Factors such as temperature, water availability, and nutrient levels play a crucial role in the size and quality of the fruit. A well-pollinated flower will produce a symmetrical fruit with uniform skin color and shape.

Breeding Methods in Pumpkin

Pumpkin breeding focuses on improving yield, disease resistance, fruit quality, and adaptability to different environmental conditions. Breeding programs use both conventional and modern techniques to develop new varieties that meet the needs of farmers and consumers.

Conventional Breeding Methods

1. **Introduction:**

   » Introduction involves the evaluation of pumpkin varieties introduced from different regions or countries. This method is used to identify varieties that perform well under local environmental conditions and are resistant to local pests and diseases. Successful introductions can expand the genetic diversity of pumpkins, providing more options for farmers and consumers.

2. **Selection:**

   » Selection is a key breeding method in pumpkin cultivation. Farmers and breeders select plants with desirable traits such as higher yields, better fruit quality, and disease resistance. These selected plants are propagated and evaluated over multiple generations to develop improved varieties. Selection can be done through mass selection, where superior plants are chosen from a population, or through pedigree selection, where individual plants are selected and their offspring are evaluated.

Source Selections in Pumpkin:

S No	Variety Name	Source	Key Features
1	Jack O'Lantern	USA	Ideal for carving, medium-sized fruits
2	Big Max	USA	Large fruits, high yields
3	Musquee de Provence	France	Sweet flavor, excellent for pies
4	Buttercup	USA	Dense flesh, good storage quality
5	Atlantic Giant	USA	Extremely large fruits, competition

3. **Hybridization (Heterosis Breeding):**

 » Hybridization is used to combine the desirable traits of two or more parent plants. This method involves crossing different varieties to produce offspring that exhibit hybrid vigor, or heterosis, which results in better yield, disease resistance, and adaptability to stress. Hybrid varieties are often more uniform in their growth and fruiting patterns, making them ideal for commercial production.

4. **Backcross Breeding:**

 » Backcross breeding is used to introduce specific traits, such as disease resistance or improved fruit quality, into an established variety. This method involves crossing a hybrid plant with one of its parent plants to retain the desirable trait while maintaining the overall characteristics of the original variety.

Non-Conventional Breeding Methods

1. **Tissue Culture:**

 » Tissue culture is a technique used to rapidly propagate disease-free pumpkin plants from a small amount of tissue. This method is used in breeding programs to produce large quantities of uniform planting material, particularly for new varieties. Tissue culture is also used for the conservation of genetic material and the rapid multiplication of elite breeding lines.

2. **Genetic Engineering (Recombinant DNA Technology):**

 » Genetic engineering offers the potential to introduce new traits into pumpkin varieties, such as resistance to viruses or improved nutrient content. Although genetic engineering in pumpkins is still in its early stages, it holds promise for developing improved varieties that can withstand biotic and abiotic stresses.

Breeding for Biotic and Abiotic Stresses in Pumpkin

Pumpkins face a range of biotic stresses, including fungal diseases, insect pests, and viruses. Abiotic stresses such as drought, heat, and poor soil conditions also affect pumpkin yields. Breeding for resistance to these stresses is essential for improving pumpkin production, particularly in regions with challenging growing conditions.

Table: Biotic Stress-Resistant Varieties

S No	Disease/Pest	Source Species	Resistant Varieties
1	Powdery Mildew	*Cucurbita moschata*	Big Max, Buttercup
2	Downy Mildew	*Cucurbita pepo*	Jack O'Lantern, Musquee de Provence
3	Squash Vine Borer	*Cucurbita maxima*	Atlantic Giant, Buttercup

1. **Powdery Mildew Resistance:**
   » Powdery mildew is a common fungal disease affecting the leaves of pumpkin plants. It can reduce photosynthesis and lead to early leaf drop, ultimately decreasing yields. Breeding for powdery mildew resistance has resulted in varieties like *Big Max* and *Buttercup*, which show greater tolerance to this disease.
2. **Downy Mildew Resistance:**
   » Downy mildew is caused by the fungus-like organism *Pseudoperonosporacubensis*. It affects the foliage of pumpkins, leading to premature leaf drop and reduced yields. Varieties like *Jack O'Lantern* and *Musquee de Provence* have been bred for resistance to downy mildew, allowing for more reliable production in areas where this disease is prevalent.
3. **Squash Vine Borer Resistance:**
   » The squash vine borer is a significant pest of pumpkins, boring into the stems and causing the plant to wilt and die. Resistant varieties like *Atlantic Giant* and *Buttercup* have been developed to withstand this pest, reducing the need for chemical control measures.

Post-Harvest Management of Pumpkin

Proper post-harvest management is crucial for maintaining the quality of pumpkins and extending their shelf life. Pumpkins are highly perishable, and improper handling can result in significant post-harvest losses due to rot, bruising, and shrinkage. Post-harvest management practices such as proper curing, storage, and transportation are essential for ensuring that pumpkins reach consumers in optimal condition.

1. **Harvesting Techniques**:
   » Pumpkins should be harvested when the fruits are fully mature, indicated by a hard, smooth skin and a deep, uniform color. The stems should be cut several inches from the fruit to prevent decay at the stem end. It is important to handle pumpkins carefully during harvest to avoid bruising or damaging the skin, which can lead to post-harvest rot.
2. **Curing and Drying:**
   » After harvesting, pumpkins should be cured by placing them in a warm, dry area with good air circulation for 10-14 days. Curing allows the skin to harden and the

wounds to heal, reducing the risk of rot during storage. Curing also helps to enhance the flavor and sweetness of the fruit by reducing moisture content in the flesh.

3. **Storage:**

   » Pumpkins should be stored in a cool, dry environment with good ventilation. The ideal storage temperature is between 10°C and 15°C, with relative humidity levels of around 60-70%. Under these conditions, pumpkins can be stored for several months without significant loss of quality. Proper storage helps to prevent sprouting, rot, and shrinkage, ensuring that the pumpkins remain fresh for an extended period.

Table: Post-Harvest Management Techniques for Pumpkin

S No	Post-Harvest Stage	Best Practices	Impact on Quality
1	Harvesting	Harvest at full maturity, handle carefully	Preserves fruit quality, reduces damage
2	Curing and Drying	Cure in a warm, dry area for 10-14 days	Enhances sweetness, reduces spoilage
3	Storage	Store in cool, dry conditions with ventilation	Extends shelf life, reduces rot

B. Bitter Gourd (*Momordica charantia*)

Family: Cucurbitaceae

Chromosome Number: 2n = 2x = 22

Bitter gourd, also known as bitter melon, is a highly nutritious and medicinal crop grown primarily in tropical and subtropical regions. It is valued for its unique bitter taste and high medicinal properties, including its potential to regulate blood sugar levels, making it popular in traditional medicine for treating diabetes. Bitter gourd is also rich in vitamins A, C, and K, and is a good source of dietary fiber, iron, and potassium. It is widely used in Asian, African, and Caribbean cuisines, often in stir-fries, soups, and curries.This crop thrives in warm, humid climates and grows as a climbing vine. Its fruits are oblong and vary in color from light green to dark green. The surface of the fruit is covered in wart-like bumps, and the fruit's interior contains seeds that become orange-red when ripe.

Origin and Domestication

Bitter gourd is native to tropical Asia, particularly India and southern China. Its domestication is believed to have occurred over 2,000 years ago when it was first cultivated for its medicinal properties. Over time, various landraces and cultivars were developed, exhibiting different fruit shapes, sizes, and degrees of bitterness.The crop spread throughout Southeast Asia, Africa, and the Caribbean through trade and migration. Traditional breeding practices in these regions resulted in the development of local varieties adapted to specific climates and uses. Domestication focused on selecting varieties with manageable bitterness levels, larger fruit sizes, and enhanced medicinal benefits.

Botanical Varieties of Bitter Gourd

Bitter gourd varieties can be classified based on their fruit shape, size, and surface texture. Some varieties are small and spiny, while others are large and smooth. Different varieties are used for distinct culinary and medicinal purposes.

Table: Major Types of Bitter Gourd on Basis of Fruit Characteristics and Use

S No	Variety Name	Fruit Characteristics	Region Grown	Main Use
1	Pusa Do Mausami	Small to medium, spiny, dark green	India, Southeast Asia	Fresh consumption, stir-fries
2	Arka Harit	Large, smooth, light green	India, Bangladesh	Fresh consumption, curries
3	Long Green	Elongated, smooth, pale green	China, Japan, Taiwan	Cooking, pickling

4	White Pearl	Small to medium, smooth, white	Southeast Asia, Caribbean	Soups, medicinal use
5	Baby Bitter Gourd	Small, spiny, deep green	Southeast Asia	Medicinal, pickling

1. **Pusa Do Mausami:**

   » This variety produces medium-sized fruits with a spiny surface and dark green skin. It is widely grown in India and other parts of Southeast Asia. Known for its robust flavor and moderate bitterness, *Pusa Do Mausami* is often used in stir-fries and curries. It is a highly adaptable variety that performs well under diverse environmental conditions.

2. **Arka Harit:**

   » *Arka Harit* is known for its large, smooth, light-green fruits. This variety is popular in India and Bangladesh and is prized for its mild bitterness and high yield. It is commonly used in various dishes, including curries and stir-fries. The smooth-skinned fruits are easier to clean and prepare, making them a popular choice in markets.

3. **Long Green:**

   » *Long Green* is an elongated, smooth variety that is less bitter than traditional spiny varieties. It is widely cultivated in China, Japan, and Taiwan, where it is used in soups, stir-fries, and pickling. The variety is known for its delicate flavor and tender texture, making it ideal for fresh consumption.

4. **White Pearl:**

   » *White Pearl* is a smooth-skinned, white-fruited variety of bitter gourd. It is commonly grown in Southeast Asia and the Caribbean, where it is used in traditional soups and medicinal preparations. The white fruits are less bitter than green varieties and are often used in mild-flavored dishes.

5. **Baby Bitter Gourd:**

   » *Baby Bitter Gourd* produces small, spiny fruits that are popular for pickling and medicinal use. It is commonly grown in Southeast Asia and is known for its intense bitterness and strong medicinal properties. The small size of the fruit makes it ideal for quick preparation and consumption.

Growth Habit of Bitter Gourd

Bitter gourd is a fast-growing, climbing vine that requires support to grow vertically. The plant produces tendrils that allow it to attach to trellises or other structures. The growth habit of bitter gourd makes it well-suited for vertical farming or growing in small spaces, where the vines can spread along trellises to maximize space utilization.

1. **Vine Growth:**

   » The vine can grow up to 5 meters in length, producing long, trailing stems that require support to grow upward. The plant's vigorous growth habit allows it to cover large areas quickly, making it an ideal crop for vertical or trellis farming. The vines are covered with large, deeply lobed leaves that help shade the fruit and protect it from sunscald.

2. **Root System:**

   » Bitter gourd has a deep taproot system that allows it to access water and nutrients from deep within the soil. The extensive root system makes the plant relatively drought-tolerant, though consistent moisture is necessary for optimal growth. The root system also helps anchor the plant as it climbs and spreads.

3. **Flowering and Pollination:**

   » Bitter gourd is monoecious, meaning it produces separate male and female flowers on the same plant. The large, bright yellow flowers are attractive to pollinators such as bees, which play a crucial role in transferring pollen from the male to the female flowers. Inadequate pollination can result in poor fruit set, making pollinator activity critical for successful production.

4. **Fruit Development:**

   » After pollination, the female flower develops into a fruit. The size, shape, and bitterness of the fruit depend on the variety and growing conditions. Bitter gourd fruits are typically harvested when they are still immature, as fully mature fruits become overly bitter and tough. The fruit can vary from 10 to 50 cm in length, depending on the variety.

Floral Biology of Bitter Gourd

Understanding the floral biology of bitter gourd is essential for optimizing fruit set and improving yields. The crop's monoecious nature, combined with its reliance on insect pollination, makes proper flower management important for achieving good yields.

1. **Male and Female Flowers**:

   » Bitter gourd produces separate male and female flowers on the same plant. Male flowers appear first, followed by female flowers. Female flowers can be distinguished by the small, bulbous ovary at their base, which develops into the fruit after successful pollination. Male flowers contain pollen-bearing stamens, while female flowers contain the ovary, style, and stigma.

2. **Pollination:**

   » Pollination is mainly carried out by bees, which transfer pollen from male flowers to female flowers. Cross-pollination is important for fruit development, and the presence of pollinators is crucial for achieving high fruit set. In commercial farming, beehives may be introduced to the fields to ensure adequate pollination.

3. **Fruit Set and Development:**

   » Once pollinated, the female flower's ovary begins to swell and develop into a fruit. The time from pollination to fruit maturity is generally 2-3 weeks, depending on environmental conditions. Proper fruit set requires consistent pollination, and poor pollination can lead to misshapen or underdeveloped fruits.

Breeding Methods in Bitter Gourd

Breeding programs for bitter gourd focus on improving yield, fruit quality, disease resistance, and tolerance to environmental stresses. Both conventional and non-conventional breeding methods are used to develop new varieties that meet the needs of farmers and consumers.

Conventional Breeding Methods

1. **Introduction:**

   » The introduction of bitter gourd varieties from different regions allows breeders to evaluate their performance under local conditions. This method helps expand the genetic base of bitter gourd and identifies varieties that perform well in specific environments.

2. **Selection:**

   » Selection is a common breeding method used to improve bitter gourd varieties. Superior plants are selected based on traits such as yield, fruit size, bitterness level, and disease resistance. Selection can be done through mass selection, pedigree selection, or recurrent selection, depending on the breeding goals.

Source Selections in Bitter Gourd:

S No	Variety Name	Source	Key Features
1	Pusa Vishesh	India	High yield, dark green fruits
2	Coimbatore Long	India	Long fruits, mild bitterness
3	Long Green	China	Long, smooth fruits
4	Arka Harit	India	Large, smooth fruits
5	Baby Bitter Gourd	Southeast Asia	Small, spiny fruits, medicinal use

3. **Hybridization (Heterosis Breeding):**

   » Hybridization is used to combine desirable traits from different parent plants. This method often leads to hybrids that exhibit hybrid vigor, resulting in higher yields, better disease resistance, and improved fruit quality. Hybrid varieties are commonly developed for commercial farming due to their uniformity and consistent performance.

4. **Backcross Breeding:**

   » Backcross breeding is used to introduce specific traits, such as disease resistance, into an established variety. The hybrid offspring is crossed back with one of its parents to retain the desired trait while maintaining the overall characteristics of the original variety.

Non-Conventional Breeding Methods

1. **Tissue Culture:**

   » Tissue culture techniques are used to propagate disease-free planting material and maintain genetic integrity in breeding programs. This method is particularly important for producing virus-free plants and ensuring high-quality seeds for commercial production.

2. **Genetic Engineering (Recombinant DNA Technology):**

   » Genetic engineering is being explored to develop bitter gourd varieties with enhanced traits such as disease resistance, improved nutritional content, and tolerance to abiotic stresses. Although still in the experimental phase, genetic engineering holds promise for future bitter gourd breeding programs.

Breeding for Biotic and Abiotic Stresses in Bitter Gourd

Bitter gourd is susceptible to various biotic stresses, including fungal diseases, viral infections, and insect pests. Abiotic stresses such as drought, salinity, and heat also affect bitter gourd production, particularly in regions with erratic rainfall or poor soils. Breeding for resistance to these stresses is essential for improving the productivity and resilience of bitter gourd crops.

Table: Biotic Stress-Resistant Varieties

S No	Disease/Pest	Source Species	Resistant Varieties
1	Powdery Mildew	*Momordica charantia*	Pusa Vishesh, Arka Harit
2	Downy Mildew	*Momordica charantia*	Coimbatore Long, Pusa Do Mausami
3	Fruit Fly	*Momordica charantia*	Long Green, White Pearl

1. **Powdery Mildew Resistance:**

   » Powdery mildew is a common fungal disease that affects the leaves of bitter gourd plants, leading to reduced photosynthesis and premature leaf drop. Resistant varieties like *Pusa Vishesh* and *Arka Harit* are developed to minimize the impact of this disease, allowing for higher yields and better fruit quality.

2. **Downy Mildew Resistance:**

   » Downy mildew is caused by the pathogen *Pseudoperonosporacubensis* and primarily affects the foliage of bitter gourd plants. This disease can lead to significant yield losses if not properly managed. Varieties such as *Coimbatore Long* and *Pusa Do Mausami* have shown resistance to downy mildew, making them suitable for areas where this disease is prevalent.

3. **Fruit Fly Resistance:**

   » Fruit flies are a major pest affecting bitter gourd production, causing damage to the developing fruits, and leading to poor fruit quality. Resistant varieties like *Long Green* and *White Pearl* have been developed to reduce the damage caused by fruit flies, improving both yield and fruit quality.

Post-Harvest Management of Bitter Gourd

Proper post-harvest management is crucial for maintaining the quality and shelf life of bitter gourd fruits. The fruits are highly perishable and require careful handling to avoid damage and spoilage. Post-harvest losses can be minimized through proper harvesting, storage, and transportation practices.

1. **Harvesting Techniques:**

   » Bitter gourd fruits should be harvested while they are still immature and green. Harvesting at the right stage of maturity ensures that the fruits retain their tenderness and have a balanced bitterness. The fruits should be carefully cut from the vine to avoid damaging the plant or bruising the fruit, which can lead to post-harvest spoilage.

2. **Curing and Drying:**

   » Curing is not commonly practiced for bitter gourd as the fruits are typically consumed fresh. However, in some regions, drying techniques are used to preserve the fruits for medicinal purposes. Dried bitter gourd slices are often used in herbal teas or as a supplement in traditional medicine.

3. **Storage:**

   » Bitter gourd fruits should be stored at low temperatures (10-12°C) to slow down respiration and delay spoilage. Proper ventilation is also important to prevent the accumulation of moisture, which can lead to fungal growth. Under optimal storage conditions, bitter gourd fruits can be stored for 7-10 days without significant loss of quality.

Table: Post-Harvest Management Techniques for Bitter Gourd

S No	Post-Harvest Stage	Best Practices	Impact on Quality
1	Harvesting	Harvest at immature stage, handle carefully	Retains fruit tenderness, reduces damage
2	Storage	Store at 10-12°C with good ventilation	Extends shelf life, reduces spoilage
3	Transportation	Pack in ventilated crates, avoid bruising	Maintains quality during transit

C. Parval (*Trichosanthes dioica*)

Family: Cucurbitaceae

Chromosome Number: 2n = 2x = 22

Parval, or pointed gourd, is an important vegetable crop cultivated mainly in tropical and subtropical regions of South Asia, particularly in India, Bangladesh, and Nepal. The fruit is highly nutritious, low in calories, and rich in dietary fiber, making it an ideal component of a healthy diet. It is commonly used in cooking, where it is consumed in fresh, fried, or curried forms. Parval is valued for its medicinal properties as well, particularly in the management of digestive disorders, diabetes, and weight loss.The crop thrives in warm, humid climates with well-drained loamy or sandy soils. Parval is a perennial climbing vine that requires proper support for optimal growth. It is propagated mainly through vine cuttings, and vegetative propagation ensures the genetic uniformity of plants. Pointed gourd requires careful management of nutrients, water, and pests to ensure high productivity and fruit quality.

Origin and Domestication

Pointed gourd is believed to have originated in the Indian subcontinent, where it has been cultivated for centuries as a staple vegetable. It is a member of the Cucurbitaceae family, which includes other important crops such as cucumbers, gourds, and melons. Parval's domestication focused on improving fruit size, yield, and resistance to pests and diseases. Over time, traditional landraces of pointed gourd have been selected for their desirable traits, such as fruit shape, size, and flavor, and modern breeding techniques have further enhanced these characteristics.

Botanical Varieties of Parval

Pointed gourd varieties are typically categorized based on fruit shape, size, and flavor. The fruits may be small to medium-sized, and the color can vary from light green to dark green, often with white or pale green stripes. The fruits are cylindrical or slightly tapering at both ends, and the flesh is soft with small, edible seeds.

Table: Major Types of Parval on Basis of Fruit Characteristics and Use

S No	Variety Name	Fruit Characteristics	Region Grown	Main Use
1	Rajendra Parwal-1	Medium-sized, dark green	India	Fresh consumption
2	Swarna Alaukik	Large-sized, smooth green fruit	India	Fresh consumption
3	Kalyanpur Long Green	Long, green, smooth	Northern India	Fresh consumption
4	Arka Neelachal	Oval-shaped, green with white stripes	Odisha, India	Fresh consumption, pickling
5	BCS Parwal-1	Medium-sized, dark green	Bangladesh	Fresh consumption

1. **Rajendra Parwal-1:**

   » *Rajendra Parwal-1* is a medium-sized pointed gourd variety with dark green skin. It is popular in India for its uniform fruit size and high yield potential. The fruits are ideal for fresh consumption and have a smooth texture and mild flavor.

2. **Swarna Alaukik:**

   » This variety produces large-sized fruits with smooth green skin. *Swarna Alaukik* is known for its high productivity and resistance to common pests and diseases. It is widely grown in India and is favored for fresh consumption due to its tender, juicy flesh.

3. **Kalyanpur Long Green:**

   » *Kalyanpur Long Green* is a long, smooth-skinned variety that is popular in Northern India. The fruits are cylindrical and dark green, making them ideal for fresh consumption. This variety is known for its good shelf life and resistance to common diseases.

4. **Arka Neelachal:**

   » *Arka Neelachal* is an oval-shaped variety with green skin and white stripes. It is cultivated primarily in Odisha, India, and is suitable for both fresh consumption and pickling. The variety is appreciated for its firm flesh and resistance to common fungal diseases.

5. **BCS Parwal-1:**

   » *BCS Parwal-1* is a medium-sized pointed gourd variety with dark green skin. It is widely grown in Bangladesh and is known for its high yield and good fruit quality. The fruits are consumed fresh and are commonly used in traditional dishes.

Growth Habit of Parval

Parval is a perennial, climbing vine that requires support for optimal growth. The plant has vigorous vegetative growth and produces long, trailing vines that can reach up to 3-5 meters in length. The crop is typically grown on trellises or stakes to allow the vines to climb and spread, which improves air circulation and sunlight exposure for better fruit development.

1. **Vine Growth:**

   » The vines of pointed gourd grow rapidly and produce large, lobed leaves that shade the fruits and help prevent sunscald. The vigorous growth of the vines requires regular pruning to maintain plant health and optimize fruit production.

2. **Root System:**

   » Parval has a fibrous root system that helps it absorb water and nutrients from the soil. It is a relatively drought-tolerant crop, but consistent irrigation is necessary during the fruiting stage to ensure good yields and high-quality fruits.

3. **Flowering and Pollination:**

   » Pointed gourd produces separate male and female flowers on the same plant (monoecious). The male flowers are more numerous and produce pollen, while the female flowers develop into fruits after successful pollination. Pollination is primarily carried out by insects such as bees and ants.

4. **Fruit Development:**

   » After pollination, the female flowers develop into fruits. Parval fruits are typically harvested when they are tender and immature. Harvesting at the right stage ensures that the fruits retain their mild flavor and tender texture.

Floral Biology of Parval

The floral biology of the pointed gourd is crucial for optimizing fruit set and improving yields. Proper pollination management is essential for ensuring successful fruit development and high-quality yields.

1. **Male and Female Flowers:**

   » Pointed gourd produces separate male and female flowers. Male flowers are borne in clusters and appear first, followed by female flowers, which have a small, bulbous ovary at the base that develops into the fruit after pollination.

2. **Pollination:**

   » Pollination is essential for fruit development in pointed gourd. The primary pollinators are insects such as bees, which transfer pollen from male to female flowers. In regions with low pollinator activity, hand pollination may be required to ensure good fruit set.

3. **Fruit Set and Development:**

   » Once pollinated, the ovary of the female flower swells and develops into a fruit. Fruits take about 60-90 days from planting to maturity, depending on the variety and environmental conditions. Fruits should be harvested while still tender to ensure the best flavor and texture.

Breeding Methods in Parval

Breeding efforts in pointed gourd focus on improving traits such as yield, fruit quality, disease resistance, and adaptability to various growing conditions. Both conventional and non-conventional breeding methods are used to develop improved varieties.

Conventional Breeding Methods

1. **Introduction:**

   » Introduction of pointed gourd varieties from different regions allows breeders to evaluate their performance in local environments. This method helps expand the genetic diversity of pointed gourd and identify varieties that perform well under specific growing conditions.

2. **Selection:**

   » Selection is a key breeding method used to improve pointed gourd varieties. Farmers and breeders select plants with desirable traits such as higher yields, disease resistance, and improved fruit quality. These selected plants are propagated and evaluated over multiple generations to develop superior varieties.

Source Selections in Parval:

S No	Variety Name	Source	Key Features
1	Rajendra Parwal-1	India	Uniform fruit size, high yield
2	Swarna Alaukik	India	Large-sized fruits, disease resistance
3	Kalyanpur Long Green	Northern India	Long, smooth fruits
4	Arka Neelachal	Odisha, India	Firm flesh, oval shape
5	BCS Parwal-1	Bangladesh	High yield, dark green fruits

3. **Hybridization (Heterosis Breeding):**

   » Hybridization is used to combine desirable traits from different parent plants. This method leads to hybrid vigor, resulting in higher yields, better fruit quality, and improved disease resistance. Hybrid pointed gourd varieties are developed to enhance productivity in commercial farming systems.

4. **Backcross Breeding**:

   » Backcross breeding is employed to introduce specific traits, such as disease resistance or improved fruit size, into an established variety. The hybrid offspring is crossed back with one of its parents to retain the desired trait while maintaining the overall characteristics of the original variety.

Non-Conventional Breeding Methods

1. **Tissue Culture:**

   » Tissue culture techniques are used to propagate disease-free planting material and maintain genetic uniformity in pointed gourd breeding programs. This method is particularly important for producing high-quality seeds and ensuring the genetic purity of elite varieties.

2. **Genetic Engineering (Recombinant DNA Technology)**:

   » Genetic engineering is being explored to develop pointed gourd varieties with enhanced traits such as disease resistance, improved nutritional content, and tolerance to abiotic stresses. Although still in the experimental phase, genetic engineering holds promise for future breeding programs in pointed gourd.

Breeding for Biotic and Abiotic Stresses in Parval

Pointed gourd is susceptible to various biotic stresses, including fungal diseases, viral infections, and insect pests. Abiotic stresses such as drought, salinity, and heat also affect pointed gourd production. Breeding for resistance to these stresses is essential for improving the productivity and resilience of the crop.

Table: Biotic Stress-Resistant Varieties

S No	Disease/Pest	Source Species	Resistant Varieties
1	Powdery Mildew	*Trichosanthes dioica*	Arka Neelachal, Rajendra Parwal-1
2	Downy Mildew	*Trichosanthes dioica*	Swarna Alaukik, Kalyanpur Long Green
3	Fruit Fly	*Trichosanthes dioica*	BCS Parwal-1, Arka Neelachal

1. **Powdery Mildew Resistance:**
   » Powdery mildew is a common fungal disease affecting pointed gourd plants, leading to reduced photosynthesis and premature leaf drop. Resistant varieties like *Arka Neelachal* and *Rajendra Parwal-1* have been developed to minimize the impact of this disease, ensuring higher yields and improved fruit quality.
2. **Downy Mildew Resistance**:
   » Downy mildew is caused by the pathogen *Pseudoperonosporacubensis* and affects the leaves of pointed gourd plants. This disease can cause significant yield losses if not managed. Varieties such as *SwarnaAlaukik* and *Kalyanpur Long Green* have shown resistance to downy mildew, making them suitable for cultivation in regions where this disease is prevalent.
3. **Fruit Fly Resistance:**
   » Fruit flies are a major pest affecting pointed gourd fruits. Resistant varieties like *BCS Parwal-1* and *Arka Neelachal* help reduce the damage caused by fruit flies, improving both yield and fruit quality.

Post-Harvest Management of Parval

Proper post-harvest management is essential for maintaining the quality and shelf life of pointed gourd fruits. Since parval is a perishable crop, it requires careful handling during and after harvest to avoid damage and spoilage.

1. **Harvesting Techniques:**
   » Parval fruits should be harvested while they are still tender and immature. Overripe fruits can become tough and lose their flavor. The fruits should be cut from the vine carefully to avoid bruising, as damaged fruits are more prone to spoilage.

2. **Storage:**

   » Pointed gourd should be stored at temperatures between 10-12°C to slow down respiration and delay spoilage. Proper ventilation is important to prevent moisture buildup, which can lead to fungal growth. Under optimal conditions, parval can be stored for 7-10 days without significant loss of quality.

3. **Transportation:**

   » Parval fruits are transported in ventilated crates or containers to prevent bruising during transit. Proper cushioning is used to protect the fruits from mechanical damage, ensuring that they arrive at the market in good condition.

Table: Post-Harvest Management Techniques for Parval

S No	Post-Harvest Stage	Best Practices	Impact on Quality
1	Harvesting	Harvest while tender, handle carefully	Retains flavor and texture
2	Storage	Store at 10-12°C with good ventilation	Extends shelf life, reduces spoilage
3	Transportation	Pack in ventilated crates, cushion properly	Maintains quality during transi

D. Ridge Gourd (*Luffa acutangula*)

Family: Cucurbitaceae

Chromosome Number: 2n = 2x = 26

Ridge gourd, also known as angled luffa or sponge gourd, is an important vegetable crop grown for its immature fruits. It is widely consumed in tropical and subtropical regions for its tender, fibrous texture and mild flavor. Ridge gourd is a versatile crop used in various culinary applications, including soups, stir-fries, and curries. Apart from its culinary use, ridge gourd is also valued for its medicinal properties, such as aiding digestion, promoting skin health, and acting as a natural detoxifier.The crop is well-suited to warm and humid climates, with a long growing season. Ridge gourd is commonly grown in countries like India, China, Thailand, and other parts of Southeast Asia, where it is an integral part of the diet.

Origin and Domestication

Ridge gourd is believed to have originated in South and Southeast Asia, where it has been cultivated for centuries. Historical records suggest that it was domesticated in India, and from there, it spread to other parts of Asia and Africa. The domestication of ridge gourd focused on selecting varieties with tender fruits, low bitterness, and high productivity.The plant's adaptability to diverse climatic conditions allowed it to be cultivated in various regions. Traditional farmers practiced seed saving and selection, leading to the development of local varieties suited to specific growing conditions and culinary preferences.

Botanical Varieties of Ridge Gourd

Ridge gourd varieties can vary in fruit size, shape, and ribbing pattern. The most common varieties have elongated, ribbed fruits, which give the vegetable its distinctive appearance. Depending on the growing region and market demands, different varieties are cultivated for specific uses, such as fresh consumption or processing.

Table: Major Botanical Varieties of Ridge Gourd

S No	Variety Name	Fruit Characteristics	Region Grown	Main Use
1	Pusa Nasdar	Elongated, ridged, green fruits	India, Southeast Asia	Fresh consumption, cooking
2	Arka Prasan	Large, smooth, dark green fruits	India, Bangladesh	Fresh consumption, stir-fries
3	Coimbatore Long	Long, light green fruits	India, Southeast Asia	Cooking, medicinal use
4	Jaipur Long	Slender, deeply ribbed fruits	India	Culinary uses, pickling
5	White Beauty	Small to medium, white fruits	China, Southeast Asia	Fresh consumption

1. **Pusa Nasdar:**
   - This variety is known for its long, green, ridged fruits that are commonly used in Indian cuisine. It is a high-yielding variety with tender, non-bitter fruits, making it ideal for stir-fries, soups, and curries. *Pusa Nasdar* is widely grown in India and other parts of Southeast Asia, where it is valued for its consistent productivity and adaptability to various climates.
2. **Arka Prasan:**
   - *Arka Prasan* is a high-yielding variety with smooth, dark green fruits that are larger and thicker than other varieties. It is popular in Bangladesh and India for its uniform fruit size and mild flavor. The fruits are often used in curries and stir-fries and have a long shelf life, making them suitable for commercial markets.
3. **Coimbatore Long:**
   - *Coimbatore Long* produces elongated, light green fruits with tender flesh and a mild flavor. This variety is commonly grown in South India and is used in a variety of traditional dishes. Its tender fruits are highly regarded for their culinary versatility and ease of preparation.
4. **Jaipur Long:**
   - *Jaipur Long* is a slender, deeply ribbed variety that is highly prized for its aesthetic appearance and taste. It is commonly grown in North India and is used in pickling, curries, and stir-fries. This variety is known for its distinctively long fruits and adaptability to local growing conditions.
5. **White Beauty:**
   - *White Beauty* is a rare variety of ridge gourd that produces small to medium-sized, smooth, white fruits. It is commonly grown in China and Southeast Asia and is used for fresh consumption. The white fruits are mild in flavor and are often included in soups and stir-fries.

Growth Habit of Ridge Gourd

Ridge gourd is a vigorous, climbing vine that requires support to grow optimally. It is grown as an annual crop in warm climates, where it thrives under full sunlight and well-drained soils. Ridge gourd's growth habit makes it ideal for trellising, which enhances fruit quality and improves yields.

1. **Vine Growth:**
   - The plant produces long, trailing vines that can extend up to 5 meters. The vines are supported by tendrils that help the plant climb trellises or other support structures. Ridge gourd's vigorous growth habit allows it to cover large areas quickly, making it an efficient crop for vertical farming or intercropping with other crops.

2. **Root System:**

    » Ridge gourd has a deep taproot system that supports its rapid growth and allows it to access water and nutrients from deeper soil layers. The deep roots help the plant tolerate drought conditions, although consistent moisture is required for optimal fruit development. The root system also helps anchor the plant, making it more stable as it grows vertically.

3. **Flowering and Pollination:**

    » Ridge gourd produces large, yellow flowers that are either male or female. The plant is monoecious, meaning that both male and female flowers are produced on the same plant. Pollination is carried out by insects, mainly bees, which transfer pollen from male flowers to female flowers. Adequate pollinator activity is essential for good fruit set.

4. **Fruit Development:**

    » After successful pollination, the female flower develops into a fruit. Ridge gourd fruits are long and ridged, with a tender texture when harvested at the immature stage. The fruits can grow to lengths of 30 to 60 cm, depending on the variety. Ridge gourd is typically harvested before the fruit becomes fully mature, as overripe fruits can become tough and bitter.

Floral Biology of Ridge Gourd

Understanding the floral biology of ridge gourd is important for optimizing fruit set and improving yields. Proper pollination management is key to ensuring successful fruit development and high-quality yields.

1. **Male and Female Flowers:**

    » Ridge gourd produces separate male and female flowers on the same plant. Male flowers appear first and are more numerous than female flowers. Female flowers can be distinguished by the small ovary at their base, which develops into the fruit after pollination. Male flowers produce pollen that is transferred to female flowers by pollinators, primarily bees.

2. **Pollination:**

    » Pollination is a critical factor in ridge gourd production. Poor pollination can lead to misshapen or underdeveloped fruits. To ensure successful pollination, growers may introduce beehives to the fields or plant pollinator-attracting flowers nearby. Hand pollination can also be practiced in small-scale farming to improve fruit set.

3. **Fruit Set and Development:**

    » Following successful pollination, the ovary of the female flower begins to swell and develop into a fruit. Ridge gourd fruits develop rapidly, typically reaching harvestable size within 10-14 days after pollination. The fruits should be harvested while they are still tender to ensure the best flavor and texture.

Breeding Methods in Ridge Gourd

Ridge gourd breeding focuses on improving traits such as yield, fruit quality, disease resistance, and tolerance to environmental stresses. Both conventional and non-conventional breeding methods are used to develop improved varieties.

Conventional Breeding Methods

1. **Introduction:**

   » Introduction of ridge gourd varieties from different regions allows breeders to evaluate their performance under local conditions. This method helps identify varieties that are well-suited to specific climates, soils, and market demands.

2. **Selection:**

   » Selection is a key breeding method for improving ridge gourd varieties. Farmers and breeders select plants with desirable traits such as higher yields, better fruit quality, and disease resistance. These selected plants are then propagated and evaluated over multiple generations to develop improved varieties.

Source Selections in Ridge Gourd:

S No	Variety Name	Source	Key Features
1	Pusa Nasdar	India	High yield, tender fruits
2	Coimbatore Long	India	Long fruits, mild flavor
3	Arka Prasan	India	High yielding, uniform fruits
4	Jaipur Long	India	Long, ribbed fruits
5	White Beauty	China	Small, white fruits

3. **Hybridization (Heterosis Breeding):**

   » Hybridization is used to combine desirable traits from different parent plants. This method often leads to hybrids that exhibit hybrid vigor, resulting in higher yields, better disease resistance, and improved fruit quality. Hybrid varieties are commonly developed for commercial farming due to their uniformity and consistent performance.

4. Ba**ckcross Breeding:**

   » Backcross breeding is used to introduce specific traits, such as disease resistance or improved fruit quality, into an established variety. The hybrid offspring is crossed back with one of its parents to retain the desired trait while maintaining the overall characteristics of the original variety.

Non-Conventional Breeding Methods

1. **Tissue Culture:**

   » Tissue culture techniques are used to propagate disease-free planting material and maintain genetic integrity in breeding programs. This method is particularly important for producing virus-free plants and ensuring high-quality seeds for commercial production.

2. **Genetic Engineering (Recombinant DNA Technology):**

   » Genetic engineering is being explored to develop ridge gourd varieties with enhanced traits such as disease resistance, improved nutritional content, and tolerance to abiotic stresses. Although still in the experimental phase, genetic engineering holds promise for future ridge gourd breeding programs.

Breeding for Biotic and Abiotic Stresses in Ridge Gourd

Ridge gourd is susceptible to a range of biotic stresses, including fungal diseases, viral infections, and insect pests. Abiotic stresses such as drought, salinity, and high temperatures also affect ridge gourd production. Breeding for resistance to these stresses is essential for improving the productivity and resilience of ridge gourd crops.

Table: Biotic Stress-Resistant Varieties

S No	Disease/Pest	Source Species	Resistant Varieties
1	Powdery Mildew	*Luffa acutangula*	Pusa Nasdar, Arka Prasan
2	Downy Mildew	*Luffa acutangula*	Coimbatore Long, Jaipur Long
3	Fruit Fly	*Luffa acutangula*	White Beauty, Arka Prasan

1. **Powdery Mildew Resistance:**

   » Powdery mildew is a common fungal disease affecting the leaves of ridge gourd plants, leading to reduced photosynthesis and premature leaf drop. Resistant varieties like *Pusa Nasdar* and *Arka Prasan* are developed to minimize the impact of this disease, allowing for higher yields and better fruit quality.

2. **Downy Mildew Resistance:**

   » Downy mildew is caused by the pathogen *Pseudoperonosporacubensis* and primarily affects the foliage of ridge gourd plants. This disease can lead to significant yield losses if not properly managed. Varieties such as *Coimbatore Long* and *Jaipur Long* have shown resistance to downy mildew, making them suitable for areas where this disease is prevalent.

3. **Fruit Fly Resistance:**

   » Fruit flies are a major pest affecting ridge gourd production, causing damage to the developing fruits, and leading to poor fruit quality. Resistant varieties like *White*

Beauty and *Arka Prasan* have been developed to reduce the damage caused by fruit flies, improving both yield and fruit quality.

Post-Harvest Management of Ridge Gourd

Proper post-harvest management is crucial for maintaining the quality and shelf life of ridge gourd fruits. The fruits are highly perishable and require careful handling to avoid damage and spoilage. Post-harvest losses can be minimized through proper harvesting, storage, and transportation practices.

1. **Harvesting Techniques:**
   » Ridge gourd fruits should be harvested while they are still immature and tender. Harvesting at the right stage ensures that the fruits retain their mild flavor and tender texture. The fruits should be carefully cut from the vine to avoid damaging the plant or bruising the fruit, which can lead to post-harvest spoilage.
2. **Curing and Drying:**
   » Curing is not commonly practiced for ridge gourd as the fruits are typically consumed fresh. However, in some regions, drying techniques are used to preserve the fruits for medicinal purposes. Dried ridge gourd is often used in traditional medicine or as a supplement in herbal preparations.
3. **Storage:**
   » Ridge gourd fruits should be stored at low temperatures (10-12°C) to slow down respiration and delay spoilage. Proper ventilation is also important to prevent the accumulation of moisture, which can lead to fungal growth. Under optimal storage conditions, ridge gourd fruits can be stored for 7-10 days without significant loss of quality.

Table: Post-Harvest Management Techniques for Ridge Gourd

S No	Post-Harvest Stage	Best Practices	Impact on Quality
1	Harvesting	Harvest at immature stage, handle carefully	Retains fruit tenderness, reduces damage
2	Storage	Store at 10-12°C with good ventilation	Extends shelf life, reduces spoilage
3	Transportation	Pack in ventilated crates, avoid bruising	Maintains quality during transit

C. Round Gourd (*Praecitrullusfistulosus*)

Family: Cucurbitaceae

Chromosome Number: 2n = 24

Round gourd, also known as Indian squash, apple gourd, or *tinda*, is an important vegetable crop cultivated in South Asia, particularly in India and Pakistan. It is primarily grown for its small, round, and green fruits, which are consumed while they are tender. The fruits have a mild, slightly sweet flavor and are used in various culinary preparations, including stir-fries, curries, and soups. Round gourd is also valued for its medicinal properties, including its digestive benefits and ability to help regulate blood sugar levels. A round gourd is a warm-season crop that thrives in well-drained soils and requires a tropical or subtropical climate for optimal growth. The plant is relatively easy to grow and can tolerate a wide range of soil conditions, making it a popular crop among small-scale and commercial farmers.

Origin and Domestication

The round gourd is native to India and has been cultivated for centuries. It is an important part of traditional agriculture in South Asia, where it is grown for its edible fruits. Domestication of round gourd likely occurred alongside other cucurbit species, with farmers selecting plants with desirable traits such as high yields, uniform fruit size, and improved taste. The crop spread to neighboring countries, including Pakistan and Bangladesh, where it became a staple vegetable. Traditional breeding methods, such as seed saving and selection, led to the development of landraces and local varieties that were well-adapted to different growing conditions and consumer preferences.

Botanical Varieties of Round Gourd

Round gourd varieties are primarily classified based on fruit size, shape, and skin color. The fruits are typically small and round, with a green or pale green skin. Different varieties are cultivated for specific purposes, such as fresh consumption or processing.

Table: Major Botanical Varieties of Round Gourd

S No	Variety Name	Fruit Characteristics	Region Grown	Main Use
1	Pusa Swarnima	Small, round, green fruits	India	Fresh consumption, curries
2	Arka Tinda	Medium-sized, round, smooth fruits	India	Fresh consumption
3	Tinda F1	Uniform, small, green fruits	India, Pakistan	Fresh consumption, processing
4	Punjab Tinda	Large, round, light green fruits	India, Pakistan	Culinary uses
5	Coimbatore Round	Small, round, pale green fruits	India	Fresh consumption

1. **Pusa Swarnima:**

   » This is a high-yielding variety with small, round fruits that have a deep green skin. The fruits are tender and mildly flavored, making them ideal for curries and stir-fries. *Pusa Swarnima* is widely grown in India for its consistent productivity and resistance to common diseases.

2. **Arka Tinda:**

   » *Arka Tinda* produces medium-sized, round fruits with smooth skin. It is popular in both commercial and home gardens due to its high yield and uniform fruit size. The variety is valued for its tender texture and mild taste, which makes it suitable for various culinary preparations.

3. **Tinda F1:**

   » *Tinda F1* is a hybrid variety known for its uniform fruit size and shape. It produces small, round fruits that are highly preferred for fresh consumption and processing. This variety is popular in both India and Pakistan for its high yield and resistance to biotic stresses.

4. **Punjab Tinda:**

   » *Punjab Tinda* is a large-fruited variety that produces round, light green fruits. It is well-suited for cultivation in the Punjab region of India and Pakistan, where it is widely used in traditional dishes. The variety is known for its high yield and adaptability to local growing conditions.

5. **Coimbatore Round:**

   » *Coimbatore Round* is a pale green, small-fruited variety that is primarily grown in South India. The fruits are tender and mildly flavored, making them ideal for fresh consumption. The variety is popular in local markets due to its consistent fruit size and quality.

Growth Habit of Round Gourd

Round gourd is a climbing vine that requires support to grow optimally. It is an annual plant that thrives in warm, humid conditions and is commonly grown on trellises to improve fruit quality and ease of harvest. The plant's growth habit makes it well-suited for vertical farming systems.

1. **Vine Growth:**

   » Round gourd produces long, trailing vines that can extend up to 3-4 meters in length. The vines are supported by tendrils that help the plant climb trellises or other support structures. The vigorous growth of round gourd allows it to cover large areas quickly, making it ideal for intercropping or vertical farming systems.

2. **Root System:**
   - » The plant has a deep taproot system that allows it to access water and nutrients from deeper soil layers. The root system helps the plant tolerate drought conditions, though consistent moisture is necessary for optimal growth and fruit development. The deep roots also help anchor the plant as it grows vertically.
3. **Flowering and Pollination:**
   - » Round gourd produces large, yellow flowers that are either male or female. The plant is monoecious, meaning both male and female flowers are produced on the same plant. Pollination is carried out by insects, mainly bees, which transfer pollen from male flowers to female flowers. Adequate pollinator activity is essential for good fruit set.
4. **Fruit Development:**
   - » After pollination, the female flower develops into a round fruit. Round gourd fruits are typically small, tender, and have a mild flavor when harvested at the immature stage. The fruits should be harvested before they become fully mature, as overripe fruits can become tough and lose their flavor.

Floral Biology of Round Gourd

The floral biology of the round gourd is essential for optimizing fruit set and improving yields. Proper pollination management is key to ensuring successful fruit development and high-quality yields.

1. **Male and Female Flowers:**
   - » Round gourd produces separate male and female flowers on the same plant. Male flowers appear first and are more numerous than female flowers. Female flowers can be distinguished by the small ovary at their base, which develops into the fruit after pollination. Male flowers produce pollen that is transferred to female flowers by pollinators, primarily bees.
2. **Pollination:**
   - » Pollination is a critical factor in round gourd production. Poor pollination can lead to misshapen or underdeveloped fruits. To ensure successful pollination, growers may introduce beehives to the fields or plant pollinator-attracting flowers nearby. Hand pollination can also be practiced in small-scale farming to improve fruit set.
3. **Fruit Set and Development:**
   - » Following successful pollination, the ovary of the female flower begins to swell and develop into a fruit. Round gourd fruits develop rapidly, typically reaching harvestable size within 10-14 days after pollination. The fruits should be harvested while they are still tender to ensure the best flavor and texture.

Breeding Methods in Round Gourd

Breeding programs for round gourd focus on improving traits such as yield, fruit quality, disease resistance, and tolerance to environmental stresses. Both conventional and non-conventional breeding methods are used to develop improved varieties.

Conventional Breeding Methods

1. **Introduction:**

   » The introduction of round gourd varieties from different regions allows breeders to evaluate their performance under local conditions. This method helps expand the genetic diversity of round gourds and identify varieties that perform well in specific environments.

2. **Selection:**

   » Selection is a key breeding method for improving round gourd varieties. Farmers and breeders select plants with desirable traits such as higher yields, better fruit quality, and disease resistance. These selected plants are then propagated and evaluated over multiple generations to develop improved varieties.

Source Selections in Round Gourd:

S No	Variety Name	Source	Key Features
1	Pusa Swarnima	India	High yield, tender fruits
2	Coimbatore Round	India	Small, uniform fruits
3	Arka Tinda	India	High yielding, uniform fruits
4	Punjab Tinda	India	Large, round fruits
5	Tinda F1	India	Hybrid, uniform fruit size

3. **Hybridization (Heterosis Breeding):**

   » Hybridization is used to combine desirable traits from different parent plants. This method often leads to hybrids that exhibit hybrid vigor, resulting in higher yields, better disease resistance, and improved fruit quality. Hybrid varieties are commonly developed for commercial farming due to their uniformity and consistent performance.

4. **Backcross Breeding:**

   » Backcross breeding is used to introduce specific traits, such as disease resistance or improved fruit quality, into an established variety. The hybrid offspring is crossed back with one of its parents to retain the desired trait while maintaining the overall characteristics of the original variety.

Non-Conventional Breeding Methods

1. **Tissue Culture:**

    » Tissue culture techniques are used to propagate disease-free planting material and maintain genetic integrity in breeding programs. This method is particularly important for producing virus-free plants and ensuring high-quality seeds for commercial production.

2. **Genetic Engineering (Recombinant DNA Technology):**

    » Genetic engineering is being explored to develop round gourd varieties with enhanced traits such as disease resistance, improved nutritional content, and tolerance to abiotic stresses. Although still in the experimental phase, genetic engineering holds promise for future round gourd breeding programs.

Breeding for Biotic and Abiotic Stresses in Round Gourd

Round gourd is susceptible to various biotic stresses, including fungal diseases, viral infections, and insect pests. Abiotic stresses such as drought, heat, and salinity also affect round gourd production. Breeding for resistance to these stresses is essential for improving the productivity and resilience of round gourd crops.

Table: Biotic Stress-Resistant Varieties

S No	Disease/Pest	Source Species	Resistant Varieties
1	Powdery Mildew	*Praecitrullusfistulosus*	Pusa Swarnima, Arka Tinda
2	Downy Mildew	*Praecitrullusfistulosus*	Coimbatore Round, Punjab Tinda
3	Fruit Fly	*Praecitrullusfistulosus*	Tinda F1, Arka Tinda

1. **Powdery Mildew Resistance:**

    » Powdery mildew is a common fungal disease affecting the leaves of round gourd plants, leading to reduced photosynthesis and premature leaf drop. Resistant varieties like *Pusa Swarnima* and *Arka Tinda* have been developed to minimize the impact of this disease, allowing for higher yields and better fruit quality.

2. **Downy Mildew Resistance:**

    » Downy mildew is caused by the pathogen *Pseudoperonosporacubensis* and primarily affects the foliage of round gourd plants. This disease can lead to significant yield losses if not properly managed. Varieties such as *Coimbatore Round* and *Punjab Tinda* have shown resistance to downy mildew, making them suitable for areas where this disease is prevalent.

3. **Fruit Fly Resistance:**

    » Fruit flies are a major pest affecting round gourd production, causing damage to the developing fruits and leading to poor fruit quality. Resistant varieties like *Tinda*

F1 and *Arka Tinda* have been developed to reduce the damage caused by fruit flies, improving both yield and fruit quality.

Post-Harvest Management of Round Gourd

Proper post-harvest management is crucial for maintaining the quality and shelf life of round gourd fruits. The fruits are highly perishable and require careful handling to avoid damage and spoilage. Post-harvest losses can be minimized through proper harvesting, storage, and transportation practices.

1. **Harvesting Techniques:**
   » Round gourd fruits should be harvested while they are still immature and tender. Harvesting at the right stage ensures that the fruits retain their mild flavor and tender texture. The fruits should be carefully cut from the vine to avoid damaging the plant or bruising the fruit, which can lead to post-harvest spoilage.
2. **Curing and Drying:**
   » Curing is not commonly practiced for round gourd, as the fruits are typically consumed fresh. However, in some regions, drying techniques are used to preserve the fruits for medicinal purposes. Dried round gourd is often used in traditional medicine or as a supplement in herbal preparations.
3. **Storage:**
   » Round gourd fruits should be stored at low temperatures (10-12°C) to slow down respiration and delay spoilage. Proper ventilation is also important to prevent the accumulation of moisture, which can lead to fungal growth. Under optimal storage conditions, round gourd fruits can be stored for 7-10 days without significant loss of quality.

Table: Post-Harvest Management Techniques for Round Gourd

S No	Post-Harvest Stage	Best Practices	Impact on Quality
1	Harvesting	Harvest at immature stage, handle carefully	Retains fruit tenderness, reduces damage
2	Storage	Store at 10-12°C with good ventilation	Extends shelf life, reduces spoilage
3	Transportation	Pack in ventilated crates, avoid bruising	Maintains quality during transit

D. Watermelon (*Citrullus lanatus*)

Family: Cucurbitaceae

Chromosome Number: 2n = 2x = 22

Watermelon is a popular fruit grown in warm climates around the world. Known for its sweet, juicy flesh and refreshing qualities, watermelon is widely consumed in fresh form, particularly during the summer months. The crop is an excellent source of vitamins A, C, and B6, as well as antioxidants like lycopene, which has been associated with numerous health benefits, including heart health and cancer prevention. Watermelon has a high water content, making it an important hydrating food. Watermelon is cultivated primarily in tropical and subtropical regions, with major producers including China, India, the United States, Turkey, and Brazil. The crop is grown both for fresh consumption and processed into juice, snacks, and other products. It thrives in hot, dry climates with well-drained soils and requires careful management of water and nutrients to achieve optimal yields.

Origin and Domestication

Watermelon originated in Africa, particularly in the dry regions of the Kalahari Desert in Southern Africa. It is believed to have been domesticated over 4,000 years ago, with early evidence of cultivation found in ancient Egypt. Wild watermelons, which are smaller and more bitter than modern varieties, were likely selected for their increased sweetness and larger fruit size. The crop spread through trade routes to the Mediterranean, Europe, and Asia, where it became a widely cultivated fruit. Over centuries of cultivation, numerous landraces and varieties were developed, each adapted to local climates and preferences. The introduction of modern breeding techniques has led to the development of hybrid varieties that are higher-yielding, more uniform, and resistant to diseases.

Botanical Varieties of Watermelon

Watermelon varieties are classified based on fruit characteristics such as shape, size, flesh color, rind thickness, and seed presence. Varieties may have red, yellow, or orange flesh, and they range from small, round fruits to large, oblong ones. Modern varieties include both seeded and seedless types, with seedless watermelons being highly popular in the market.

Table: Major Botanical Varieties of Watermelon

S No	Variety Name	Fruit Characteristics	Region Grown	Main Use
1	Sugar Baby	Small, round, deep red flesh	India, USA, Mediterranean	Fresh consumption
2	Crimson Sweet	Large, oblong, bright red flesh	USA, India	Fresh consumption, processing

3	Charleston Gray	Large, oblong, pale green rind	USA, China	Fresh consumption, juice
4	Yellow Crimson	Medium-sized, yellow flesh	China, Southeast Asia	Fresh consumption
5	Seedless 313	Medium-sized, seedless red flesh	USA, Europe	Fresh consumption

1. **Sugar Baby:**

» *Sugar Baby* is a small, round variety with deep red flesh and a sweet flavor. It is widely grown in India, the United States, and Mediterranean countries. The variety is popular for its early maturity and high sugar content, making it ideal for fresh consumption during the summer months.

2. **Crimson Sweet:**

» *Crimson Sweet* is a large, oblong variety with bright red flesh and a thick, green rind. It is one of the most popular watermelon varieties in the United States and India. The fruit is known for its high sugar content, crisp texture, and excellent storage quality. *Crimson Sweet* is used both for fresh consumption and processing.

3. **Charleston Gray:**

» *Charleston Gray* is a large, oblong watermelon with a pale green rind and deep red flesh. It is grown extensively in the United States and China. The variety is valued for its high yield, disease resistance, and suitability for transportation. It is commonly used for fresh consumption and juice production.

4. **Yellow Crimson:**

» *Yellow Crimson* is a medium-sized variety with yellow flesh. It is widely cultivated in China and Southeast Asia for its unique flavor and color. The fruit has a mild sweetness and is often used in fresh salads and desserts.

5. **Seedless 313:**

» *Seedless 313* is a medium-sized variety with seedless, bright red flesh. Seedless watermelons have become highly popular in markets worldwide due to their convenience and sweetness. This variety is grown in the United States and Europe, where it is favored for fresh consumption.

Growth Habit of Watermelon

Watermelon is a sprawling vine that grows best in warm, sunny environments. The plant produces long, trailing stems that spread along the ground. Its growth habit requires a large amount of space, making it well-suited for open fields or large gardens. Watermelon plants produce large leaves that help shade the fruits from direct sunlight, preventing sunscald.

1. **Vine Growth:**

   » The vines can grow up to 3-4 meters in length and require ample space to spread. Watermelon plants are typically grown on the ground, though they can also be trained to grow on trellises in smaller gardens. The plant produces tendrils that help anchor it to the soil or other structures.

2. **Root System:**

   » Watermelon has a deep taproot system that helps it access water and nutrients from deeper soil layers. This root system makes the plant relatively drought-tolerant, though consistent moisture is necessary for optimal fruit development. The roots also help stabilize the plant as it spreads along the ground.

3. **Flowering and Pollination:**

   » Watermelon produces separate male and female flowers on the same plant (monoecious). Male flowers appear first, followed by female flowers. Pollination is carried out by insects, particularly bees, which transfer pollen from male to female flowers. Adequate pollination is essential for good fruit set.

4. **Fruit Development:**

   » After pollination, the female flower develops into a fruit. Watermelon fruits vary in size, shape, and color, depending on the variety. Fruits can take 75-90 days from planting to reach maturity. The fruits should be harvested when fully ripe, as unripe watermelons have a bland taste and lack sweetness.

Floral Biology of Watermelon

Understanding the floral biology of watermelon is essential for optimizing fruit set and improving yields. Proper pollination management is key to ensuring successful fruit development and high-quality yields.

1. **Male and Female Flowers:**

   » Watermelon produces separate male and female flowers on the same plant. Male flowers appear in clusters and are more numerous than female flowers. Female flowers can be distinguished by the small, bulbous ovary at their base, which develops into the fruit after pollination.

2. **Pollination:**

   » Pollination is critical for watermelon fruit development. Inadequate pollination can lead to poor fruit set or misshapen fruits. Bees and other pollinators are vital for transferring pollen from male flowers to female flowers. In commercial production, farmers may introduce beehives to ensure adequate pollination.

3. **Fruit Set and Development:**

   » Once pollinated, the ovary of the female flower begins to swell and develop into a fruit. Fruit development occurs over 75-90 days, depending on the variety and environmental conditions. Fruits should be harvested when they have reached full maturity, as indicated by a hollow sound when tapped and a dry tendril near the fruit's stem.

Breeding Methods in Watermelon

Watermelon breeding focuses on improving traits such as yield, fruit size, sweetness, disease resistance, and seedlessness. Both conventional and non-conventional breeding methods are used to develop improved varieties.

Conventional Breeding Methods

1. **Introduction:**

   » Introduction of watermelon varieties from different regions allows breeders to evaluate their performance under local conditions. This method helps expand the genetic base of watermelon and identify varieties that perform well in specific environments.

2. **Selection:**

   » Selection is a key breeding method for improving watermelon varieties. Farmers and breeders select plants with desirable traits such as higher yields, better fruit quality, and disease resistance. These selected plants are then propagated and evaluated over multiple generations to develop improved varieties.

Source Selections in Watermelon:

S No	Variety Name	Source	Key Features
1	Sugar Baby	USA	Small, round, sweet fruits
2	Crimson Sweet	USA	High yield, bright red flesh
3	Charleston Gray	USA, China	Large, disease-resistant
4	Yellow Crimson	China	Unique yellow flesh
5	Seedless 313	USA, Europe	Seedless, sweet fruits

3. **Hybridization (Heterosis Breeding):**

   » Hybridization is used to combine desirable traits from different parent plants. This method often leads to hybrids that exhibit hybrid vigor, resulting in higher yields, better disease resistance, and improved fruit quality. Seedless watermelons are often produced through hybridization, using a sterile triploid plant to produce seedless fruits.

4. **Backcross Breeding:**

   » Backcross breeding is used to introduce specific traits, such as disease resistance or improved fruit quality, into an established variety. The hybrid offspring is crossed back with one of its parents to retain the desired trait while maintaining the overall characteristics of the original variety.

Non-Conventional Breeding Methods

1. **Tissue Culture:**

    » Tissue culture techniques are used to propagate disease-free planting material and maintain genetic integrity in breeding programs. This method is particularly important for producing virus-free plants and ensuring high-quality seeds for commercial production.

2. **Genetic Engineering (Recombinant DNA Technology):**

    » Genetic engineering is being explored to develop watermelon varieties with enhanced traits such as disease resistance, improved nutritional content, and tolerance to abiotic stresses. Although still in the experimental phase, genetic engineering holds promise for future watermelon breeding programs.

Breeding for Biotic and Abiotic Stresses in Watermelon

Watermelon is susceptible to a variety of biotic stresses, including fungal diseases, viral infections, and insect pests. Abiotic stresses such as drought, heat, and salinity also affect watermelon production. Breeding for resistance to these stresses is essential for improving the productivity and resilience of watermelon crops.

Table: Biotic Stress-Resistant Varieties

S No	Disease/Pest	Source Species	Resistant Varieties
1	Fusarium Wilt	*Citrullus lanatus*	Charleston Gray, Crimson Sweet
2	Downy Mildew	*Citrullus lanatus*	Sugar Baby, Arka Manik
3	Anthracnose	*Citrullus lanatus*	Pusa Bedana, Arka Jyoti

1. **Fusarium Wilt Resistance:**

    » Fusarium wilt is a soil-borne fungal disease that affects watermelon plants, causing wilting and eventual death of the plant. Resistant varieties such as *Charleston Gray* and *Crimson Sweet* have been developed to minimize the impact of this disease, allowing for higher yields and improved plant health.

2. **Downy Mildew Resistance:**

    » Downy mildew is caused by the pathogen *Pseudoperonosporacubensis* and primarily affects the leaves of watermelon plants. This disease can lead to significant yield losses if not properly managed. Varieties such as *Sugar Baby* and *Arka Manik* have shown resistance to downy mildew, making them suitable for areas where this disease is prevalent.

3. Anthracnose Resistance:

- » Anthracnose is a fungal disease that causes dark lesions on the leaves, stems, and fruits of watermelon plants. Resistant varieties like *Pusa Bedana* and *Arka Jyoti* have been developed to reduce the damage caused by anthracnose, improving both yield and fruit quality.

Post-Harvest Management of Watermelon

Proper post-harvest management is crucial for maintaining the quality and shelf life of watermelon fruits. The fruits are large and heavy, requiring careful handling to avoid damage. Post-harvest losses can be minimized through proper harvesting, storage, and transportation practices.

1. Harvesting Techniques:

- » Watermelons should be harvested when fully ripe to ensure maximum sweetness and flavor. The maturity of the fruit can be determined by observing the tendril near the fruit's stem, which dries up when the fruit is ripe. Harvesting should be done carefully to avoid bruising or cracking the fruit, as damaged fruits are more prone to spoilage.

2. Storage:

- » Watermelons should be stored at temperatures between 10-15°C to slow down respiration and delay spoilage. Under proper storage conditions, watermelons can be stored for 2-3 weeks without significant loss of quality. Fruits should not be stacked too high, as their weight can cause bruising and damage.

3. Transportation:

- » Watermelons are transported in ventilated crates or containers to prevent moisture buildup, which can lead to fungal growth. Proper cushioning is used to prevent bruising during transit, ensuring that the fruits reach the market in good condition.

Table: Post-Harvest Management Techniques for Watermelon

S No	Post-Harvest Stage	Best Practices	Impact on Quality
1	Harvesting	Harvest at full maturity, handle carefully	Ensures sweetness, reduces damage
2	Storage	Store at 10-15°C with good ventilation	Extends shelf life, reduces spoilage
3	Transportation	Pack in ventilated crates, cushion properly	Maintains quality during transit

E. Muskmelon (*Cucumis melo*)

Family: Cucurbitaceae

Chromosome Number: 2n = 24

Muskmelon, also known as cantaloupe, is a popular fruit grown primarily in warm, arid regions. It is prized for its sweet, aromatic flesh, which can vary in color from pale green to deep orange. The fruit is typically consumed fresh, but it is also used in juices, salads, and desserts. Muskmelon is a rich source of vitamins A and C, potassium, and dietary fiber, making it a nutritious addition to a balanced diet. Muskmelon is widely cultivated in countries like India, the United States, China, and Turkey. The crop thrives in well-drained, sandy loam soils and requires a long growing season with plenty of sunlight. Muskmelon is sensitive to frost and high humidity, making it better suited for regions with dry, hot climates.

Origin and Domestication

Muskmelon is believed to have originated in Africa and parts of southwest Asia. Evidence suggests that it was domesticated over 4,000 years ago. The fruit was introduced to the Mediterranean region and later spread to Europe, where it became a widely cultivated crop. Today, muskmelon is grown in many parts of the world, with numerous varieties developed to suit different climates and consumer preferences. The domestication of muskmelon has focused on improving fruit size, sweetness, aroma, and shelf life. Traditional landraces of muskmelon were selected for these traits, and modern breeding techniques have further refined them, resulting in the development of high-yielding hybrid varieties.

Botanical Varieties of Muskmelon

Muskmelon varieties are classified based on fruit characteristics such as skin texture, flesh color, and aroma. The most common varieties include netted muskmelons, smooth-skinned types, and winter melons. Varieties with orange flesh tend to be sweeter, while those with green flesh have a milder flavor.

Table: Major Botanical Varieties of Muskmelon

S No	Variety Name	Fruit Characteristics	Region Grown	Main Use
1	Hales Best Jumbo	Netted skin, orange flesh	USA, India	Fresh consumption
2	Punjab Sunehri	Netted skin, firm orange flesh	India	Fresh consumption
3	Arka Rajhans	Smooth skin, white flesh	India	Fresh consumption
4	Honey Dew	Smooth skin, pale green flesh	USA, China	Fresh consumption
5	Cantaloupe	Netted skin, firm orange flesh	Mediterranean, USA	Fresh consumption

1. **Hales Best Jumbo:**

    » *Hales Best Jumbo* is a netted muskmelon variety with orange flesh and a sweet aroma. It is widely grown in the United States and India. The fruit is medium to large in size and has a firm texture, making it ideal for fresh consumption.

2. **Punjab Sunehri:**

    » This is a popular Indian variety with netted skin and firm, orange flesh. *Punjab Sunehri* is known for its sweetness and excellent shelf life, making it a preferred choice for both local markets and long-distance transportation.

3. **Arka Rajhans:**

    » *Arka Rajhans* is a smooth-skinned muskmelon variety with white flesh. It is cultivated primarily in India and is known for its mild flavor and refreshing qualities. The variety is well-suited for regions with dry, hot climates.

4. **Honey Dew:**

    » *Honey Dew* is a smooth-skinned muskmelon with pale green flesh. It is grown in the United States, China, and other parts of the world. The fruit has a mildly sweet flavor and is popular for its juiciness and refreshing taste.

5. **Cantaloupe:**

    » *Cantaloupe* is a well-known variety with netted skin and firm, orange flesh. It is widely cultivated in Mediterranean countries and the USA. The variety is known for its rich aroma and sweet flavor, making it a favorite for fresh consumption.

Growth Habit of Muskmelon

Muskmelon is a vigorous, trailing vine that grows well in warm, sunny environments. The plant produces long, sprawling stems that spread along the ground. It is well-suited for open fields and can also be grown on trellises to improve fruit quality and ease of harvest.

1. **Vine Growth:**

    » Muskmelon vines can grow up to 3 meters in length and require ample space to spread. The plant produces tendrils that help it anchor to the soil or support structures. The vigorous growth habit of muskmelon allows it to cover large areas quickly, making it ideal for open-field cultivation.

2. **Root System:**

    » Muskmelon has a deep taproot system that allows it to access water and nutrients from deeper soil layers. This root system makes the plant relatively drought-tolerant, though consistent moisture is necessary for optimal fruit development. The roots also help stabilize the plant as it spreads along the ground.

3. **Flowering and Pollination:**

   » Muskmelon produces separate male and female flowers on the same plant (monoecious). Male flowers appear first, followed by female flowers. Pollination is carried out by insects, primarily bees, which transfer pollen from male to female flowers. Adequate pollination is essential for good fruit set.

4. **Fruit Development:**

   » After pollination, the female flower develops into a fruit. Muskmelon fruits vary in size, shape, and color, depending on the variety. Fruits take 75-90 days from planting to reach maturity. Harvesting should be done when the fruit has developed its characteristic aroma and the skin begins to change color.

Floral Biology of Muskmelon

Understanding the floral biology of muskmelon is essential for optimizing fruit set and improving yields. Proper pollination management is key to ensuring successful fruit development and high-quality yields.

1. **Male and Female Flowers:**

   » Muskmelon produces separate male and female flowers on the same plant. Male flowers appear first and are more numerous than female flowers. Female flowers can be distinguished by the small, bulbous ovary at their base, which develops into the fruit after pollination.

2. **Pollination:**

   » Pollination is a critical factor in muskmelon production. Poor pollination can lead to small or misshapen fruits. Bees and other pollinators are vital for transferring pollen from male flowers to female flowers. In commercial production, growers may introduce beehives to ensure adequate pollination.

3. **Fruit Set and Development:**

   » Once pollinated, the ovary of the female flower begins to swell and develop into a fruit. Fruit development occurs over 75-90 days, depending on the variety and environmental conditions. Fruits should be harvested when they emit a sweet aroma and the skin changes color, indicating full maturity.

Breeding Methods in Muskmelon

Muskmelon breeding focuses on improving traits such as yield, fruit size, sweetness, disease resistance, and shelf life. Both conventional and non-conventional breeding methods are used to develop improved varieties.

Conventional Breeding Methods

1. Introduction:

» Introduction of muskmelon varieties from different regions allows breeders to evaluate their performance under local conditions. This method helps expand the genetic base of muskmelon and identify varieties that perform well in specific environments.

2. Selection:

» Selection is a key breeding method for improving muskmelon varieties. Farmers and breeders select plants with desirable traits such as higher yields, better fruit quality, and disease resistance. These selected plants are then propagated and evaluated over multiple generations to develop improved varieties.

Source Selections in Muskmelon:

S No	Variety Name	Source	Key Features
1	Hales Best Jumbo	USA	High yield, sweet, aromatic fruits
2	Punjab Sunehri	India	High yield, firm flesh
3	Arka Rajhans	India	Mild flavor, smooth skin
4	Honey Dew	USA, China	Juicy, pale green flesh
5	Cantaloupe	Mediterranean	Rich aroma, orange flesh

3. Hybridization (Heterosis Breeding):

» Hybridization is used to combine desirable traits from different parent plants. This method often leads to hybrids that exhibit hybrid vigor, resulting in higher yields, better disease resistance, and improved fruit quality. Hybrid muskmelons are popular in commercial farming for their uniformity and consistent performance.

4. Backcross Breeding:

» Backcross breeding is used to introduce specific traits, such as disease resistance or improved fruit quality, into an established variety. The hybrid offspring is crossed back with one of its parents to retain the desired trait while maintaining the overall characteristics of the original variety.

Non-Conventional Breeding Methods

1. Tissue Culture:

» Tissue culture techniques are used to propagate disease-free planting material and maintain genetic integrity in breeding programs. This method is particularly important for producing virus-free plants and ensuring high-quality seeds for commercial production.

2. Genetic Engineering (Recombinant DNA Technology):

» Genetic engineering is being explored to develop muskmelon varieties with enhanced traits such as disease resistance, improved nutritional content, and tolerance to abiotic stresses. Although still in the experimental phase, genetic engineering holds promise for future muskmelon breeding programs.

Breeding for Biotic and Abiotic Stresses in Muskmelon

Muskmelon is susceptible to a variety of biotic stresses, including fungal diseases, viral infections, and insect pests. Abiotic stresses such as drought, salinity, and heat also affect muskmelon production. Breeding for resistance to these stresses is essential for improving the productivity and resilience of muskmelon crops.

Table: Biotic Stress-Resistant Varieties

S No	Disease/Pest	Source Species	Resistant Varieties
1	Powdery Mildew	*Cucumis melo*	Arka Rajhans, Punjab Sunehri
2	Downy Mildew	*Cucumis melo*	Hales Best Jumbo, Honey Dew
3	Aphids	*Cucumis melo*	Arka Rajhans, Cantaloupe

1. Mildew Resistance:

» Powdery mildew is a common fungal disease affecting the leaves of muskmelon plants, leading to reduced photosynthesis and premature leaf drop. Resistant varieties like *Arka Rajhans* and *Punjab Sunehri* have been developed to minimize the impact of this disease, allowing for higher yields and better fruit quality.

2. Downy Mildew Resistance:

» Downy mildew is caused by the pathogen *Pseudoperonosporacubensis* and primarily affects the foliage of muskmelon plants. This disease can lead to significant yield losses if not properly managed. Varieties such as *Hales Best Jumbo* and *Honey Dew* have shown resistance to downy mildew, making them suitable for areas where this disease is prevalent.

3. Aphid Resistance:

» Aphids are a major pest affecting muskmelon plants, causing damage to the leaves and spreading viral diseases. Resistant varieties like *Arka Rajhans* and *Cantaloupe* have been developed to reduce the damage caused by aphids, improving both yield and fruit quality.

Post-Harvest Management of Muskmelon

Proper post-harvest management is crucial for maintaining the quality and shelf life of muskmelon fruits. The fruits are highly perishable and require careful handling to avoid damage. Post-harvest losses can be minimized through proper harvesting, storage, and transportation practices.

1. Harvesting Techniques:

» Muskmelons should be harvested when they emit a sweet aroma and the skin begins to change color, indicating full maturity. Harvesting should be done carefully to avoid bruising the fruit, as damaged fruits are more prone to spoilage. Muskmelons should be cut from the vine rather than pulled to avoid damage to the fruit and plant.

2. Storage:

» Muskmelons should be stored at temperatures between 5-10°C to slow down respiration and delay spoilage. Under proper storage conditions, muskmelons can be stored for 2-3 weeks without significant loss of quality. The fruits should be handled carefully to avoid bruising and should not be stacked too high, as their weight can cause damage.

3. Transportation:

» Muskmelons are transported in ventilated crates or containers to prevent moisture buildup, which can lead to fungal growth. Proper cushioning is used to prevent bruising during transit, ensuring that the fruits reach the market in good condition.

Table: Post-Harvest Management Techniques for Muskmelon

S No	Post-Harvest Stage	Best Practices	Impact on Quality
1	Harvesting	Harvest at full maturity, handle carefully	Ensures sweetness, reduces damage
2	Storage	Store at 5-10°C with good ventilation	Extends shelf life, reduces spoilage
3	Transportation	Pack in ventilated crates, cushion properly	Maintains quality during transit

F. Kheera (*Cucumis sativus*)

Family: Cucurbitaceae

Chromosome Number: 2n = 14

Cucumber, or kheera, is one of the most widely cultivated vegetable crops around the world. It is prized for its refreshing and crunchy texture, often consumed raw in salads, sandwiches, or pickled. Cucumbers are also a great source of hydration due to their high water content, making them popular during the summer. In addition to their culinary uses, cucumbers are known for their medicinal properties, particularly in promoting skin health and digestion. They are rich in vitamins C and K, potassium, and fiber, making them a valuable addition to a healthy diet. Cucumber cultivation is common in tropical and subtropical regions, and it is grown both in open fields and under protected cultivation in greenhouses. It thrives in warm, sunny conditions and well-drained soils with a pH of 6.0-7.0. Cucumber plants are relatively fast-growing and require careful management of water, nutrients, and pests to achieve optimal yields.

Origin and Domestication

Cucumber is believed to have originated in South Asia, with evidence of its cultivation dating back over 3,000 years. It spread to Europe and the Mediterranean region through trade routes, where it became a popular vegetable in ancient civilizations, including Greece and Rome. Over time, cucumber was introduced to other parts of the world, including North America, where it became a staple in gardens and farms. Domestication efforts focused on improving fruit size, flavor, and texture. Traditional landraces were selected for these traits, and modern breeding techniques have further enhanced the crop, resulting in high-yielding, disease-resistant varieties that are well-suited for different climates and consumer preferences.

Botanical Varieties of Cucumber

Cucumber varieties are classified based on their intended use, such as slicing cucumbers, pickling cucumbers, and seedless (burpless) cucumbers. The fruit shape, skin texture, and size can vary greatly between varieties, with some producing long, smooth fruits and others developing short, spiny ones.

Table: Major Botanical Varieties of Cucumber

S No	Variety Name	Fruit Characteristics	Region Grown	Main Use
1	Pusa Uday	Medium-sized, dark green, smooth	India	Fresh consumption
2	Poinsett 76	Long, slender, smooth	USA, India	Fresh consumption
3	Beit Alpha	Seedless, smooth skin	Israel, USA	Greenhouse cultivation
4	Gherkin	Short, spiny	Europe, India	Pickling
5	Chinese Long	Long, thin, smooth	China	Fresh consumption

1. **Pusa Uday:**
   » *Pusa Uday* is a medium-sized cucumber variety with dark green, smooth skin. It is widely grown in India for fresh consumption and is known for its crisp texture and mild flavor. The variety is suitable for both open-field and protected cultivation.
2. **Poinsett 76:**
   » This is a long, slender cucumber variety with smooth skin. *Poinsett 76* is popular in the USA and India for its excellent yield and adaptability to different growing conditions. The fruits are ideal for slicing and are consumed fresh.
3. **Beit Alpha:**
   » *Beit Alpha* cucumbers are seedless and have smooth, thin skin. This variety is primarily grown in greenhouses and is valued for its sweet flavor and lack of bitterness. It is a popular choice for fresh consumption, especially in Israel and the USA.
4. **Gherkin:**
   » *Gherkin* is a short, spiny cucumber variety grown primarily for pickling. The small, firm fruits are ideal for preserving in brine or vinegar. Gherkin varieties are widely grown in Europe and India for the production of pickles.
5. **Chinese Long:**
   » *Chinese Long* cucumbers are characterized by their long, thin fruits with smooth skin. This variety is grown extensively in China for fresh consumption. The fruits are crisp and mild in flavor, making them ideal for salads and other fresh preparations.

Growth Habit of Cucumber

Cucumber is a climbing vine that produces long, trailing stems. The plant can be grown on the ground or trained to climb trellises for better air circulation and improved fruit quality. Cucumber plants produce large, lobed leaves that provide shade for the developing fruits, helping to prevent sunscald.

1. **Vine Growth:**
   » Cucumber vines can grow up to 2-3 meters in length, making them well-suited for vertical gardening and trellising. The plant produces tendrils that help it climb, and this growth habit allows for efficient use of space in small gardens or greenhouse systems.
2. **Root System:**
   » Cucumber has a shallow root system, which makes it sensitive to moisture stress. Consistent irrigation is necessary to support the plant's fast growth and ensure proper fruit development. Mulching is often used to retain soil moisture and regulate soil temperature.

3. **Flowering and Pollination:**

   » Cucumber produces separate male and female flowers on the same plant (monoecious). Male flowers appear first and produce pollen, which is transferred to female flowers by insects, primarily bees. In greenhouse production, hand pollination or the introduction of pollinators may be required to ensure a good fruit set.

4. **Fruit Development:**

   » After pollination, the female flowers develop into fruits. Cucumber fruits are typically harvested while still immature, as overripe cucumbers can become bitter. The fruits can vary in size and shape, depending on the variety and growing conditions.

Floral Biology of Cucumber

The floral biology of cucumber is essential for optimizing fruit sets and improving yields. Proper pollination management ensures successful fruit development and high-quality yields.

1. **Male and Female Flowers:**

   » Cucumber produces separate male and female flowers on the same plant. Male flowers are more numerous and appear first, followed by female flowers, which can be distinguished by the small ovary at their base. Pollination occurs when pollen from the male flowers is transferred to the female flowers, leading to fruit development.

2. **Pollination:**

   » Pollination is critical for cucumber fruit development. Poor pollination can result in misshapen fruits or low yields. Bees and other pollinators are vital for transferring pollen from male to female flowers. In protected cultivation, hand pollination or the introduction of bumblebees may be necessary to achieve good pollination rates.

3. **Fruit Set and Development:**

   » Once pollinated, the female flowers develop into fruits. Fruit development is rapid, with cucumbers reaching harvestable size within 40-60 days of planting, depending on the variety. Regular harvesting is essential to prevent overripe fruits, which can negatively affect the plant's productivity.

Breeding Methods in Cucumber

Cucumber breeding focuses on improving traits such as yield, disease resistance, fruit quality, and adaptability to different growing conditions. Both conventional and non-conventional breeding methods are used to develop improved varieties.

Conventional Breeding Methods

1. **Introduction:**

   » Introduction involves bringing cucumber varieties from different regions to evaluate their performance in local conditions. This method helps expand the genetic diversity of cucumber and allows breeders to identify varieties that perform well under specific environmental conditions.

2. **Selection:**
   » Selection is a fundamental breeding method used to improve cucumber varieties. Farmers and breeders select plants with desirable traits, such as higher yields, disease resistance, and improved fruit quality. These selected plants are then propagated to develop improved varieties.

Source Selections in Cucumber:

S No	Variety Name	Source	Key Features
1	Pusa Uday	India	Dark green, medium-sized fruits
2	Poinsett 76	USA	Long, smooth fruits
3	Beit Alpha	Israel	Seedless, smooth-skinned fruits
4	Gherkin	Europe	Small, spiny fruits
5	Chinese Long	China	Long, thin, smooth fruits

3. **Hybridization (Heterosis Breeding):**
   » Hybridization involves crossing two different parent plants to combine desirable traits. This method leads to hybrid vigor, resulting in higher yields, better disease resistance, and improved fruit quality. Hybrid cucumber varieties are popular in commercial farming due to their uniformity and consistent performance.

4. **Backcross Breeding:**
   » Backcross breeding is used to introduce specific traits, such as disease resistance, into an established variety. The hybrid offspring is crossed back with one of its parents to retain the desired trait while maintaining the overall characteristics of the original variety.

Non-Conventional Breeding Methods

1. **Tissue Culture:**
   » Tissue culture techniques are used to propagate disease-free planting material and maintain genetic integrity in cucumber breeding programs. This method is especially important for producing virus-free plants and ensuring high-quality seeds for commercial production.

2. **Genetic Engineering (Recombinant DNA Technology):**
   » Genetic engineering is being explored to develop cucumber varieties with enhanced traits such as disease resistance, improved nutritional content, and tolerance to abiotic stresses. Although still in the experimental phase, genetic engineering holds promise for future cucumber breeding programs.

Breeding for Biotic and Abiotic Stresses in Cucumber

Cucumber is vulnerable to various biotic stresses, including fungal diseases, viral infections, and insect pests. Abiotic stresses such as drought, heat, and salinity also affect cucumber production. Breeding for resistance to these stresses is crucial for improving the productivity and resilience of cucumber crops.

Table: Biotic Stress-Resistant Varieties

S No	Disease/Pest	Source Species	Resistant Varieties
1	Powdery Mildew	*Cucumis sativus*	Poinsett 76, Arka Anmol
2	Downy Mildew	*Cucumis sativus*	Pusa Uday, Beit Alpha
3	Aphids	*Cucumis sativus*	Chinese Long, Poinsett 76

1. **Powdery Mildew Resistance:**
   » Powdery mildew is a common fungal disease that affects cucumber plants, causing white, powdery spots on the leaves. Resistant varieties like *Poinsett 76* and *Arka Anmol* have been developed to minimize the impact of this disease, leading to healthier plants and higher yields.
2. **Downy Mildew Resistance:**
   » Downy mildew is a major disease caused by the pathogen *Pseudoperonosporacubensis*. It affects the leaves and can cause significant yield losses if not managed. Varieties such as *Pusa Uday* and *Beit Alpha* have shown resistance to downy mildew, making them ideal for regions where this disease is prevalent.
3. **Aphid Resistance:**
   » Aphids are a common pest in cucumber crops, feeding on the plant sap and transmitting viral diseases. Resistant varieties like *Chinese Long* and *Poinsett 76* help reduce the damage caused by aphids and improve crop yields.

Post-Harvest Management of Cucumber

Proper post-harvest management is essential for maintaining the quality and shelf life of cucumber fruits. Since cucumbers are highly perishable, they require careful handling during and after harvest to avoid damage and spoilage.

1. **Harvesting Techniques:**
   » Cucumbers should be harvested while they are still immature and tender. Overripe cucumbers tend to become bitter and lose their crisp texture. The fruits should be cut from the vine carefully to avoid bruising, as damaged fruits are more prone to spoilage.
2. **Storage:**
   » Cucumbers should be stored at temperatures between 10-12°C to slow down

respiration and delay spoilage. Proper ventilation is important to prevent moisture buildup, which can lead to fungal growth. Under optimal conditions, cucumbers can be stored for 7-10 days without significant loss of quality.

3. **Transportation:**

   » Cucumbers are transported in ventilated crates or containers to prevent bruising during transit. Proper cushioning is used to protect the fruits from mechanical damage, ensuring that they arrive at the market in good condition.

Table: Post-Harvest Management Techniques for Cucumber

S No	Post-Harvest Stage	Best Practices	Impact on Quality
1	Harvesting	Harvest while tender, handle carefully	Retains crisp texture, reduces damage
2	Storage	Store at 10-12°C with good ventilation	Extends shelf life, reduces spoilage
3	Transportation	Pack in ventilated crates, cushion properly	Maintains quality during transit

Chapter - 7

Vegetable Breeder's Kit

Vegetable Breeder's Kit

In the field of vegetable breeding, precision, efficiency, and attention to detail are key to successfully developing new varieties. A vegetable breeder's kit comprises various tools and materials specifically designed to carry out breeding tasks such as pollination, emasculation, and plant selection. The effectiveness of these tools is critical for achieving desired genetic outcomes in crops.

Materials Required

The basic materials required for successful plant breeding include forceps, scissors, alcohol, tags, pencils, and butter paper bags. These materials form the backbone of the breeding process, ensuring the breeder can handle delicate plant tissues and keep track of their work. Forceps and scissors are used for handling plant parts, while alcohol is crucial for sterilization, preventing contamination during the breeding process. Tags and pencils ensure that breeders can accurately record their observations and identify specific plants, while butter paper bags serve to protect flowers from unintended pollination. The meticulous use of these items ensures that cross-pollination and emasculation are carried out with precision.

In addition to these basic materials, modern breeding kits often include digital tools such as electronic tagging devices and data management systems. These tools streamline record-keeping, making it easier for breeders to track genetic traits, manage large datasets, and make informed decisions based on recorded information. The combination of traditional and modern tools enhances the breeder's ability to produce high-quality, genetically superior plant varieties.

DOI: 10.1201/9781003790976-7

Equipment and Their Uses

1. Scissors:

Scissors are essential for removing unwanted leaves, buds, and other plant parts that may interfere with the pollination process. In breeding, it is vital to clear the plant of any unnecessary growth to expose the desired flowers or buds, making it easier to control pollination. Scissors allow breeders to make clean cuts, preventing damage to the plant and ensuring healthy flower development. High-quality, stainless-steel scissors with a pointed tip are preferred for accuracy in cutting delicate plant parts without crushing or bruising them.In some advanced kits, micro-scissors with ultra-fine blades are used for handling very small or delicate flowers. These specialized scissors allow for precision cuts in flowers with minute parts, such as those found in some legumes or cruciferous crops. The use of these fine tools ensures that the breeder can access flower organs without damaging them, which is crucial for achieving controlled pollination.

2. Needle (Long and Pointed):

A long and pointed needle is indispensable for opening small flower buds during the emasculation process. Emasculation, which involves removing the male reproductive organs (anthers) from the flower, requires delicate handling, especially with small or immature buds. The pointed needle allows breeders to carefully open the flower without causing damage to the other parts, such as the petals or the stigma. Proper emasculation ensures that only selected pollen grains will fertilize the flower, which is a fundamental step in controlled breeding programs.Advanced breeding kits may also include microsurgical tools or fine-tipped needles with ergonomic grips to facilitate more comfortable and efficient handling during extended periods of work. These tools reduce hand fatigue and increase precision, particularly when working with crops that have small or intricate flower structures.

3. Forceps/Tweezers:

Fine-tipped forceps or tweezers with long points are crucial for removing anthers during emasculation and transferring pollen grains from the male flowers to the stigma of emasculated female flowers. The use of forceps ensures that the breeder can handle the fragile anthers and pollen without damaging them or contaminating the flower. Proper use of forceps is key to ensuring successful pollination and ultimately leads to the desired crossbreed. Modern breeding kits may include forceps made from anti-magnetic and corrosion-resistant stainless steel, which prevents static buildup and contamination from chemical exposure. Some advanced models are designed with angled tips for better access to hard-to-reach flower parts, ensuring maximum efficiency in pollen transfer.

4. Brush (Hair Brush):

A soft hairbrush is typically used for collecting pollen grains and dusting them onto the stigma of emasculated female flowers. Brushes with fine, soft bristles are preferred to prevent damage to the delicate stigma and to ensure even distribution of pollen. The use of a brush allows for careful handling of pollen, which can be very sensitive to external conditions such as temperature and humidity. In some breeding kits, the brush is designed with an adjustable handle for easier maneu verability. In more sophisticated kits, mechanical pollen applicators or electrostatic pollen dusters are employed. These devices use an electrical charge to attract and transfer pollen, ensuring a more controlled and uniform application to the stigma. This technology improves the success rate of pollination and is particularly useful in large-scale breeding programs.

5. Spirit Bottle:

A small bottle of alcohol or spirit is essential for sterilizing scissors, needles, forceps, and brushes, as well as the breeder's hands, to prevent cross-contamination during the breeding process. Proper sterilization is critical for maintaining the integrity of the breeding program, as contamination from foreign pollen can lead to unintended genetic results. Sterilization also prevents the spread of diseases that could compromise the health of the plants.In addition to alcohol, some kits include UV sterilizers or portable autoclaves for quick and efficient sterilization of tools between breeding tasks. These modern sterilization methods are particularly useful in large breeding operations, where speed and efficiency are paramount.

6. Magnifying Lens:

A magnifying lens is essential for emasculating small flower buds and observing the stigma of emasculated flowers to confirm that it is free of any anthers or pollen grains. The magnifying lens allows the breeder to work with greater precision, especially when dealing with very small or intricate flowers, where naked-eye inspection may not suffice. More advanced kits may include digital microscopes with high-resolution imaging capabilities. These tools allow breeders to capture detailed images of flower parts, which can be analyzed and shared with other breeders or researchers. Such tools are invaluable for documenting breeding progress and for educating others in breeding techniques.

7. Bags:

Various types of bags—such as brown paper bags, muslin cloth bags, and parchment paper bags—are used to cover flowers or inflorescences before and after pollination. These bags protect the flowers from unwanted pollination by external agents like wind, insects, or accidental contact. The choice of bag depends on the size and type of the crop, ensuring that the flowers are securely covered while still allowing for proper ventilation and growth.

In some modern setups, specially designed pollination tents or isolation chambers are used. These chambers provide a controlled environment where temperature, humidity, and light can be regulated to optimize the breeding process.

8. Labels:

Paper or aluminum labels are used to record essential information about the breeding process, such as the date of pollination, the parent plants involved, and any observations made during the process. Accurate labeling is crucial for tracking the progress of breeding experiments and for ensuring that the results are correctly attributed to the appropriate cross-pollinations. In advanced breeding programs, digital tagging systems with QR codes or RFID chips are becoming more common. These systems allow breeders to scan plants with a device and instantly access all relevant data, streamlining the process of data collection and analysis.

9. U-Pins or Wire Rings:

U-pins or wire rings are used to hold the paper bags securely over the flowers or inflorescences. These tools ensure that the bags remain in place throughout the pollination process, protecting the flowers from unintended exposure to external pollen. The use of sturdy, corrosion-resistant pins ensures that they can be reused across multiple breeding cycles.

10. Field Notebook:

A field notebook is indispensable for recording all important observations during the breeding process. This includes details about the number of crosses made, the dates of emasculation or selfing, the number of seeds set in crossed flowers, and any other relevant information. A well-maintained field notebook provides a comprehensive record of the breeding process, which is essential for evaluating the success of the breeding program.In modern breeding programs, digital field notebooks or mobile applications are used for real-time data entry. These tools allow breeders to collect, store, and analyze data more efficiently, ensuring that breeding decisions are based on accurate and up-to-date information.By using a combination of traditional tools and modern technology, vegetable breeders can achieve greater precision, efficiency, and success in their breeding programs.

References

Acquaah, G. (2012). Principles of Plant Genetics and Breeding (2nd ed.). Wiley-Blackwell.

Allard RW. 1960. Principle of plant breeding. John Willey and Sons, USA.

Ashraf, M., & Harris, P. J. C. (2013). Abiotic Stress in Plants: Mechanisms and Adaptations. CRC Press.

Bassett, M. J. (Ed.). (2007). Vegetable Genetics and Breeding. CABI.

Bose, T. K., & Som, M. G. (1986). Vegetable Crops in India (1st ed.). Naya Prokash.

Frankel, R., & Galun, E. (2012). Pollination Mechanisms, Reproduction and Plant Breeding. Springer.

George, R. A. T. (2009). Vegetable Seed Production (3rd ed.). CABI Publishing.

Hazra, P., & Banerjee, M. K. (2010). Breeding and Biotechnology of Vegetables. Kalyani Publishers.

Kalloo G. 1988. Vegetable breeding (Vol. I, II, III). CRC Press, Fl, USA.

Kole CR. 2007. Genome mapping and molecular breeding in plants-vegetables. Springer, USA.

Peter KVand Pradeep Kumar T. 1998. Genetics and breeding of vegetables. ICAR, New Delhi, p. 488.

Peter, K. V., & Hazra, P. (2012). Breeding of Horticultural Crops: Principles and Practices. New India Publishing Agency.

Prohens J and Nuez F. 2007. Handbook of plant breeding-vegetables (Vol I and II). Springer, USA.

Prohens, J., & Nuez, F. (2007). Handbook of Plant Breeding: Vegetables I. Springer.

Prohens, J., & Nuez, F. (Eds.). (2008). Vegetables I: Asteraceae, Brassicaceae, Chenopodicaceae, and Cucurbitaceae. Springer.

Prohens, J., & Nuez, F. (Eds.). (2008). Vegetables II: Fabaceae, Liliaceae, Solanaceae, and Umbelliferae. Springer.

Simmonds, N. W., & Smartt, J. (1999). Principles of Crop Improvement (2nd ed.). John Wiley & Sons.

Singh Ram J. 2007. Genetic resources, chromosome engineering, and crop improvement-vegetable crops (Vol. 3). CRC Press, Fl, USA

Singh, B. D. (2019). Plant Breeding: Principles and Methods (11th ed.). Kalyani Publishers.

Singh, B. D., & Singh, A. K. (2015). Marker-Assisted Plant Breeding: Principles and Practices. Springer.

For Product Safety Concerns and Information please contact our EU representative GPSR@taylorandfrancis.com Taylor & Francis Verlag GmbH, Kaufingerstraße 24, 80331 München, Germany